Ozone Reaction Kinetics for Water and Wastewater Systems

FERNANDO J. BELTRÁN

LEWIS PUBLISHERS

A CRC Press Company
Boca Raton   London   New York   Washington, D.C.

Library of Congress Cataloging-in-Publication Data

Beltrán, Fernando J., 1955-
  Ozone reaction kinetics for water and wastewater systems / Fernando J. Beltrán.
     p. cm.
  Includes bibliographical references and index.
  ISBN 1-56670-629-7 (alk. paper)
  1. Water—Purification—Ozonization. 2. Sewage—Purification—Ozonization. I. Title.

TD461.B45 2003
628.1'662—dc22   2003060323

This book contains information obtained from authentic and highly regarded sources. Reprinted material is quoted with permission, and sources are indicated. A wide variety of references are listed. Reasonable efforts have been made to publish reliable data and information, but the author and the publisher cannot assume responsibility for the validity of all materials or for the consequences of their use.

Neither this book nor any part may be reproduced or transmitted in any form or by any means, electronic or mechanical, including photocopying, microfilming, and recording, or by any information storage or retrieval system, without prior permission in writing from the publisher.

The consent of CRC Press LLC does not extend to copying for general distribution, for promotion, for creating new works, or for resale. Specific permission must be obtained in writing from CRC Press LLC for such copying.

Direct all inquiries to CRC Press LLC, 2000 N.W. Corporate Blvd., Boca Raton, Florida 33431.

Trademark Notice: Product or corporate names may be trademarks or registered trademarks, and are used only for identification and explanation, without intent to infringe.

Visit the CRC Press Web site at www.crcpress.com

© 2004 by CRC Press LLC
Lewis Publishers is an imprint of CRC Press LLC

No claim to original U.S. Government works
International Standard Book Number 1-56670-629-7
Library of Congress Card Number 2003060323
Printed in the United States of America 1 2 3 4 5 6 7 8 9 0
Printed on acid-free paper

*To my wife, Rosa Maria,
and to my son, Fernando*

To my parents

Acknowledgments

I am very grateful to my colleagues in the Department of Ingeniería Química at the University of Extremadura for their help in conducting the many laboratory experiments I used to study the ozonation kinetics of compounds in water and wastewater. I am especially grateful to Juan Fernando García-Araya, Francisco J. Rivas, Pedro M. Álvarez, Benito Acedo, Jose M. Encinar, Manuel González, and many others who wrote their doctoral dissertations on this challenging subject under my supervision.

I acknowledge the research grants from the CICYT of the Spanish Ministry of Science and Technology, the European FEDER funds, and the Junta of Extremadura, which have enabled me to conduct ozonation kinetic studies for more than 15 years.

I also acknowledge Christine Andreasen, my CRC project editor, for her invaluable help editing, and at times virtually translating, my "Spanish-English" manuscript.

Finally, I express my deep appreciation to my wife and son for their patience and support during the many hours I spent preparing this book and conducting my research.

Preface

Today ozone is considered an alternative oxidant-disinfectant agent with multiple possible applications in water, air pollution, medicine, etc. In water treatment, in particular, ozone has the ability to disinfect, oxidize, or to be used in combination with other technologies and reagents. Much of the information about these general aspects of ozone has been reported in excellent works, such as Langlais et al. (1991).[1] There is another aspect, however, that the literature has not dealt with sufficiently — the ozonation kinetics of compounds in water, especially those organic compounds usually considered water pollutants. In contrast, many works published in scientific journals, such as *Ozone Science and Engineering*, *Water Research*, *Industrial and Engineering Chemistry Research*, and the like, present simple examples of the multiple possibilities of ozone in water and the kinetics of wastewater treatment. I thought that this wide variety of ozone kinetic information should be published in a unique book that examined the many aspects of this subject and provided a general overview that would facilitate a better understanding of the fundamentals.

For more than 20 years I have worked on the use of ozone to oxidize organic compounds, both in organic and, especially, aqueous media. The results of my research have generated more than 100 papers in scientific journals and several doctoral theses on the ozonation of dyes, phenols, herbicides, polynuclear aromatic hydrocarbons, and wastewater. For many years I have lectured on ozonation kinetics in graduate courses at the University of Extremadura (Badajoz, Spain). As a result of this accumulated experience, I can confirm that the numerous possible applications of ozone in water and wastewater treatment make the study of ozonation kinetics a challenging subject in which theory and practice can be examined simultaneously. The work presented here is a compilation of my years of study in this field.

This book is intended for both undergraduate, graduate, and postgraduate students, and for teachers and professionals involved with water and wastewater treatment. Students who want to become involved with ozone applications in water must be familiar with the many aspects of the subject covered here, including absorption or solubility of ozone, stability or decomposition, reactivity, kinetic regime of absorption, ozonation kinetics, and reactor modeling. Practicing professionals in ozone water treatment, that is, professionals in the ozonation processing field, can augment their fund of knowledge with the advanced information in this book. Finally, this book can also be used as a teaching tool for verifying the fundamentals of chemistry, reaction mechanisms, and, particularly, chemical engineering kinetics and heterogeneous kinetics by examining the results of the ozonation of organic compounds in water.

The subjects that affect ozone kinetics in water are detailed in 11 chapters. Chapter 1 presents a short history of naturally occurring ozone and explains the electronic structure of the ozone molecule, which is responsible for ozone reactivity.

Chapter 2 reviews the chemistry of ozone reactions in water by studying direct and indirect or free radical reaction types. Chapter 3 focuses on the kinetics of direct ozone reactions and explains that these studies can be developed through experimental homogeneous and heterogeneous ozone reactions. Chapters 4 and 5 continue with studies on direct ozone reaction kinetics, but they deal exclusively with heterogeneous gas–liquid reaction kinetics, which represents the way ozone is applied in water and wastewater treatment — that is, in gas form. Chapter 4 presents the fundamentals of the kinetics of these reactions and includes detailed explanations of the kinetic equations of gas–liquid reactions, which are later applied to ozone direct reaction kinetic studies in Chapter 5. Chapter 5 discusses examples of kinetic works on ozone gas–water reactions, starting with the fundamental tools to accomplish this task: the properties of ozone in water, such as solubility and diffusivity. The ozone kinetic studies are presented according to the kinetic regimes of ozone absorption that, once established, allow the rate constant and mass transfer coefficients to be determined. Chapter 6 focuses on wastewater ozonation reactions, including classification of wastewater according to its reactivity with ozone, characterizing parameters, the importance of pH, and the influence of ozonation on biological processes. Chapter 6 also addresses the kinetics of wastewater ozone reactions and provides insight into experimental studies in this field.

Chapters 7 through 9 examine the kinetics of indirect ozone reactions that can also be considered advanced oxidation reactions involving ozone: ozone alone and ozone combined with hydrogen peroxide and UV radiation. Chapter 7 discusses indirect reactions that result from the decomposition of ozone (without the addition of hydrogen peroxide or UV radiation). Chapter 7 begins with a study of the relative importance of ozone direct and decomposition reactions whose results are fundamental to establishing the overall kinetics of any ozone–compound B reaction. Chapter 7 also explores methods to determine the rate constant of the reactions between the hydroxyl free radical and any compound B, and the characteristic relationships of natural water to ozone reactivity. Chapter 8 explains the kinetic study of ozone–hydrogen peroxide processes, including those aspects related to the rate constant determination, kinetic regimes, and competition with direct ozone reactions. Chapter 9 focuses on the UV radiation/ozone processes: the direct photolytic and UV radiation/hydrogen peroxide processes. The latter process is also important because it is present when ozone and UV radiation are simultaneously applied. Chapter 9 includes methods to determine quantum yields, rate constants of hydroxyl radical reactions, and multiple aspects of the relative importance of different reactions; ozone direct reactions, ozone–peroxide reactions, and ozone direct photolysis, among other subjects.

Chapter 10 discusses the state of the art of heterogeneous catalytic ozonation. Although this field dates from the 1970s, the past decade has witnessed a considerable increase in work on heterogeneous catalytic ozonation. Chapter 10 details the fundamentals of the kinetics of these gas–liquid–solid catalytic reactions, followed by applications to the catalytic ozonation of compounds in water. An extensive, annotated list of published studies on this ozone action is provided in table format. Chapter 11 presents the kinetic modeling of ozone reactions, beginning with a detailed classification of possible ozone kinetic modeling based on the different kinetic regimes of ozone absorption. Mathematical models are presented together with the ways in which they can be solved, together with examples from the literature on ozone. The focus is on studies of ozone reactions on model compounds, which are more related to drinking water treatment and wastewater ozonation. The appendices provide mathematical tools, concepts on ideal reactors and actinometry, and nonideal flow studies needed to solve and understand the ozonation kinetic examples previously developed.

About the Author

Fernando Juan Beltrán Novillo, Ph.D., received his doctorate in chemistry in 1982 from the University of Extremadura in Badajoz, Spain. In 1986, he became Professor Titular in Chemical Engineering at the University of Extremadura. In 1985 and 1986, he did postdoctoral work at the Laboratoire de Chimie de l'eau et de Nuisances at the University of Poitiers (France), where he worked with Professors Marcel Doré, Bernard Legube, and Jean-Philippe Croué on the ozonation of natural fulvic substances and its effect on trihalomethane formation. In 1988 and 1989, he researched the catalytic combustion of PCBs and catalytic wet air oxidation with Professors Stan Kolaczkowski and Barry Crittenden at the School of Chemical Engineering, University of Bath (U.K.). He did further research with Professor William H. Glaze on the UV radiation/hydrogen peroxide oxidation system in the Department of Environmental Science and Engineering at the University of North Carolina in 1991.

Dr. Beltrán became Catedratico (Professor) in Chemical Engineering at the University of Extremadura in 1992. In 1993, he was a Visiting Professor at the University of Bath.

Dr. Beltrán has published more than 100 papers on ozonation, most of them on kinetics. He has co-supervised 13 doctoral theses, primarily on the ozonation kinetics of model compounds and wastewaters.

Dr. Beltrán is a member of the International Ozone Association and a member of the editorial board of *Ozone Science and Engineering* and *International Water Quality*. He has collaborated in the peer-review process of many scientific and engineering journals, such as *Ozone Science and Engineering, Industrial Engineering Chemistry Research, Environmental Science and Technology, Water Research*, and *Applied Catalysis B*.

Dr. Beltrán teaches courses on chemical reaction engineering to undergraduate students and ozone reaction kinetics in water to postgraduate students at the University of Extremadura, where he is also director of a research group on water treatment.

Nomenclature

a	Specific interfacial area in gas–liquid systems, s^{-1}
a_c	External surface area per unit of catalyst mass, m^2g^{-1}
A	Absorbance, dimensionless
Acc	Accumulation rate term, $mols^{-1}$, see Equation (5.32)
Alk	Alkalinity of any surface water, mgL^{-1} $CaCO_3$, see Equations (7.31) to (7.34)
BOD	Biological oxygen demand, mgL^{-1}
C	Concentration, M or mgL^1
COD	Chemical oxygen demand, mgL^{-1}
cosh(x)	Hyperbolic cosine of x, dimensionless, see Appendix A2
D	Molecular diffusivity, m^2s^{-1}, or axial dispersion coefficient, m^2s^{-1}
D_{eA}	Effective diffusivity, m^2s^{-1}, defined in Equation (10.21)
Dam	Damkohler number, dimensionless, defined in Equation (A3.18)
DF	Depletion factor, dimensionless, defined in Equation (11.22)
DOC	Dissolved organic carbon, mgL^{-1}
E	Reaction factor, dimensionless, defined in Equation (4.31), energy of radiation, J, or residence time distribution function, s^{-1}, defined in Equation (A3.2)
E_i	Instantaneous reaction factor, dimensionless, defined in Equation (4.46), (4.67), or (4.68)
E_0	Radiant energy of the lamp, $Einstein.cm^{-1}s^{-1}$
f	Fugacity, defined in Equation (5.15) or (5.16)
F	Molar rate, $mols^{-1}$, or fraction of absorbed radiation, dimensionless, defined in Equation (9.12), or F function of a distribution, dimensionless, defined in Equation (A3.3)
g	Gravity constant, m^2s^{-1}
G	Gibbs free energy, J, or generation rate term, Ms^{-1}
h	Height of a column, m, or salting-out coefficient of an ionic species, M^{-1}, see Equation (5.22)
h_G	Salting-out coefficient of gas species, M^{-1}, defined in Equation (5.23)
h_T	Parameter defined in Equation (5.23), $M^{-1}K^{-1}$
H	Total height of a column, m
H_A	Heat of absorption of a gas, $Jmol^{-1}$
Ha_1	Hatta number of a first-order gas–liquid reaction, dimensionless, defined in Equation (4.20)

Ha_2	Hatta number of a second-order gas–liquid reaction, dimensionless, defined in Equation (4.40)
Ha_s	Modified Hatta number for series-parallel gas–liquid reactions, dimensionless, defined in Equation (4.54)
He	Henry constant, PaM^{-1}, see Equation (4.78)
He_{ap}	Apparent Henry constant, PaM^{-1}, defined in Equation (5.19)
HeCO	Heterogeneous catalytic ozonation
HoCO	Homogeneous catalytic ozonation
I	Ionic strength, M^1, defined in Equation (5.21)
I_a	Local rate of absorbance radiation, Einstein $L^{-1}s^{-1}$
I_0	Intensity of incident radiation, Einstein $L^{-1}s^{-1}$
IC	Inorganic carbon, mgL^{-1}
k	Chemical reaction rate constant, s^{-1} or $M^{-1}s^{-1}$
k_c	Individual liquid–solid coefficient, ms^{-1}, see Equation (10.16)
k_G	Gas phase individual mass transfer coefficient, $mol.s^{-1}m^{-2}Pa^{-1}$
k_L	Liquid phase individual mass transfer coefficient, ms^{-1}
$k_L a$	Liquid phase volumetric mass transfer coefficient, s^{-1}
k_v	Volatility coefficient, s^{-1}
K_S	Sechenov constant, M^{-1}, see Equation (5.19)
L	Effective path of radiation through a photoreactor, m
M	molar rate, $mols^{-1}$
M_1	Maximum physical absorption rate, $mols^{-1}m^{-2}$, defined in Equation (4.22)
M_2	Maximum physical diffusion through film layer, $mols^{-1}m^{-2}$, defined in Equation (4.43)
MCL	Maximum contaminant level
MOC	Mean oxidation number of carbon, dimensionless
MW	Molecular weight, $gmol^{-1}$, see Equation (5.1)
n	Molar amount, mol
N	Absorption rate or flux of a component, $mols^{-1}m^{-2}$
N_{AV}	Avogadro's number, molecules mol^{-1}
N_D	Dispersion number, dimensionless, defined in Equation (11.60)
NOM	Natural organic matter
NPOC	Nonpurgeable organic carbon, mgL^{-1}
P	Pressure, Pa
Pe	Peclet number, dimensionless, defined in Equation (11.60)
POC	Purgeable organic carbon, mgL^{-1}
q	Density flux of radiation, Einstein $m^{-2}s^{-1}$
q_0	Density flux of radiation at the internal wall of photoreactor, Einstein $m^{-2}s^{-1}$

r	Chemical reaction rate, Ms^{-1}
R	Gas perfect constant, $Pa m^3 K^{-1} mol^{-1}$, or catalyst particle radius, m
R_{b1}	Maximum chemical reaction rate in bulk liquid for a first-order reaction, $mol s^{-1} m^{-2}$, defined in Equation (4.24)
R_{b2}	Maximum chemical reaction rate in bulk liquid for a second-order reaction, $mol s^{-1} m^{-2}$, defined in Equation (4.41)
R_{CT}	Coefficient defined in Equation (7.61), dimensionless
R_F	Maximum chemical reaction rate in bulk liquid, $mol s^{-1} m^{-2}$
R_0	Internal wall radius of photoreactor, m
s	Surface renewal velocity, s^{-1}, see Equation (4.12)
S	Entropy, JK^{-1}, defined in Equation (5.7) or surface section of a column, m^2, or solubility ratio for ozone–water equilibrium, dimensionless, see Equation (5.24)
Sc	Schmidt number, dimensionless, defined in Equation (5.40)
S_g	Internal surface area of a porous catalyst, $m^2 g^{-1}$
sinh(x)	Hyperbolic sine of x, dimensionless, see Appendix A2
SOC	Suspended or particulate organic carbon, mgL^{-1}
SS	Suspended solids, mgL^{-1}
t	Reaction time, s or min
t_D	Diffusion time, s, defined in Equation (4.84)
t_i	Time needed to reach steady-state conditions, s, see Equation (5.81)
t_m	Mean residence time of a distribution function, s, defined in (A3.4)
t_R	Reaction time, s, defined in Equation (4.85) or (4.86)
T	Temperature, K
tanh(x)	Hyperbolic tangent of x, dimensionless, see Appendix A2
ThOD	Theoretical oxygen demand, mgL^{-1}
TOC	Total organic carbon, mgL^{-1}
U or u	Superficial velocity in a column, ms^{-1}
v	Flow rate, $m^3 s^{-1}$
V	Reaction volume, m^3
V_A	Molar volume of diffusing solute, $cm^3 mol^{-1}$, see Equation (5.1)
w	Parameter defined in Equation (7.22), dimensionless, or catalyst concentration, mgL^{-1}
x	Depth of liquid penetration from the gas–liquid interface, m, or liquid molar fraction, dimensionless
y	Gas molar fraction, dimensionless
z	Stoichiometric coefficient, dimensionless, or valency of an ionic species, dimensionless

GREEK LETTERS

α	Degree of dissociation, dimensionless, defined in Equation (3.21), or parameter defined in Equation (5.52), dimensionless
β	Liquid holdup, dimensionless, defined in Equation (5.43) or (5.44) or Bunsen coefficient for ozonewater equilibrium, see Equation (5.24)
χ	Parameter defined in Equation (4.63) for surface renewal theory, dimensionless,
δ	Phase film, m, see Equation (4.7)
ε	Extinction coefficient, base 10, $M^{-1}cm^{-1}$
ε_{O3}	Rate coefficient of wastewater ozonation, $Lmg^{-1}s^{-1}$, defined in Equation (6.7)
ε_p	Catalyst particle porosity, dimensionless
φ	Dimensionless concentration in a porous catalyst, defined in Equation (10.25)
ϕ_1	Thiele number for a first-order fluid–solid catalytic reaction, dimensionless, defined in Equation (10.26)
ϕ_s	Association parameter of solvent, dimensionless, see Equation (5.1)
Φ	Quantum yield, mol Einstein^{-1}, defined in Equation (9.14)
γ	Activity coefficient, M, see Equation (5.17)
Γ	Global effectiveness factor for a fluid–solid catalytic reaction, dimensionless, defined in Equation (10.31)
η	Effectiveness factor for a fluidsolid catalytic reaction, dimensionless, defined in Equation (10.29)
ς	Parameter defined in Equation (6.23), $(mgm)^{-1/2}$
λ	Dimensionless distance defined in Equation (11.66) or (A3.17)
μ	Attenuation coefficient, cm^{-1}, defined in Equation (9.7)
μ_L	Liquid viscosity, $kgm^{-1}s^{-1}$, see Equation (5.40)
μ_s	Solvent viscosity, poise, see Equation (5.1)
μ_i^{phase}	Chemical potential of the i component in a given phase, defined in Equation (5.10), Pam^3mol^{-1}
ν	Fugacity coefficient, dimensionless, see Equation (5.15)
θ	Dimensionless reaction time, defined in Equation (11.66)
ρ_L	Liquid density, kgm^{-3}
ρ_p	Apparent density of a catalyst particle, kgm^{-3}
ρ_b	Bulk density of a catalyst bed, kgm^{-3}
σ_L	Surface tension of liquid, kgm^3s^{-1}
σ^2	Standard deviation of a distribution function, s^2, defined in Equation (A3.5)
σ^2_θ	Dimensionless standard deviation of a distribution function, defined in Equation (A3.11)
τ	Hydraulic residence time, s
τ_p	Catalyst particle tortuosity, dimensionless

ϑ	Parameter defined in Equation (5.50), dimensionless
ψ	Dimensionless concentration defined in Equation (11.66)
ψ(t)	Surface renewal distribution function, s^{-1}, defined in Equation (4.11)
Ω	Oxidation competition coefficient, mgL^{-1}, defined in Equation (7.45)

SUPERINDEXES

g	Refers to the gas phase
l	Refers to the liquid phase
m	Reaction order, dimensionless
n	Reaction order, dimensionless
*	Refers to gas–liquid equilibrium conditions

SUBINDEXES

A	Refers to any compound A
ap	Refers to an apparent value of a given parameter
b	Refers to bulk phase
B	Refers to any compound B
bg	Refers to band gap in semiconductor photocatalysis
c, c1, c2	Refers to Equation (7.18) and Reactions (7.16) and (7.17) between carbonate species and the hydroxyl radical
cb	Refers to conduction band in semiconductor photocalysis
CH, CH1, CH2	Refers to Equation (7.23) and reactions between hydrogen peroxide species and the carbonate ion radical
CM	Refers to any reaction between the carbonate ion radical and any substance present in water but hydrogen peroxide
D	Refers to direct ozone reaction
Dd	Refers to the direct reaction between the dissociated form of a given compound and ozone, see Equation (8.20)
Dn	Refers to the direct reaction between the nondissociated form of a given compound and ozone, see Equation (8.20)
g	Refers to the gas phase
G	Refers to a global value of a given parameter or to the gas phase
HCO3t	Refers to total bicarbonate
HO	Refers to hydroxyl radicals
HOB	Refers to the reaction between hydroxyl radicals and a compound B
H2O2t	Refers to total hydrogen peroxide
i	Refers to any component of water or to gas–liquid interface conditions or to reactor inlet conditions
i.S	Refers to an adsorbed i species on a catalyst surface

Ii	Refers to any compound that initiates the decomposition of ozone in water, see Equation (2.70)
L	Refers to the liquid phase
Mi	Refers to any compound that directly reacts with ozone, see Equation (2.70)
o	Refers to reactor outlet conditions
O_3	Refers to ozone
O_{3_l}	Refers to ozone in water
O_{3_g}	Refers to gaseous ozone
P	Refers to any products from ozone direct reactions
Pi	Refers to any compound that promotes the decomposition of ozone in water, see Equation (2.70)
Rad	Refers to free radical reactions
rel	Refers to a relative value between parameters, see Equation (3.16)
S, s	Refers to any scavenger of hydroxyl radicals
Si	Refers to any compound that inhibits the decomposition of ozone in water, see Equation (2.70)
t	Refers to total active centers of a catalyst surface, see Equation (10.11)
T	Refers to a tracer compound for nonideal flow studies (see Appendix A3) or total conditions
UV	Refers to UV radiation
v	Refers to free active centers of a catalyst surface
vb	Refers to valence band in semiconductor photocatalysis
vgi	Refers to any i volatile compound in the gas phase
vi	Refers to any i volatile compound dissolved in water
0	Refers to initial conditions or conditions at reactor inlet

Contents

Chapter 1 Introduction 1
1.1 Ozone in Nature 2
1.2 The Ozone Molecule 3
References 5

Chapter 2 Reactions of Ozone in Water 7
2.1 Oxidation–Reduction Reactions 7
2.2 Cycloaddition Reactions 9
2.3 Electrophilic Substitution Reactions 11
2.4 Nucleophilic Reactions 13
2.5 Indirect Reactions of Ozone 14
     2.5.1 The Ozone Decomposition Reaction 19
References 26

Chapter 3 Kinetics of the Direct Ozone Reactions 31
3.1 Homogeneous Ozonation Kinetics 33
     3.1.1 Batch Reactor Kinetics 33
     3.1.2 Flow Reactor Kinetics 39
     3.1.3 Influence of pH on Direct Ozone Rate Constants 40
     3.1.4 Determination of the Stoichiometry 42
3.2 Heterogeneous Kinetics 43
     3.2.1 Determination of the Stoichiometry 44
References 44

Chapter 4 Fundamentals of Gas–Liquid Reaction Kinetics 47
4.1 Physical Absorption 47
     4.1.1 The Film Theory 48
     4.1.2 Surface Renewal Theories 50
4.2 Chemical Absorption 50
     4.2.1 Film Theory 51
            4.2.1.1 Irreversible First-Order or Pseudo First-Order Reactions 51
            4.2.1.2 Irreversible Second-Order Reactions 54
            4.2.1.3 Series-Parallel Reactions 58
     4.2.2 Danckwerts Surface Renewal Theory 62
            4.2.2.1 First-Order or Pseudo First-Order Reactions 62
            4.2.2.2 Irreversible Second-Order Reactions 62
            4.2.2.3 Series-Parallel Reactions 65

4.2.3　Influence of Gas Phase Resistance 65
　　　　　　4.2.3.1　Slow Kinetic Regime 66
　　　　　　4.2.3.2　Fast Kinetic Regime 66
　　　4.2.4　Diffusion and Reaction Times 67
References 68

Chapter 5　Kinetic Regimes in Direct Ozonation Reactions 69

5.1　Determination of Ozone Properties in Water 69
　　　5.1.1　Diffusivity 69
　　　5.1.2　Ozone Solubility: The Ozone–Water Equilibrium System 71
5.2　Kinetic Regimes of the Ozone Decomposition Reaction 80
5.3　Kinetic Regimes of Direct Ozonation Reactions 83
　　　5.3.1　Checking Secondary Reactions 84
　　　5.3.2　Some Common Features of the Kinetic Studies 84
　　　　　　5.3.2.1　The Ozone Solubility 87
　　　　　　5.3.2.2　The Individual Liquid Phase Mass-Transfer Coefficient,
　　　　　　　　　　　k_L 88
　　　5.3.3　Instantaneous Kinetic Regime 89
　　　5.3.4　Fast Kinetic Regime 92
　　　5.3.5　Moderate Kinetic Regime 101
　　　　　　5.3.5.1　Case of No Dissolved Ozone 102
　　　　　　5.3.5.2　Case of Pseudo First-Order Reaction with Moderate
　　　　　　　　　　　Kinetic Regime 102
　　　5.3.6　Slow Kinetic Regime 102
　　　　　　5.3.6.1　The Slow Diffusional Kinetic Regime 105
　　　　　　5.3.6.2　Very Slow Kinetic Regime 105
5.4　Changes of the Kinetic Regimes during Direct Ozonation Reactions 107
5.5　Comparison between Absorption Theories in Ozonation Reactions 107
References 109

Chapter 6　Kinetics of the Ozonation of Wastewaters 113

6.1　Reactivity of Ozone in Wastewater 118
6.2　Critical Concentration of Wastewater 120
6.3　Characterization of Wastewater 121
　　　6.3.1　The Chemical Oxygen Demand 122
　　　6.3.2　The Biological Oxygen Demand 123
　　　6.3.3　Total Organic Carbon 123
　　　6.3.4　Absorptivity at 254 nm (A254) 124
　　　6.3.5　Mean Oxidation Number of Carbon 124
6.4　Importance of pH in Wastewater Ozonation 125
6.5　Chemical Biological Processes 129
　　　6.5.1　Biodegradability 130
　　　6.5.2　Sludge Settling 132
　　　6.5.3　Sludge Production 132
6.6　Kinetic Study of the Ozonation of Wastewaters 133

        6.6.1   Establishment of the Kinetic Regime of Ozone Absorption 134
        6.6.2   Determination of Ozone Properties for the Ozonation Kinetics
                of Wastewater 136
        6.6.3   Determination of Rate Coefficients for the Ozonation Kinetics
                of Wastewater 140
                6.6.3.1   Fast Kinetic Regime (High COD) 141
                6.6.3.2   Slow Kinetic Regime (Low COD) 143
References 145

Chapter 7   Kinetics of Indirect Reactions of Ozone in Water 151

7.1   Relative Importance of the Direct Ozone–Compound B Reaction and the
      Ozone Decomposition Reaction 152
      7.1.1   Application of Diffusion and Reaction Time Concepts 152
7.2   Relative Rates of the Oxidation of a Given Compound 154
7.3   Kinetic Parameters 156
      7.3.1   The Ozone Decomposition Rate Constant 157
              7.3.1.1   Influence of Alkalinity 159
      7.3.2   Determination of the Rate Constant of the OH–Compound B
              Reaction 160
              7.3.2.1   The Absolute Method 161
              7.3.2.2   The Competitive Method 162
7.4   Characterization of Natural Waters Regarding Ozone Reactivity 163
      7.4.1   Dissolved Organic Carbon, pH, and Alkalinity 163
      7.4.2   The Oxidation–Competition Value 164
      7.4.3   The R_{CT} Concept 170
References 172

Chapter 8   Kinetics of the Ozone/Hydrogen Peroxide System 175

8.1   The Kinetic Regime of the O_3/H_2O_2 Process 176
      8.1.1   Slow Kinetic Regime 177
      8.1.2   Fast-Moderate Kinetic Regime 177
      8.1.3   Critical Hydrogen Peroxide Concentration 178
8.2   Determination of Kinetic Parameters 180
      8.2.1   The Absolute Method 180
      8.2.2   The Competitive Method 181
      8.2.3   The Effect of Natural Substances on the Inhibition of Free
              Radical Ozone Decomposition 181
8.3   The Ozone/Hydrogen Peroxide Oxidation of Volatile Compounds 182
8.4   The Competition of the Direct Reaction 183
      8.4.1   Comparison between the Kinetic Regimes of the Ozone–
              Compound B and Ozone–Hydrogen Peroxide Reactions 183
      8.4.2   Comparison between the Rates of the Ozone–Compound B and
              Hydroxyl Radical–Compound B Reactions 186
      8.4.3   Relative Rates of the Oxidation of a Given Compound 190
References 191

Chapter 9 Kinetics of the Ozone–UV Radiation System 193

9.1 Kinetics of the UV Radiation for the Removal of Contaminants from Water 193
    9.1.1 The Molar Absorptivity 194
    9.1.2 The Quantum Yield 194
    9.1.3 Kinetic Equations for the Direct Photolysis Process 195
    9.1.4 Determination of Photolytic Kinetic Parameters: The Quantum Yield 199
        9.1.4.1 The Absolute Method 199
        9.1.4.2 The Competitive Method 201
    9.1.5 Quantum Yield for Ozone Photolysis 201
        9.1.5.1 The Ozone Quantum Yield in the Gas Phase 204
        9.1.5.2 The Ozone Quantum Yield in Water 205
9.2 Kinetics of the UV/H_2O_2 System 206
    9.2.1 Determination of Kinetic Parameters 206
        9.2.1.1 The Absolute Method 207
        9.2.1.2 The Competitive Method 208
    9.2.2 Contribution of Direct Photolysis and Free Radical Oxidation in the UV/H_2O_2 Oxidation System 209
9.3 Comparison between the Kinetic Regimes of the Ozone–Compound B and Ozone–UV Photolysis Reactions 211
    9.3.1 Comparison between Ozone Direct Photolysis and the Ozone Direct Reaction with a Compound B through Reaction and Diffusion Times 211
    9.3.2 Contributions of Direct Photolysis and Direct Ozone Reaction to the Ozone Absorption Rate 214
        9.3.2.1 Strong UV Absorption Exclusively due to Dissolved Ozone 215
        9.3.2.2 Strong UV Absorption due to Dissolved Ozone and a Compound B 216
        9.3.2.3 Weak UV Absorption 216
    9.3.3 Contributions of the Direct Ozone and Free Radical Reactions to the Oxidation of a Given Compound B 216
        9.3.3.1 Strong UV Absorption Exclusively due to Dissolved Ozone 217
        9.3.3.2 Strong UV Absorption due to Dissolved Ozone and a Compound B 218
        9.3.3.2 Weak UV Absorption 218
    9.3.4 Estimation of the Relative Importance of the Rates of the Direct Photolysis/Direct Ozonation and Free Radical Oxidation of a Compound B 218
        9.3.4.1 Relative Importance of Free Radical Initiation Reactions in the UV/O_3 Oxidation System 219
        9.3.4.2 Relative Importance of the Direct Reactions and Free Radical Oxidation Rates of Compound B 221
References 224

Chapter 10 Heterogeneous Catalytic Ozonation 227

10.1 Fundamentals of Gas–Liquid–Solid Catalytic Reaction Kinetics 241
    10.1.1 Slow Kinetic Regime 242
    10.1.2 Fast Kinetic Regime or External Diffusion Kinetic Regime 245
    10.1.3 Internal Diffusion Kinetic Regime 246
    10.1.4 General Kinetic Equation for Gas–Liquid–Solid Catalytic
           Reactions 249
    10.1.5 Criteria for Kinetic Regimes 250
10.2 Kinetics of Heterogeneous Catalytic Ozone Decomposition in
     Water 251
10.3 Kinetics of Heterogeneous Catalytic Ozonation of Compounds in
     Water 258
    10.3.1 The Slow Kinetic Regime 259
    10.3.2 External Mass Transfer Kinetic Regime 261
        10.3.2.1 Catalyst in Powder Form 262
        10.3.2.2 Catalyst in Pellet Form 263
    10.3.3 Internal Diffusion Kinetic Regime 263
        10.3.3.1 Determination of the Effective Diffusivity and
                 Tortuosity Factor of the Porous Catalyst 264
        10.3.3.2 Determination of the Rate Constant of the Catalytic
                 Reaction 264
10.4 Kinetics of Semiconductor Photocatalytic Processes 265
    10.4.1 Mechanism of TiO_2 Semiconductor Photocatalysis 267
    10.4.2 Langmuir–Hinshelwood Kinetics of Semiconductor
           Photocatalysis 268
    10.4.3 Mechanism and Kinetics of Photocatalytic Ozonation 269
References 271

Chapter 11 Kinetic Modeling of Ozone Processes 277

11.1 Case of Slow Kinetic Regime of Ozone Absorption 279
11.2 Case of Fast Kinetic Regime of Ozone Absorption 281
11.3 Case of Intermediate or Moderate Kinetic Regime of Ozone
     Absorption 283
11.4 Time Regimes in Ozonation 285
11.5 Influence of the Type of Water and Gas Flows 286
11.6 Mathematical Models 288
    11.6.1 Slow Kinetic Regime 289
        11.6.1.1 Both Gas and Water Phases in Perfect Mixing Flow 289
        11.6.1.2 Both Gas and Water Phases in Plug Flow 291
        11.6.1.3 The Water Phase in Perfect Mixing Flow and the Gas
                 Phase in Plug Flow 299
        11.6.1.4 The Water Phase as N Perfectly Mixed Tanks in Series
                 and the Gas Phase in Plug Flow 300

- 11.6.1.5 Both the Gas and Water Phases as N and N' Perfectly Mixed Tanks in Series 301
- 11.6.1.6 Both the Gas and Water Phases with Axial Dispersion Flow 303
- 11.6.2 Fast Kinetic Regime 305
  - 11.6.2.1 Both the Water and Gas Phases in Perfect Mixing 307
  - 11.6.2.2 The Gas Phase in Plug Flow and the Water Phase in Perfect Mixing Flow 307
  - 11.6.2.3 Both the Gas and Water Phases in Plug Flow 308
- 11.6.3 The Moderate Kinetic Regime: A General Case 309
- 11.7 Examples of Kinetic Modeling for Model Compounds 312
- 11.8 Kinetic Modeling of Wastewater Ozonation 316
  - 11.8.1 Case of Slow Kinetic Regime: Wastewater with Low COD 317
    - 11.8.1.1 Kinetic Modeling of Wastewater Ozonation without Considering a Free Radical Mechanism 317
    - 11.8.1.2 Kinetic Modeling of Wastewater Ozonation Considering a Free Radical Mechanism 318
  - 11.8.2 Case of Fast Kinetic Regime: Wastewater with High COD 322
  - 11.8.3 A General Case of Wastewater Ozonation Kinetic Model 324
- References 326

Appendices 331
Appendix A1 Ideal Reactor Types: Design Equations 331
- A1.1 Perfectly Mixed Reactor 331
- A1.2 Plug Flow Reactor 333

Appendix A2 Useful Mathematical Functions 334
- A2.1 Hyperbolic Functions 334
- A2.2 The Error Function 335

Appendix A3 The Influence of the Type of Flow on Reactor Performance 335
- A3.1 Nonideal Flow Study 335
  - A3.1.1 Fundamentals of RTD Function 336
    - A3.1.1.1 Determination of the E Function ... 336
    - A3.1.1.2 Moments of the RTD 338
  - A3.1.2 RTD Functions of Ideal Flows through the Reactors 338
- A3.2 Some Fluid Flow Models 339
  - A3.2.1 The Perfectly Mixed Tanks in Series Model ... 340
  - A3.2.2 The Axial Dispersion Model 340
- A3.3 Ozone Gas as a Tracer 342

Appendix A4 Actinometry 342
- A4.1 Determination of Intensity of Incident Radiation 343
- A4.2 Determination of the Effective Path of Radiation 344

Appendix A5   Some Useful Numerical Procedures 345
        A5.1   The Newton–Raphson Method for a Set of Nonlinear
                Algebraic Equations 345
        A5.2   The Runge–Kutta Method for a Set of Nonlinear
                First-Order Differential Equations 348

References 349

Index 351

1 Introduction

In the late 1970s, the discovery of trihalomethanes (THM) in drinking water due to chlorination of natural substances present in the raw water[1,2] gave rise to two different research lines: the identification of the structures of these natural substances (i.e., humic substances) and the formation of organochlorine compounds from their chlorination.[3,4] The search began for alternative oxidant-disinfectants that could play the role of chlorine without generating the problem of trihalomethane formation.[5,6] This latter research line led to numerous studies on the use of ozone in drinking water treatment and the study of the kinetics of ozonation reactions in water. This research line is still productive. Recent surveys[7,8] have shown that organohalogen compounds formed in the treatment of surface waters with chlorine and other chlorine-derived oxidant-disinfectants (i.e., chloramines) yield a greater number of disinfection by-products than ozone. However, chlorine is not the only factor affecting water contamination. Other compounds are often discharged in natural waters or in soils and then migrate to underground water. The result is contamination of wells, aquifers, etc. The literature reports underground contamination from compounds such as volatile aromatics including benzene, toluene, xylenes (BTX); methyltertbutylether (MTBE); and volatile organochlorinated compounds.[9–12] Ozonation or advanced oxidation processes (AOP) or hydroxyl radical oxidant–based processes, among others, have proven to be efficient technologies for the removal of these types of pollutants from water.[13–16]

The application of ozone is not exclusive to the treatment of drinking water. Ozone also has numerous applications for the treatment of wastewater. Here, chlorine is mainly used for disinfection purposes, leading to many problems in the aquatic environment where treated wastewater is released.[17] Thus, organochlorine compounds generated from wastewater chlorination can harm aquatic organisms in receiving waters. The U.S. Environmental Protection Agency (EPA) has established a limit of less than 11 µg/L for total residual chlorine in fresh water,[18] which is usually surpassed when chlorinated wastewater is discharged.[19] Thus, wastewater treatment plant operators must often balance two contradictory aspects: the use of chlorine for wastewater disinfection and the preservation of aquatic life. Thus, alternative oxidant-disinfectant agents are needed for wastewater treatment. As shown in Chapter 6, ozone has been used in the treatment of a variety of wastewater. It should be highlighted, however, that ozone, like other oxidants, also produces by-products such as bromate (in water containing bromide), which can be harmful.[20] The EPA promulgated the Stage 1 Disinfectants/Disinfection By-Products (D/DBP) Rule to regulate the MCL of bromate (10 µg/L), chlorite (1 mg/L), THMs (80 µg/L), and haloacetic acids (10 µg/L).[21] This rule took effect on January 1, 2002 but the EPA plans to reexamine the bromate MCL in its 6-year review process.[22] So, when using ozone in the treatment of water some care must be taken to eliminate or reduce DBPs as much as possible.

Contrary to what might be assumed from this history, the use of ozone in the treatment of drinking water was not new when THMs were discovered in chlorinated drinking water. In fact, ozone started to be used, mainly as a disinfectant, in the late 19th century in many water treatment plants in Europe.[23] The fact that chlorine was the main oxidant-disinfectant agent was due, among other reasons, to both extensive studies on its use during World War I for chemical weapons and its low cost.

Today, however, there are numerous water treatment plants, mainly for drinking water, that include some ozonation step in their treatment lines. In addition, interest in the kinetics of these processes has been growing because of the dual practical–academic aspects. Since ozonation of compounds in water is a gas–liquid heterogeneous reaction, the process is of great academic interest because it is one of the few practical cases outside the chemical industry in which different chemical reaction engineering concepts (mass transfer, chemical kinetics, reactor design, etc.) apply.

Data on ozonation and related processes (i.e., advanced oxidation processes) are also of practical interest for addressing the design of ozone reactors or contact times to achieve a given reduction in water pollution, improvement of wastewater biodegradability during conventional biological oxidation, or increased settling rate in sedimentation.

Ozone applications in the treatment of water and wastewater can be grouped into three categories: disinfectants or biocides, classical oxidants to remove organic pollutants, and pre- or posttreatment agents to aid in other unit operations (coagulation, flocculation, sedimentation, biological oxidation, carbon adsorption, etc.).[24–28]

1.1 OZONE IN NATURE

In 1785, the odor released from the electric discharges of storms led Van Mauren, a Dutch chemist, to suspect the presence of a new compound. In 1840, Christian Schonbein finally discovered ozone although its chemical structure as a triatomic oxygen molecule was not confirmed until 1872,[23] and in 1952 it was established as a hydride resonance structure.[29]

Ozone is formed naturally in the upper zones of the atmosphere (about 25 km above sea level and a few kilometers wide) where it surrounds the Earth and protects the surface of the planet from UV-B and UV-C radiation. The spontaneous generation of ozone is due to the combination of bimolecular and atomic oxygen, a reaction that starts to develop from approximately 70 km high above sea level down to about 20 km from the Earth's surface where unfavorable conditions are established. In the atmosphere close to the Earth's surface, however, ozone is a toxic compound with a maximum contaminant level of 0.1 ppm for an exposure of at least 8 h.[30] From the positive point of view, the properties of ozone, derived from its reactivity, have been applied in the treatment of water in medicine, organic chemical synthesis, etc.

Introduction

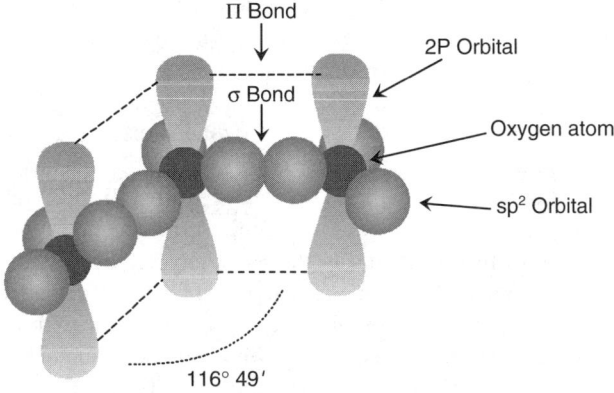

FIGURE 1.1 The molecular structure of ozone.

1.2 THE OZONE MOLECULE

Ozone reactivity is due to the structure of the molecule. The ozone molecule consists of three oxygen atoms. Each oxygen atom has the following electronic configuration surrounding the nucleus: $1s^2\ 2s^2\ 2p_x^2\ 2p_y^1\ 2p_z^1$, i.e., in its valence band it has two unpaired electrons, each one occupying one $2p$ orbital. In order to combine three oxygen atoms and yield the ozone molecule, the central oxygen rearranges in a plane sp^2 hybridization from the $2s$ and two $2p$ atomic orbitals of the valence band. With this rearrangement the three new sp^2 hybrid orbitals form an equilateral triangle with an oxygen nucleus in its center, i.e., with an angle of 120° between the orbitals. However, in the ozone molecule this angle is 116° 49′.[29] The other $2p$ orbital of the valence band stays perpendicular to the sp^2 plane, as Figure 1.1 shows, with two coupled electrons. Two of the sp^2 orbitals from the central oxygen, forming the angle indicated above, combine with one $2p$ orbital (each containing one electron) of the other two adjacent oxygen atoms in the ozone molecule, while the third sp^2 orbital has a couple of nonshared electrons. Finally, the third $2p$ orbital of each adjacent atomic oxygen, which has only one electron, combines with the remaining $2p^2$ orbital of the central oxygen to yield two π molecular orbitals that move throughout the ozone molecule. As a consequence, the ozone molecule represents a hybrid formed by the four possible structures shown in Figure 1.2. The length of the bond between

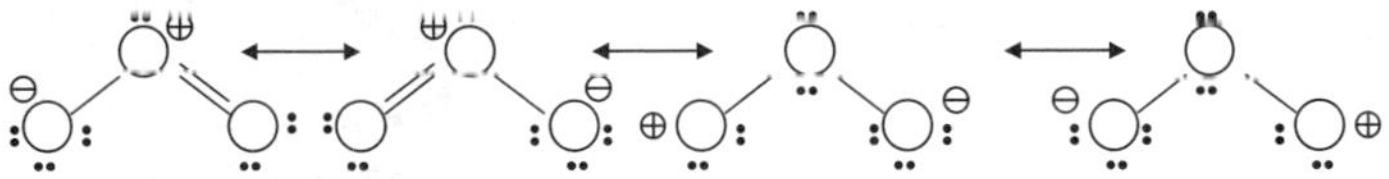

FIGURE 1.2 Resonance forms of the ozone molecule.

oxygen atoms in the ozone molecule has been found experimentally to be 1.278 Å, which is an intermediate value between the length of an oxygen double bond (1.21 Å) and that of a simple oxygen–hydrogen bond in the hydrogen peroxide molecule (1.47 Å). According to the literature[29] the calculated lengths show a 50% likelihood that the bond between oxygen atoms in the ozone molecule is a double bond. Therefore, the resonance structures I and II in Figure 1.2 basically represent the electronic structure of ozone. Nonetheless, resonance forms III and IV also contribute to some extent to the ozone molecule because the ozone angle is lower than 120° due to the attraction of positively and negatively charged adjacent oxygen atoms.

The resonance forms of the ozone molecule confer some sort of polarity. Different properties of molecules (solubility, type of reactivity of bonds, etc.) are partially due to the polarity that is measured with the dipolar momentum. The ozone molecule presents a weak polarity (0.53 D), probably due to the electronegativity of oxygen atoms and the unshared pairs of electrons in some of the orbitals that contribute to the total dipolar momentum in opposing directions.

The high reactivity of ozone can then be attributed to the electronic configuration of the molecule. Thus, the absence of electrons in one of the terminal oxygen atoms in some of the resonance structures confirms the electrophilic character of ozone. Conversely, the excess negative charge present in some other oxygen atom imparts a nucleophilic character. These properties make ozone an extremely reactive compound. Table 1.1 presents some physico-chemical properties of ozone.

TABLE 1.1
Physico-Chemical Properties of Ozone

Property	Value
Melting point, °C	–251
Boiling point, °C	–112
Critical pressure, atm	54.62
Critical temperature, °C	–12.1
Specific gravity	1.658 higher than air
	1.71 gcm^{-3} at –183°C
Critical density, kgm^{-3}	436
Heat of vaporization, calmol^{-1} [a]	2,980
Heat of formation, calmol^{-1} [b]	33,880
Free energy of formation, calmol^{-1} [b]	38,860
Oxidation potential, V[c]	2.07

[a] At the boiling point temperature. [b] At 1 atm and 25°C. [c] At pH = 0.

Source: Perry, R.H. and Green, D.W., *Perry's Chemical Engineers' Handbook*, 7th ed., McGraw-Hill, New York, 1997. With permission.

REFERENCES

1. Rook, J.J., Formation of haloforms during chlorination of natural waters, *Water Treat. Exam.*, 23, 234–243, 1974.
2. Bellar, T.A., Lichtenberg, J.J., and Kroner, R.C., The occurrence of organohalides in chlorinated drinking water, *J. Am. Water Works Assoc.*, 66, 703–706, 1974.
3. Visser, S.A., Comparative study of the elementary composition of fulvic acid and humic acid of aquatic origin and from soils and microbial substrates, *Water Res.*, 17, 1393–1396, 1983.
4. Rook, J.J., Chlorination reactions of fulvic acids in natural waters, *Environ. Sci. Technol.*, 11, 478–482, 1977.
5. Rice, R.G., The use of ozone to control trihalomethanes in drinking water treatment, *Ozone Sci. Eng.*, 2, 75–99, 1980.
6. Croué, J.P., Beltrán, F.J., Legube, B., and Doré, M., Effect of preozonation on the organic halide formation potential of an aquatic fulvic acid, *Ind. Eng. Chem. Res.*, 28, 1082–1089, 1989.
7. Hu, J.H. et al., Disinfection by-products in water produced by ozonation and chlorination, *Environ. Monit. Assess.*, 59, 81–93, 1999.
8. Richardson, S.D. et al., Identification of new drinking water disinfection by-products from ozone, chlorine dioxide, chloramine, and chlorine, *Water Air Soil Pollut.*, 123, 95–102, 2000.
9. Howard, P.H. (Ed.), *Handbook of Environmental Fate and Priority Pollutants. Vol. I, Large Production and Priority Pollutants*, Lewis Publishers, Chelsea, MI, 1989.
10. Howard, P.H. (Ed.), *Handbook of Environmental Fate and Priority Pollutants. Vol. II, Solvents*, Lewis Publishers, Chelsea, 1990.
11. Howard, P.H. (Ed.), *Handbook of Environmental Fate and Priority Pollutants. Vol. III, Pesticides*, Lewis Publishers, Chelsea, MI, 1991.
12. Einarson, M.D. and Mackay, D.M., Water contamination, *Environ. Sci. Technol.*, 35, 67A–73A, 2001.
13. Camel, V. and Bermond, A., The use of ozone and associated oxidation processes in drinking water treatment, *Water Res.*, 32, 3208–3222, 1998.
14. Peyton, G.R. et al., Destruction of pollutants in water by ozone in combination with ultraviolet radiation. 1. General principles and oxidation of tetrachloroethylene, *Environ. Sci. Technol.*, 16, 448–453, 1982.
15. Meijers, R.T. et al., Degradation of pesticides by ozonation and advanced oxidation, *Ozone Sci. Eng.*, 17, 673–686, 1995.
16. Kang, J.W. et al., Sonolytic destruction of methyl tert-butyl ether by ultrasonic irradiation: the role of O_3, H_2O_2, frequency, and power density, *Environ. Sci. Technol.*, 33, 3199–3205, 1999.
17. Lazarova, V. et al., Advanced wastewater disinfection technologies: state of the art and perspectives, *Water Sci. Technol.*, 40, 203–213, 1999.
18. U.S. Environmental Protection Agency, Ambient Water Quality Criteria for Chlorine-1984, EPA 440/5-84-030, U.S. Government Printing Office, Washington, D.C., 1985.
19. Helz, G. and Nweke, A.C., Incompleteness of wastewater dechlorination, *Environ. Sci. Technol.*, 29, 1018–1022, 1995.
20. Legube, B., A survey of bromate ion in European drinking water, *Ozone Sci. Eng.*, 18, 325–348, 1996.
21. National Primary Drinking Water Regulations: disinfectants and disinfection by-products: final Rule, *Fed. Regist.*, 63, 69,389, 1998.

22. Richardson, S.D., Simmons, J.E., and Rice, G., Disinfection by-products: the next generation, *Environ. Sci. Technol.*, 36, 198A–205A, 2002.
23. Langlais, B., Reckhow, D.A., and Brink, D.R. (Eds.), *Ozone in Water Treatment: Application and Engineering*, Lewis Publishers, Chelsea, MI, 1991.
24. Rice, R.G., Ozone in the United States of America — state of the art, *Ozone Sci. Eng.*, 21, 99–118, 1999.
25. Andreozzi, R. et al., Integrated treatment of olive oil mill effluents (OME): study of ozonation coupled with anaerobic digestion, *Water Res.*, 32, 2357–2364, 1998.
26. Boere, J.A., Combined use of ozone and granular activated carbon (GAC) in potable water treatment: effects on GAC quality after reactivation, *Ozone Sci. Eng.*, 14, 123–137, 1992.
27. Beltrán, F.J. et al., Improvement of domestic wastewater sedimentation through ozonation, *Ozone Sci. Eng.*, 21, 605–614, 1999.
28. Beltrán, F.J. et al., Wine-distillery wastewater degradation 1. Oxidative treatment using ozone and its effect on the wastewater biodegradability, *J. Agric. Food Chem.*, 47, 3911–3918, 1999.
29. Trambarulo, R. et al., The molecular structure, dipole moment, and g factor of ozone from its microwave spectrum, *J. Phys. Chem.*, 21, 851–855, 1953.
30. Nebel, C., Ozone, in *Kirk-Othmer: Encyclopedia of Chemical Technology*, 3rd ed., Vol. 16, John Wiley & Sons, New York, 1981, 683–713.
31. Perry, R.H. and Green, D.W., *Perry's Chemical Engineers' Handbook*, 7th ed., McGraw-Hill, New York, 1997.

2 Reactions of Ozone in Water

Due to its electronic configuration, ozone has different reactions in water. These reactions can be divided into three categories:

- Oxidation–reduction reactions
- Dipolar cycloaddition reactions
- Electrophilic substitution reactions

A possible fourth type of reaction could be some sort of nucleophilic addition, although this reaction has only been confirmed in nonaqueous systems.[1]

In some cases, free radicals are formed from these reactions. These free radicals propagate themselves through mechanisms of elementary steps to yield hydroxyl radicals. These hydroxyl radicals are extremely reactive with any organic (and some inorganic) matter present in water.[2] For this reason, ozone reactions in water can be classified as direct and indirect reactions. Direct reactions are the true ozone reactions, that is, the reactions the ozone molecule undergoes with any other type of chemical species (molecular products, free radicals, etc.). Indirect reactions are those between the hydroxyl radical, formed from the decomposition of ozone or from other direct ozone reactions, with compounds present in water. It can be said that a direct ozone reaction is the initiation step leading to an indirect reaction.

2.1 OXIDATION–REDUCTION REACTIONS

Redox reactions are characterized by the transfer of electrons from one species (reductor) to another one (oxidant).[3] The oxidizing or reducing character of any chemical species is given by the standard redox potential. Ozone has one of the highest standard redox potentials,[4] lower only than those of the fluorine atom, oxygen atom, and hydroxyl radical (see Table 2.1). Because of its high standard redox potential, the ozone molecule has a high capacity to react with numerous compounds by means of this reaction type. This reactivity is particularly important in the case of some inorganic species such as Fe^{2+} or I^-. However, in most of these reactions there is no explicit electron transfer, but rather an oxygen transfer from the ozone molecule to the other compound. Examples of explicit electron transfer reactions are few, but the reactions between ozone and the hydroperoxide ion and the superoxide ion radical can be classified in this group:[6]

$$O_3 + HO_2^- \longrightarrow O_3^- \bullet + HO_2 \bullet \qquad (2.1)$$

TABLE 2.1
Standard Redox Potential of Some Oxidant Species[5]

Oxidant Species	E^o, Volts	Relative Potential of Ozone
Fluorine	3.06	1.48
Hydroxyl radical	2.80	1.35
Atomic oxygen	2.42	1.17
Ozone	2.07	1.00
Hydrogen peroxide	1.77	0.85
Hydroperoxide radical	1.70	0.82
Permanganate	1.67	0.81
Chlorine dioxide	1.50	0.72
Hypochlorous acid	1.49	0.72
Chlorine	1.36	0.66
Bromine	1.09	0.53
Hydrogen peroxide	0.87	0.42
Iodine	0.54	0.26
Oxygen	0.40	0.19

$$O_3 + O_2^- \bullet \longrightarrow O_3^- \bullet + O_2 \tag{2.2}$$

In most of the cases, however, one oxygen atom is transferred as, for example, in the reaction with Fe^{2+}:

$$O_3 + Fe^{2+} \longrightarrow FeO_2^+ + O_2 \tag{2.3}$$

Nonetheless, in all these reactions, some atom of the inorganic species goes to a higher valence state, that is, it loses electrons, so that these reactions can be classified theoretically as oxidation–reduction reactions since, in an implicit way, there is an electron transfer. The reaction of ozone with nitrite is one such example. The two half-reactions are:

$$NO_2^- + H_2O - 2e^- \longrightarrow NO_3^- + 2H^+ \tag{2.4}$$

$$O_3 + 2H^+ + 2e^- \longrightarrow O_2 + H_2O \tag{2.5}$$

The standard redox potential allows us to verify the possibility that ozone reacts with a given compound through redox reactions. The main electron ion half-reactions of ozone in water are Reactions (2.5) and (2.6):

$$O_3 + H_2O + e^- \longrightarrow O_2 + 2OH^- \quad E^o = 1.24V \tag{2.6}$$

From these data the importance of pH in ozone redox reactions can be deduced. Detailed information on the standard redox potential of different substances can be obtained elsewhere.[3,4]

2.2 CYCLOADDITION REACTIONS

Addition reactions are those reactions resulting from the combination of two molecules to yield a third one.[7] One of the molecules usually has atoms sharing more than two electrons (i.e., unsaturated compounds such as olefinic compounds with a carbon double bond) and the other molecule has an electrophilic character. These unsaturated compounds present π electrons that to a lesser extent keep the carbon atoms of the double bond bonded. These π electrons are readily available to electrophilic compounds. It can also be said that an addition reaction develops between a base compound (a compound with π electrons) and an acid compound (an electrophilic compound). As a general rule, the following scheme corresponds to an addition reaction:

$$-C=C-+XY \longrightarrow -XC-CY- \qquad (2.7)$$

In practice, there could be different types of addition reactions such as those between ozone and any olefinic compound. Such a reaction follows the mechanism of Criegge[8] and constitutes an example of a cycloaddition reaction. The mechanism of Criegge has three steps, as shown in Figure 2.1. In the first step, a very unstable five-member ring or primary ozonide is formed.[9] This breaks up, in the second step, to yield a zwitterion. In the third step, this zwitterion reacts in different ways, depending on the solvent where the reaction develops, on experimental conditions, and on the nature of the olefinic compound. Thus, in a neutral solvent, it decomposes to yield another ozonide, a peroxide or ketone, and polymer substances, as shown in Figure 2.2. When the reaction is in a participating solvent (i.e., a protonic or nucleophilic solvent) some oxy-hydroperoxide species is generated (Figure 2.3). A third possibility is the so-called abnormal ozonolysis that could develop both in participating and nonparticipating solvents. In this way, some ketone, aldehyde, or carboxylic acids can be formed (Figure 2.4). The cycloaddition reaction, then, leads to the breakup of both σ and π bonds of the olefinic compound while the basic addition Reaction (2.7) leads only to the breakup of the π bond. Compounds with different double bonds (C=N or C=O) do not react with ozone through this type of reaction.[10,11] This is not the case with aromatic compounds that could also react with ozone through 1,3-cycloaddition reactions leading to the breakup of the aromatic ring. However, in these cases, the cycloaddition reaction is also less probable than the electrophilic attack of one terminal oxygen of the ozone molecule on any nucleophilic center of the aromatic compound. The reason for this is the stability of the aromatic ring that results from the resonance. Note that the cycloaddition reaction leads to the breakup of the aromatic ring, then to the loss of aromaticity, while the electrophilic reaction (as discussed later) retains the aromatic ring.

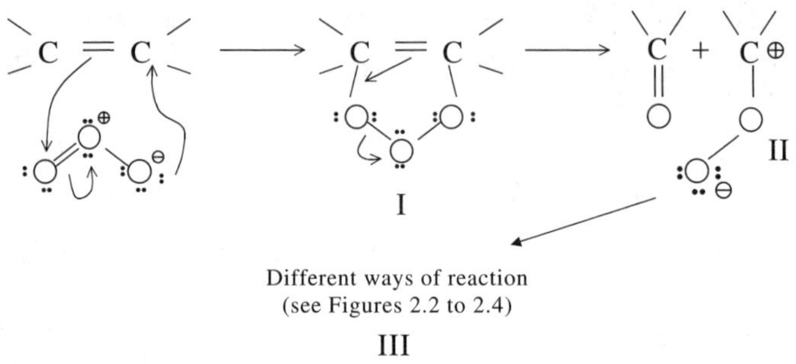

FIGURE 2.1 Criegge mechanism.

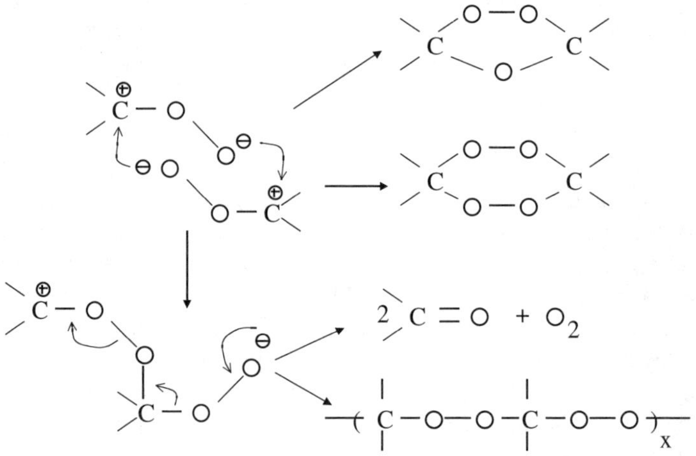

FIGURE 2.2 Steps in decomposition of primary ozonide in an inert solvent.

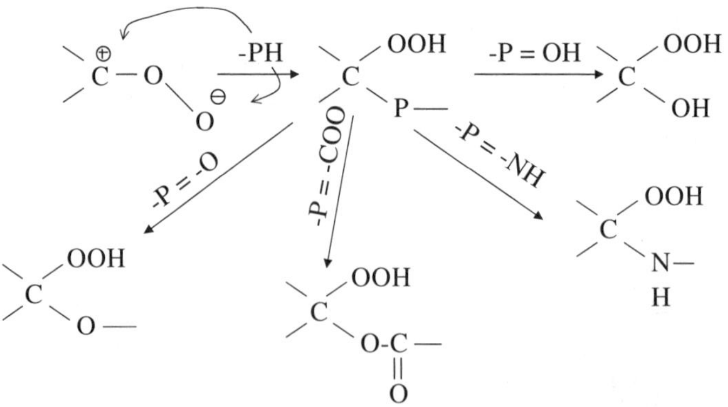

FIGURE 2.3 Steps in decomposition of primary ozonide in a participating solvent.

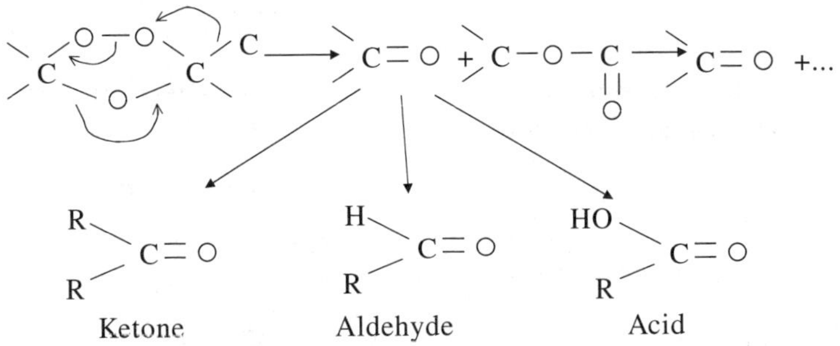

Reactions of Ozone in Water

FIGURE 2.4 Examples of abnormal ozonolysis.

2.3 ELECTROPHILIC SUBSTITUTION REACTIONS

In these reactions, one electrophilic agent (such as ozone) attacks one nucleophilic position of the organic molecule (i.e., an aromatic compound), resulting in the substitution of one part (i.e., atom, functional group, etc.) of the molecule.[7] As shown later, this type of reaction is the base of the ozonation of aromatic compounds such as phenols. Aromatic compounds are prone to undergo electrophilic substitution reactions rather than cycloaddition reactions because of the stability of the aromatic ring. For example, the benzene molecule is strongly stabilized by the resonance phenomena. The benzene molecule can be represented by different electronic structures that constitute the benzene hybrid. The difference in stability between individual structures and the hybrid is the energy of resonance. In the case of benzene, the individual structure is the cyclohexatriene, and the resonance energy is 36 kcal, that is, the energy difference between the cyclohexatriene and the benzene hybrid. It can be said that the greater the resonance energy, the stronger the aromatic properties. The reactions of aromatic compounds depend on these aromatic properties. Thus, after electrophilic substitution, the aromatic properties are still valid, and the resulting molecules have aromatic stability. This state is lost when cycloaddition takes place.

In a general way, an aromatic substitution reaction develops in two steps, as shown in Figure 2.5 for the case of benzene and one electrophilic agent YZ. In the first step, a carbocation ($C_6H_5^+HY$) is formed and, in the second step, a base compound takes a proton from the nucleophilic position.

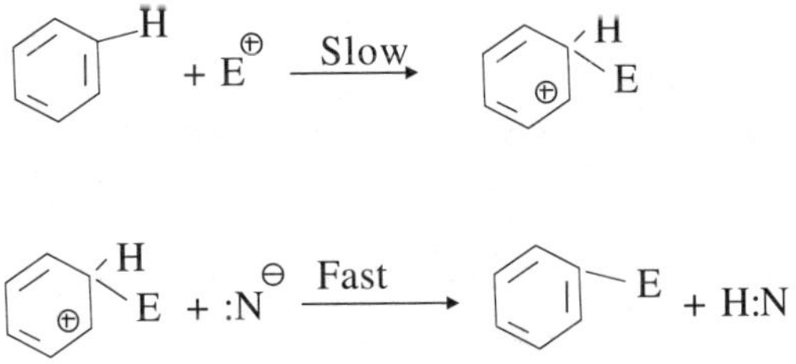

FIGURE 2.5 Basic steps of the aromatic electrophilic substitution reaction.

TABLE 2.2
Activating and Deactivating Groups of the Aromatic Electrophilic Substitution Reaction[7]

Groups	Action on Reaction	Importance
$-OH^-$, $-O^-$, $-NH_2$, $-NHR$, $-NR_2$	Activation	Strong
$-OR$, $-NHCOR$	Activation	Intermediate
$-C_6H_5$, $-Alkyl$	Activation	Weak
$-NO_2$, $-NR_3^+$	Deactivation	Strong
$-C{\equiv}N$, $-CHO$, $-COOH$	Deactivation	Intermediate
$-F$, $-Cl$, $-Br$, $-I$	Deactivation	Weak

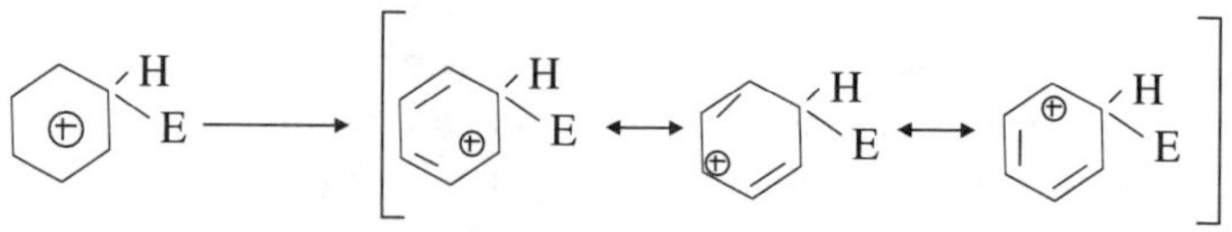

FIGURE 2.6 Resonance forms of the hybride carbocation.

Another important consideration is the presence of substituting groups in the aromatic molecule (i.e., phenols, cresols, aromatic amines, etc.). These groups strongly affect the reactivity of the aromatic ring with electrophilic agents. Thus, groups such as HO^-, NO_2^-, Cl^-, etc. activate or deactivate the aromatic ring for the electrophilic substitution reaction. Depending on the nature of the substituting group, the substitution reaction can take place in different nucleophilic points of the aromatic ring. Thus, activating groups promote the substitution of hydrogen atoms from their ortho and para positions with respect to these groups, while the deactivating groups facilitate the substitution in the meta position. Table 2.2 shows the effect of different substituting groups on the electrophilic reaction of the benzene molecule. In fact, both the resulting products of the electrophilic substitution reaction and the relative importance of the reaction rate can be predicted after considering the nature of substituting groups. Theoretically, differences in the rate of substitution reaction should be due to differences in the slow step of the process, i.e., the formation of the carbocation: the higher the stability of the carbocation, the faster the electrophilic substitution reaction rate. The carbocation is a hybrid of different possible structures where the positive charge is distributed throughout the aromatic ring, although the ortho and para positions of the substituting group position have the higher nucleophilic character. As a consequence, these positions have the highest probability of undergoing the electrophilic substitution reaction (see Figure 2.6). Factors that affect the spread of the positive charge are those that stabilize the carbocation or intermediate state.

The substituting group can increase or decrease the carbocation stability, depending on the capacity to release or take electrons. From Figure 2.6, it is evident that the stabilizing or destabilizing effect is especially important when the substituting group is bonded to the ortho or para carbon atom in the attacked nucleophilic

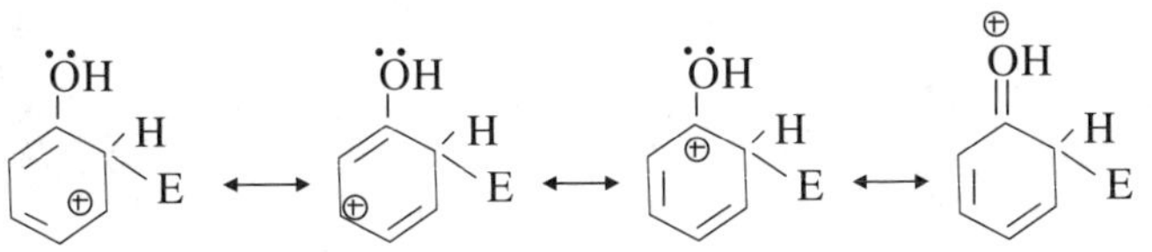

FIGURE 2.7 Resonance forms of the carbocation formed during the ozonation of phenol (attack on the ortho position).

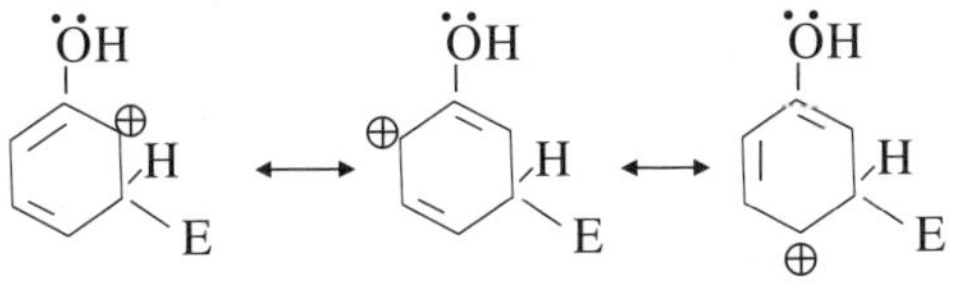

FIGURE 2.8 Resonance forms of the carbocation formed during the ozonation of phenol (attack on the meta position).

position. Groups such as alkyl radicals or –OH activate the aromatic ring because they tend to release electrons while groups such as –NO_2 deactivate the aromatic ring because they attract electrons. In the first case, the carbocation is stabilized, while in the second case it is not. For example, in the case of the ozonation of phenols, this property is particularly important due to the strong electron donor character of the hydroxyl group. In addition, the carbocation formed in the case of phenol is a hybrid formed by the contribution of structures I through III (see Figure 2.6) and also a fourth structure (see Figure 2.7), where the oxygen atom is positively charged. Structure IV is especially stable since each atom (except the hydrogen atom) has completed the orbitals (eight electrons). This carbocation is more stable than that from the electrophilic substitution in the benzene molecule (where there is no substituting group) or in the meta position of the –OH group in the phenol molecule (Figure 2.8). In these two cases, structure IV is not possible, so the ozonation of phenol is faster than that of benzene and occurs mainly at ortho and para positions of the –OH group. In fact, the literature reports kinetic studies (see Chapters 3 and 5) of the ozonation of aromatic compounds where the rate constants of the direct reactions between ozone and phenol, and ozone and benzene, were found to be 2×10^6 and 3 $M^{-1}sec^{-1}$, respectively.[12–14] It should be noted, however, that these values correspond to pH 7 and 20°C. As shown later, the rates of phenol ozonation are largely influenced by the pH of water because of the dissociating character of phenols. More information on the stability of carbocations in electrophilic substitution reactions in different aromatic structures can be obtained from organic chemistry books.[7]

In the case of the ozonation of phenol, the mechanism goes through different electrophilic substitution and cycloaddition reactions, as shown in Figure 2.9.[15–17]

2.4 NUCLEOPHILIC REACTIONS

According to the resonance structures of the ozone molecule (see Figure 1.2), there is a negative charge on one of the terminal oxygen atoms. This charge confers, at

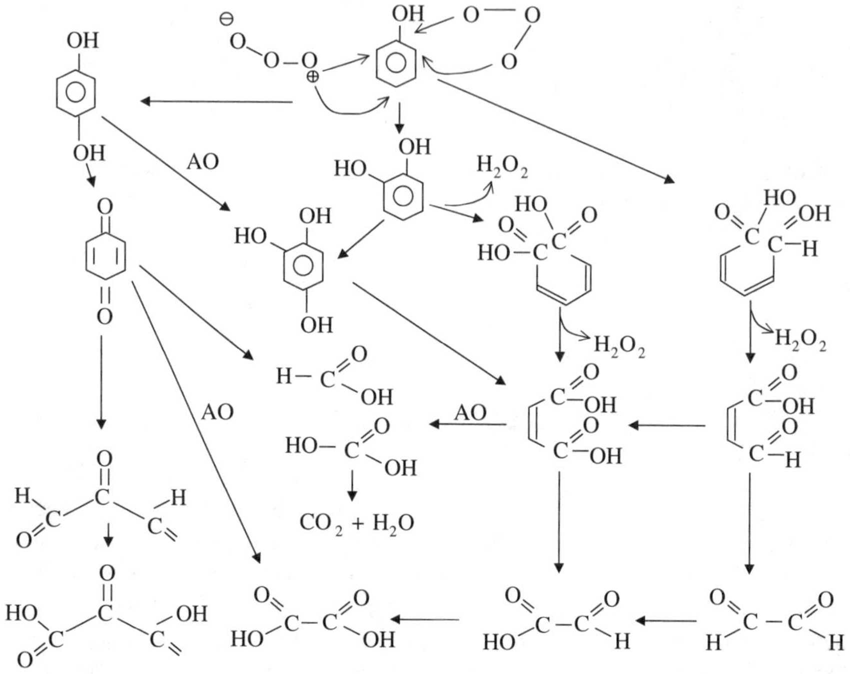

FIGURE 2.9 General mechanism of the ozonation of phenol (AO = abnormal ozonolysis).

least theoretically, a nucleophilic character on the ozone molecule. Thus, ozone could react with molecules containing electrophilic positions. These reactions are of the nucleophilic addition type, and theoretically molecules with double (and triple) bonds between atoms of different electronegativity could be involved. In the case of ozonation, the nucleophilic activity can be seen in the presence of carbonyl or double and triple carbon nitrogen bonds.[1] The following example shows two possible kinds (nucleophilic and electrophilic) of ozone attack on a ketone. For example, Figure 2.10 shows the nucleophilic reaction of ozone on Schiff bases with carbon–nitrogen double bonds. It should be noted, however, that most of the information related to the mechanism of the ozonation of organic compounds has been obtained in an organic medium, and that there is little information on this subject when water is the solvent.

2.5 INDIRECT REACTIONS OF OZONE

These reactions are due to the action of free radical species resulting from the decomposition of ozone in water. The free radical species are formed in the initiation or propagation reactions of the mechanisms of advanced oxidation processes involving ozone and other agents, such as hydrogen peroxide or UV radiation, among others.[18] An advanced oxidation process (AOP) is defined as one producing hydroxyl radicals that are strong oxidant species.[2] In the ozone decomposition mechanisms,

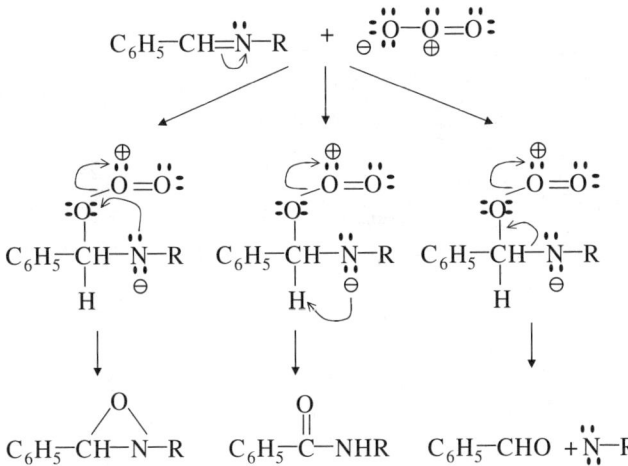

FIGURE 2.10 An ozone nucleophilic substitution reaction. (From Riebel, A.H. et al., Ozonation of carbon-nitrogen bonds. I. Nucleophilic attack of ozone, *J. Am. Chem. Soc.*, 82, 1801–1807, 1960. With permission.)

the hydroxyl radical is the main responsible species in the indirect reactions. The reaction between the hydroxyl radical and compounds (which could be called pollutants) present in water constitute the indirect ozone reactions.

Numerous studies have been developed to clarify the mechanism of decomposition of ozone in water since Weiss proposed the first model in 1934.[19] Today, the mechanism of Staehelin, Hoigné, and Buhler (SHB)[20–23] is generally accepted as the mechanism of ozone in water, although when the pH is high, the mechanism of Tomiyasu, Fukutomi, and Gordon (TFG) is considered the most representative.[24] Tables 2.3 and 2.4 show both mechanisms.

The reactions of ozone with the hydroxyl and hydroperoxide ions can be considered the main initiation reactions of the ozone decomposition mechanism in water. However, other initiation reactions develop when other agents, such as UV radiation or solid catalysts, are also present. Thus, the direct photolysis of ozone that yields hydrogen peroxide and then free radicals,[26] or the ozone adsorption and decomposition on a catalyst surface to yield active species (in some cases hydroxyl radicals,[27] as will be discussed in other chapters) are also examples of initiation reactions. The reaction of ozone and the superoxide ion radical [Reaction (2.2)] is one of the main propagating reactions of the ozone decomposition mechanism.

There are other reactions that lead to the decomposition or stabilization of ozone in water. Thus, substances of different nature can also contribute to the appearance or inhibition of free radicals. These substances are called initiators, inhibitors, and promoters of the decomposition of ozone.[21] The initiators are those substances, such as the hydroperoxide ion (the ionic form of hydrogen peroxide) mentioned above, that directly react with ozone to yield the superoxide ion radical [Reaction (2.1)]. These reactions are initiation reactions. The superoxide ion radical is the key to propagating free radical species because it rapidly reacts with ozone to yield free

TABLE 2.3
Ozone Decomposition Mechanism in Pure Water According to Staehelin, Hoigné, and Bühler[22,23]

Reaction	Rate Constant	Reaction No.
Initiation Reaction		
$O_3 + OH^- \xrightarrow{k_{i1}} HO_2\bullet + O_2^-\bullet$ [a]	70 $M^{-1}sec^{-1}$	(2.8)
Propagation Reactions		
$HO_2\bullet \xrightarrow{k_1} O_2^-\bullet + H^+$	7.9×10^5 sec^{-1} [25]	(2.9)
$O_2^-\bullet + H^+ \xrightarrow{k_1'} HO_2\bullet$	5×10^{10} $M^{-1}sec^{-1}$ [25]	(2.10)
$O_3 + O_2^-\bullet \xrightarrow{k_2} O_3^-\bullet + O_2$	1.6×10^9 $M^{-1}sec^{-1}$	(2.2)
$O_3^-\bullet + H^+ \xrightarrow{k_3} HO_3\bullet$	5.2×10^{10} $M^{-1}sec^{-1}$	(2.11)
$HO_3\bullet \xrightarrow{k_4} O_3^-\bullet + H^+$	3.3×10^2 sec^{-1}	(2.12)
$HO_3\bullet \xrightarrow{k_5} HO\bullet + O_2$	1.1×10^5 sec^{-1}	(2.13)
$O_3 + HO\bullet \xrightarrow{k_6} HO_4\bullet$	2×10^9 $M^{-1}sec^{-1}$	(2.14)
$HO_4\bullet \xrightarrow{k_7} HO_2\bullet + O_2$	2.8×10^4 sec^{-1}	(2.15)
Termination Reactions		
$HO_4\bullet + HO_4\bullet \xrightarrow{k_{T1}} H_2O_2 + 2O_3$ [b]	5×10^9 $M^{-1}sec^{-1}$ [25]	(2.16)
$HO_4\bullet + HO_3\bullet \xrightarrow{k_{T2}} H_2O_2 + O_2 + O_3$ [b]	5×10^9 $M^{-1}sec^{-1}$ [25]	(2.17)

[a] Later, Hoigné[6] suggested Reaction (2.8) should be Reaction (2.18) of the Tomiyashu et al. mechanism[24] (see Table 2.4) although the rate constant value remained the same (70 $M^{-1}sec^{-1}$). This reaction change implies that Reaction (2.1), hydrogen peroxide equilibrium Reactions (2.22) and (2.23) (see Table 2.4), and reactions between hydrogen peroxide and the hydroxyl radical (2.27) and (2.28) are also part of the mechanism.

[b] Reaction products, H_2O_2 and O_3, were tentatively proposed.

radicals, such as the ozonide ion radical [Reaction (2.2)] that eventually leads to the hydroxyl radical (see Table 2.3 or 2.4). Promoters are those species that, through their reaction with the hydroxyl radical, propagate the radical chain to yield the key free radical: the superoxide ion radical. Examples of these substances are methanol, formic acid, or some humic substances.[21] Of particular interest is the role of hydrogen peroxide in the mechanism of ozone decomposition. In fact, hydrogen peroxide is the initiating agent of ozone decomposition as proposed by Tomiyasu et al.[24] but it also acts as a promoter of ozone decomposition according to the following reactions:[28]

TABLE 2.4
Ozone Decomposition Mechanism in Pure Water at Alkaline Conditions According to Tomiyasu, Fukutomi, and Gordon[24]

Reaction	Rate Constant	Reaction No.
Initiation Reaction		
$O_3 + OH^- \xrightarrow{k_8} HO_2^- + O_2$ *	40 $M^{-1}sec^{-1}$	(2.18)
$O_3 + HO_2^- \xrightarrow{k_{12}} HO_2\bullet + O_3^-\bullet$	2.2×10^6 $M^{-1}sec^{-1}$	(2.1)
Propagation Reactions		
$HO_2\bullet \xrightarrow{k_9} O_2^-\bullet + H^+$	7.9×10^5 sec^{-1} [25]	(2.9)
$O_2^-\bullet + H^+ \xrightarrow{k'_9} HO_2\bullet$	5×10^{10} $M^{-1}sec^{-1}$ [25]	(2.10)
$O_3 + O_2^-\bullet \xrightarrow{k_2} O_3^-\bullet + O_2$	1.6×10^9 $M^{-1}sec^{-1}$	(2.2)
$O_3^-\bullet + H_2O \xrightarrow{k_{10}} HO\bullet + O_2 + OH^-$	20–30 $M^{-1}sec^{-1}$	(2.19)
$O_3^-\bullet + HO\bullet \xrightarrow{k_{11}} HO_2\bullet + O_2^-\bullet$	6×10^9 $M^{-1}sec^{-1}$	(2.20)
$O_3 + HO\bullet \xrightarrow{k_6} HO_2\bullet + O$	3×10^9 $M^{-1}sec^{-1}$	(2.21)
$HO_2^- + H^+ \xrightarrow{k_{12}} H_2O_2$	5×10^{10} $M^{-1}sec^{-1}$ [25]	(2.22)
$H_2O_2 \xrightarrow{k'_{12}} HO_2^- + H^+$	0.25 sec^{-1} [25]	(2.23)
Termination Reactions		
$O_3 + HO\bullet \xrightarrow{k_{T3}} O_3 + OH^-$	2.5×10^9 $M^{-1}sec^{-1}$	(2.24)
$HO\bullet + CO_3^= \xrightarrow{k_{C2}} OH^- + CO_3^-\bullet$ *	4.2×10^8 $M^{-1}sec^{-1}$	(2.25)
$CO_3^-\bullet + O_3 \xrightarrow{k_{T4}} (O_2 + CO_2 + O_2^-\bullet)$ *	No data given	(2.26)

* Carbonates were assumed to be present because of alkaline conditions. In fact, Reactions (2.25) and (2.26) are not true termination reactions since the superoxide ion radical, $O_2^-\bullet$, would propagate the radical chain. Reaction products O_2, CO_2, and $O_2^-\bullet$ were tentatively proposed but not confirmed. Reactions (2.27) and (2.28) (see text) have to be added to this mechanism.

$$HO\bullet + H_2O_2 \xrightarrow{k_{H1}=2.7 \times 10^7 M^{-1}s^{-1}} HO_2\bullet + H_2O \quad (2.27)$$

$$HO\bullet + HO_2^- \xrightarrow{k_{H2}=7.5 \times 10^9 M^{-1}s^{-1}} HO_2\bullet + OH^- \quad (2.28)$$

However, as shown in Chapter 8, hydrogen peroxide can also act as an indirect inhibitor of the ozone decomposition, when its concentration is so high the ozone/hydrogen peroxide reaction becomes mass transfer-controlled.

Finally, inhibitors of the ozone decomposition are those species that while reacting with the hydroxyl radical terminate the radical chain. In this group, one can cite *tert*-butanol, *p*-chlorobenzoate ion, carbonate, and bicarbonate ions or some other humic substances.[21,27] The inhibitors are also called hydroxyl-free radical scavengers because their presence limits or inhibits the action of these radicals on the target contaminants. For example, the presence of carbonates in natural water reduces the efficiency of ozonation to oxidize refractory contaminants also present in the water. Because of the importance of these three types of substances, alternative mechanisms to those of SHB or TFG have been proposed. Thus, it is particularly significant that due to their usual presence in water the mechanisms of ozone decomposition in the presence of carbonate species (carbonate and bicarbonate ions), are called *natural inhibitors*. However, carbonate ion species cannot be considered pure inhibitors of ozone decomposition. In this case, some other reactions must be added to the mechanisms shown in Tables 2.3 and 2.4, especially if hydrogen peroxide is present in significant concentration. These reactions are shown in Table 2.5. As Table 2.5 shows, the reactions of hydroxyl radicals with carbonate species yield the carbonate ion radical. In many cases, this free radical is not inactive. Instead, the carbonate ion radical is able to regenerate the carbonate ions by reacting with hydrogen peroxide (see reactions in Table 2.5). Also, the carbonate ion radical can react with some substances (i.e., phenol) and constitute another type of oxidation.[31-33] More information on the rate constant of these reactions can be obtained from other works.[2,31]

Another case, extensively studied in the literature because of its health impact, is the presence of bromide ion in ozonated water.[34,35] Ozone readily oxidizes bromide ion to yield a toxic pollutant, bromate ion. As indicated in Chapter 1, environmental agencies have established a low MCL for bromate ion in water. The reactions of bromide–ozone processes are shown in Table 2.6. It can be seen that different reactions between the species appear in this mechanism. Formation of bromate is highly dependent on the presence of other substances that consume ozone such as hydrogen peroxide or ammonia, which react with hypobromous acid to yield bromamines.[42,43]

Another important aspect often considered in the ozone decomposition mechanism in water is the presence of natural organic matter (NOM). Depending on the nature of NOM, these substances can act as promoters or inhibitors of the decomposition of ozone. For this reason, the following reaction is usually included in the mechanism of ozone decomposition when NOM is present:[44,45]

$$NOM + HO\bullet \longrightarrow \alpha O_2^- \bullet + P \qquad (2.68)$$

or

$$NOM + \alpha HO\bullet \longrightarrow O_2^- \bullet + P \qquad (2.69)$$

These reactions basically mean that a fraction of hydroxyl radical, while reacting with NOM, yields the superoxide ion radical and hence a fraction of NOM is a promoter of the ozone decomposition reaction. In this reaction, it is accepted that

TABLE 2.5
Main Reactions Involving Carbonate Species in Water during Ozonation Processes[2,25,29–31]

Reaction	Rate Constant	Reaction No.
$HCO_3^- + HO\bullet \xrightarrow{k_{C1}} HCO_3\bullet + OH^-$	$8.5 \times 10^6\ M^{-1}sec^{-1}$	(2.29)
$CO_3^= + HO\bullet \xrightarrow{k_{C2}} CO_3^-\bullet + OH^-$	$4.2 \times 10^8\ M^{-1}sec^{-1}$	(2.30)
$HCO_3^- \xrightarrow{k_{C3}} CO_3^= + H^+$	$2.2\ sec^{-1}$ *	(2.31)
$HCO_3^- + H^+ \xrightarrow{k'_{C3}} CO_3^=$	5×10^{10} *	(2.32)
$H_2CO_3 \xrightarrow{k_{C4}} HCO_3^- + H^+$	2.25×10^4 *	(2.33)
$HCO_3^- + H^+ \xrightarrow{k'_{C4}} H_2CO_3$	5×10^{10} *	(2.34)
$HCO_3\bullet \xrightarrow{k_{C5}} CO_3^-\bullet + H^+$	$500\ sec^{-1}$ *	(2.35)
$CO_3^-\bullet + H^+ \xrightarrow{k'_{C5}} HCO_3\bullet$	$5 \times 10^{10}\ M^{-1}sec^{-1}$ *	(2.36)
$CO_3^-\bullet + H_2O_2 \xrightarrow{k_{C6}} HCO_3^- + HO_2\bullet$	$4.3 \times 10^5\ M^{-1}sec^{-1}$	(2.37)
$CO_3^-\bullet + HO_2^- \xrightarrow{k_{C7}} CO_3^= + HO_2\bullet$	$5.6 \times 10^7\ M^{-1}sec^{-1}$	(2.38)
$CO_3^-\bullet + O_2^-\bullet \xrightarrow{k_{C8}} CO_3^= + O_2$	$7.5 \times 10^8\ M^{-1}sec^{-1}$	(2.39)
$CO_3^-\bullet + O_3^-\bullet \xrightarrow{k_{C9}} CO_3^= + O_3$	$6 \times 10^7\ M^{-1}sec^{-1}$	(2.40)
$CO_3^-\bullet + B \xrightarrow{k_{C10}} CO_3^= + Products$	See Reference 31	(2.41)

* For the rate constant of protonation reactions a value of 5×10^{10} [25] has been assumed considering that these reactions are diffusion controlled. The rate constant of the inverse reaction has been calculated from the corresponding pK and the indicated value for the protonation reaction.

part of the matter present in water acts as a promoter and another part acts as an inhibitor. It should be noted that the kinetic behavior of ozone in natural water is similar to that in wastewater but with the difference that there may be multiple direct ozone reactions and actions of promoters, initiators, and inhibitors due to the complexity of the organic and inorganic matrix of the wastewater. Chapter 6 discusses aspects related to the treatment of ozone kinetics of wastewater.

2.5.1 THE OZONE DECOMPOSITION REACTION

The fact that ozone once dissolved in water is unstable and decomposes is both an advantage and a drawback. On the one hand, when ozone decomposes, free radicals, particularly the hydroxyl radical, are generated and oxidation of compounds develops in a process called *advanced oxidation* (indirect reaction). On the other hand, because

TABLE 2.6
Main Reactions Involving Bromine Species during Ozonation Processes[34,36–42]

Reaction	Rate Constant	Reaction No.
$Br^- + O_3 \longrightarrow BrO^- + O_2$	50 $M^{-1}sec^{-1}$	(2.42)
$BrO^- + O_3 \longrightarrow Br^- + 2O_2$	300 $M^{-1}sec^{-1}$	(2.43)
$BrO^- + O_3 \longrightarrow BrO_2^- + O_2$	100 $M^{-1}sec^{-1}$	(2.44)
$BrO^- + H^+ \longrightarrow HBrO$	5×10^{10} $M^{-1}sec^{-1}$	(2.45)
$HBrO \longrightarrow BrO^- + H^+$	50 sec^{-1}	(2.46)
$BrO_2^- + O_3 \longrightarrow BrO_3^- + O_2$	10^5 $M^{-1}sec^{-1}$	(2.47)
$Br^- + HO\bullet \longrightarrow BrOH^-\bullet$	1.1×10^{10} $M^{-1}sec^{-1}$	(2.48)
$BrOH^-\bullet \longrightarrow Br^- + HO\bullet$	3.3×10^7 $M^{-1}sec^{-1}$	(2.49)
$BrOH^-\bullet \longrightarrow Br\bullet + OH^-$	4.2×10^6 $M^{-1}sec^{-1}$	(2.50)
$Br\bullet + Br^- \longrightarrow Br_2^-\bullet$	2×10^9 $M^{-1}sec^{-1}$	(2.51)
$Br\bullet + O_3 \longrightarrow BrO\bullet + O_2$	10^{10} $M^{-1}sec^{-1}$	(2.52)
$2Br_2^-\bullet \longrightarrow Br\bullet + Br_3^-\bullet$	2×10^9 $M^{-1}sec^{-1}$	(2.53)
$Br_2 + H_2O \longrightarrow HBrO + Br^- + H^+$	8.24 $M^{-1}sec^{-1}$	(2.54)
$BrO^- + Br_2^-\bullet \longrightarrow BrO\bullet + 2Br^-$	8×10^7 $M^{-1}sec^{-1}$	(2.55)
$BrO^- + HO\bullet \longrightarrow BrO\bullet + OH^-$	4.5×10^9 $M^{-1}sec^{-1}$	(2.56)
$HBrO + HO\bullet \longrightarrow BrO\bullet + H_2O$	2×10^9 $M^{-1}sec^{-1}$	(2.57)
$HBrO + O_2^-\bullet \longrightarrow Br\bullet + OH^- + O_2$	9.5×10^8 $M^{-1}sec^{-1}$	(2.58)
$HBrO + H_2O_2 \longrightarrow Br^- + H_2O + H^+ + O_2$	7×10^4 $M^{-1}sec^{-1}$	(2.59)
$BrO^- + H_2O_2 \longrightarrow Br^- + H_2O$	2×10^5 $M^{-1}sec^{-1}$	(2.60)
$BrO^- + Br\bullet \longrightarrow Br^- + BrO\bullet$	4.1×10^9 $M^{-1}sec^{-1}$	(2.61)
$2BrO\bullet + H_2O \longrightarrow BrO^- + BrO_2^- + 2H^+$	4.9×10^9 $M^{-1}sec^{-1}$	(2.62)
$BrO_2^- + HO\bullet \longrightarrow BrO_2\bullet + OH^-$	2×10^9 $M^{-1}sec^{-1}$	(2.63)
$2BrO_2\bullet \longrightarrow Br_2O_4$	1.4×10^9 $M^{-1}sec^{-1}$	(2.64)
$Br_2O_4 \longrightarrow 2BrO_2\bullet$	7×10^9 sec^{-1}	(2.65)
$Br_2O_4 + OH^- \longrightarrow BrO_3^- + BrO_2^- + H^+$	7×10^8 $M^{-1}sec^{-1}$	(2.66)
$CO_3^-\bullet + BrO^- \longrightarrow CO_3^= + BrO\bullet$	4.3×10^7 $M^{-1}sec^{-1}$	(2.67)

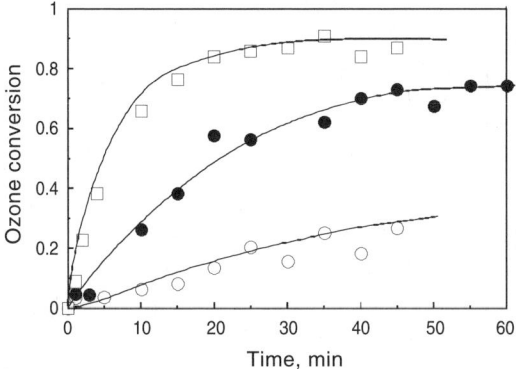

FIGURE 2.11 Decomposition of ozone in buffered distilled water at different pH. Conditions: 17°C, buffered water with phosphates; ○ = pH 2, C_{O30} = 8.1 mgL^{-1}; ● = pH 7, C_{O30} = 4.2 mgL^{-1}; □ = pH 8.5, C_{O30} = 3.7 mgL^{-1}.

of its instability, ozone cannot be used in practice as a final disinfectant of water. The kinetic study of the decomposition of ozone in water is one of the necessary steps to ascertain whether or not ozone is able to remove some given compounds from water through direct or indirect reactions.

The ozone decomposition rate, as can be deduced from the previous section, depends greatly on the nature of substances present in water. For example, Figure 2.11 presents data on the decomposition of ozone in buffered distilled water at three pH values. As deduced from Figure 2.11 (and also from the mechanisms of Tables 2.3 and 2.4), pH is also one of the main factors that influence the decomposition of ozone in water. As a general rule, for pH < 7 this variable has a slight effect on the ozone decomposition,[46] but at higher pH, the rate increases significantly. For example, Figure 2.12 shows the variation of the apparent rate constant of the decomposition of ozone with pH in a study where first-order kinetics was considered.[46]

Because of the importance of hydroxyl radical oxidation, the decomposition of ozone in water has been the subject of numerous works. Since the first work on this matter by Weiss,[19] numerous researchers have studied ozone decomposition kinetics of reaction mechanisms, based on experimental facts. The rate equations are finally fitted to the experimental data of ozone concentration time obtained in homogeneous ozone decomposition reactions in water. Thus, rate equations of different complexity with one or two terms, first-, second-, or third-, half-order reactions to the ozone, with the rate constant expressed as a function of temperature and pH, have been calculated and checked in the literature (see Table 2.7). As observed from Table 2.7, the reaction order and the influence of pH differ in many cases. For the SHB mechanism, initiators, promoters, and inhibitors of the decomposition of ozone are considered the substances responsible for the kinetic differences between rate equations reported in Table 2.7. In naturally occurring water, application of the SHB mechanism yields the following ozone decomposition rate equation:

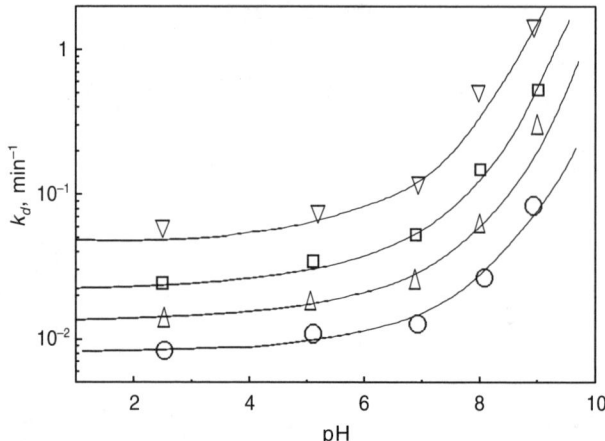

FIGURE 2.12 Variation of the apparent pseudo first-order rate constant of ozone decomposition in buffered distilled water with pH. T, °C: $\circ$ = 10; Δ = 20; $\square$ = 30; ∇ = 40. (From Sotelo, J.L. et al., Ozone decomposition in water: kinetic study, *Ind. Eng. Chem. Res.*, 26, 39–43, 1987. With permission.)

$$-r_{O3} = k_d C_{O3} = \left[\sum k_{Di} C_{Mi} + \left\{ 3 k_i C_{OH^-} + \sum k_{Ii} C_{Ii} \right\} \left(1 + \frac{\sum k_{Pi} C_{Pi}}{\sum k_{Si} C_{Si}} \right) \right] C_{O3} \quad (2.70)$$

where C_{Mi}, C_{Ii}, C_{Pi}, and C_{Si} are the concentrations of compounds that react directly with ozone, initiator, promoter, and inhibitor compounds, respectively, and C_{O3} is the concentration of dissolved ozone.

Notwithstanding the works of Hoigné et al.,[20–23] studies on this matter still continue, as shown in Table 2.7. The determination of the rate constant and reaction order was usually carried out by following the decrease in the ozone concentration in water with time in an agitated tank through homogeneous experiments (see Chapter 3). Because of the effects of natural organic matter in water, when planning any ozonation study where the decomposition of ozone is considered important, it is recommended to determine the specific rate law of ozone decomposition if the water matrix is different from that used in the studies reported previously. In fact, there are two clear ozone decomposition periods when treating natural waters.[74] The first one, called fast ozone demand, varies just from a few seconds to as much as approximately 1 or 2 min. During this period, an instantaneous or very fast consumption of ozone takes place. During the second or long ozone decomposition period, ozone slowly decomposes. It is this latter period that is the subject of kinetics studies: ozone decomposition in natural waters. The initial ozone fast-demand period is due to the presence of substances that readily react with ozone through direct reactions. Once these substances have disappeared or their concentrations decrease the longer ozone decomposition period starts. More information on ozone decomposition kinetics is given in Chapter 7.

TABLE 2.7
Works on Aqueous Ozone Decomposition Kinetics

Reacting System and Operating Conditions	Main Observations	Reference No. (Year)
Batch reactor, iodometric analysis 0°C, pH 2–4, H_2SO_4 or HNO_3	Rate equation: $-r_{O3} = A(a-C_{O3})^2 + B(a-C_{O3})$ Data on k determined; no influence of pH	47 (1913)
Batch reactor, iodometric analysis 0°C, pH 5–8, phosphate buffers	$-r_{O3} = k10^{0.36(pH\ 14)}C_{O3}$ Data on k determined	48 (1933)
Data from other works at 0°C and acid pH	Chain mechanism, $-r_{O3} = k_1 10^{pH\ 14} C_{O3} + k_2 10^{1/2(pH\ 14)} C_{O3}^{1.5}$; data on k determined Reaction H_2O_2–O_3 also studied	19 (1935)
Batch reactor, spectrophotometric, and iodometric analysis 0 and 27°C, pH 0.7–2.8, $HClO_4$	Chain mechanism, $-r_{O3} = k10^{1/2(pH\ 14)} C_{O3}$	49 (1950)
Batch reactor, pH 7.6–10.4, T = 1.2–19.8°C	$-r_{O3} = kC_{O3}$; data on k determined	50 (1954)
Batch reactor, cell, saturation reactor, spectrophotometric, iodometric, manometric analysis, 25°C, pH 0.2–10, $HClO_4$ (10^{-5}–0.96 M), H_2SO_4, NaOH, Cu(II), Fe(II) of perchlorate, glycine, phosphate, arseniate buffers	Influence of different acids and salts Kinetics varies depending on pH values: for acid-neutral pH: $-r_{O3} = kC_{O3}^{3/2}$, for pH > 7: $-r_{O3} = kC_{O3}^2$ Henry constant also determined	51 (1956)
Semibatch bubble column; iodometric analysis. 10–40°C, pH 2–4 (H_2SO_4), pH 6 (phosphate buffer), pH 8 (borate buffer)	Kinetics depends on pH: pH 2–4: $-r_{O3} = kC_{O3}^2$, pH 6: $-r_{O3} = kC_{O3}^{2-1.5}$ pH 8: $-r_{O3} = kC_{O3}$ Ozone–secondary wastewater also studied	52 (1971)
Semibatch bubble column until saturation pH 0.22–1.9, T = 5–40°C	$-r_{O3} = kC_{O3}$ or $-r_{O3} = kC_{O3}^2$; chain mechanism; data on k determined	53 (1971)
Batch reactors, spectrophotometric analysis 25°C, pH 2–11, different organics and salts	Influence of organics Relative rates of organic decomposition Two methods of ozone decomposition, critical pH	54 (1976)
Semibatch packed column (95 cm, 3.7 cm diameter); packed: glass rings; iodometric analysis, 27°C, pH 8.5–13.5, KOH,	$k_L a$ and a determined from CO_2 absorption in arseniate solutions $-r_{O3} = k10^{pH\ 14} C_{O3}$,	55 (1976)
3.2-L batch reactor, iodometric analysis, 3.5–60°C, pH 0.45–12, phosphate buffers	$-r_{O3} = k10^{0.123(pH\ 14)} C_{O3}$	56 (1979)
Cell reactors, spectrophotometric analysis, 25°C, pH 1–3 (H_2SO_4), 4–5 (acetate buffer), 7–9.5 (phosphate-borate buffer), 10–11.5 (carbonate buffer), 12–13.5 NaOH	For pH < 8 no sufficient reliable data were obtained to fix the ozone reaction order For pH > 8, $-r_{O3} = k10^{0.88(pH\ 14)} C_{O3}$ Reactions O_3-CN^- and O_3 dyes were also studied	57 (1981)
Cell reactor, spectrophotometric analysis, 20°C, pH 3, 7, and 9, H_2SO_4, and NaOH	Kinetics depends on pH: pH 3: $-r_{O3} = kC_{O3}^{3/2}$, pH 7: $-r_{O3} = kC_{O3}^{3/2\ to\ 1}$ pH 9: $-r_{O3} = kC_{O3}$ Kinetics of O_3-bromide studied	58 (1981)

TABLE 2.7 (continued)
Works on Aqueous Ozone Decomposition Kinetics

Reacting System and Operating Conditions	Main Observations	Reference No. (Year)
2.8-L Continuous stirred batch reactor, spectrophotometric analysis, 20°C, tap water (pH not given)	Mass transfer and chemical reaction model proposed and tested. Observed zero-order kinetics for ozone. Mass transfer coefficient also determined	59 (1982)
Spectrophotometer cell, spectrophotometric analysis. 20°C, pH 11–13, NaOH, carbonate and acetate (in some cases)	Acetate and carbonate inhibitors. Hydrogen peroxide promoter. Mechanism, $-r_{O3} = k10^{(pH\ 14)} C_{O3}$. $O_3^{\bullet-}$ and $CO_3^{\bullet-}$ identified	60 (1982)
Cell reactor, spectrophotometric analysis, 20°C, pH 8–10. Phosphate buffer, carbonate (in some cases)	Carbonate and phosphate inhibitors, hydrogen peroxide also investigated. Mechanism and kinetics: $-r_{O3} = k10^{(pH\ 14)} C_{O3}$	20 (1982)
Batch reactor for k determination; 20°C, pH < 4 H_2SO_4, pH 6–9 phosphate buffer, pH 7–10 borate buffer, colorimetric analysis; semibatch bubble column for checking results	Dependency on pH: pH < 4, $-r_{O3} = kC_{O3}^2$, pH > 4, $-r_{O3} = k10^{0.55(pH\ 14)} C_{O3}^2$. Kinetics checked from O_3 absorption in water in bubble column; Henry constant and mass transfer coefficient also determined	61 (1982)
Comparison of results from previous works 3.5–60°C, pH 0.45–10.2	Different ozone reaction orders were tested; no statistical differences between ozone reaction orders (first, second, and 3/2); data can be fitted with an ozone first-order kinetics; potential effects of impurities and buffer systems used	62 (1983)
Pulse radiolysis experiments, 21°C, pH 6.3–7.9, phosphate buffers adjusted with NaOH and $HClO_4$	Studies of free radical formation and decays ($O_3^{\bullet-}$, $HO_3\bullet$); initiation and propagation steps of ozone decomposition; influence of phosphate	22 (1984)
Second part of Reference 22	Studies of termination chain reactions; the role of $HO\bullet$ and $HO_4\bullet$ radicals; full mechanism proposed; recombination of $2HO_4\bullet$ most probable termination reaction	23 (1984)
Batch reactors, conditions as in Reference 52; spectrophotometric and colorimetric analysis	Definition of promoters, initiators, and inhibitors of ozone decomposition in water; first-order reaction; full mechanism proposed (see Table 2.3); full kinetic expression [see (Equation 2.68)]	(21) 1985
Stopped flow cell; spectrophotometric analysis; 20°C, pH 12, NaOH, Na_2CO_3 effect	Kinetics at alkaline conditions, pH 12: $-r_{O3} = k_1 C_{O3} + k_2 C_{O3}^2$, pH > 8: $-r_{O3} = k C_{O3}$ in the presence of carbonates; full mechanism proposed; formation of HO_2^- and O_2 in the initiation reaction	24 (1985)

TABLE 2.7 (continued)
Works on Aqueous Ozone Decomposition Kinetics

Reacting System and Operating Conditions	Main Observations	Reference No. (Year)
750-mL batch stirred reactor, spectrophotometric analysis; 10–40°C, pH 2.5–9, phosphate buffers	Mechanism proposed and tested Kinetics: $-r_{O3} = k_1 C_{O3} + k_2 10^{1/2 \text{(pH 14)}} C_{O3}^{1.5}$ Determination of rate constants and energy of activation	46 (1987)
Pulse radiolysis experiments; 20°C, pH 4.5–9, acetate buffers	Acetate acts as inhibitor of ozone decomposition Identification of acetate, ozonide free radicals Mechanism proposed; determination of rate constants	63 (1987)
Treatment of data from previous works	Discussion on the chain termination reaction Termination proposed: $O_3^{-\bullet} + O_2^{-\bullet}$ Kinetics: $3 < \text{pH} < 7$: $-r_{O3} = k_1 10^{1/2(\text{pH-14})} C_{O3}^{3/2}$, pH = 7: $-r_{O3} = k_1 10^{(\text{pH-14})} C_{O3}^{3/2}$, $7 < \text{pH} < 12$: $-r_{O3} = k_1 10^{1/2(\text{pH-14})} C_{O3}$	64 (1988)
Batch stirred reactor; colorimetric analysis; 10–30°C, pH 2–8.5: phosphate buffer, pH 6: carbonates, chlorides, sulphates, and salt-free water	Full mechanism proposed; dependency on pH, nature of salts, and ionic strength; ozone reaction orders vary from first, half, 3/2, and second Rate constants are determined from integral methods	65 (1989)
2-cm path quartz cell reactor; spectrophotometric analysis; pH > 12, 20°C, NaOH; carbonates: 0–0.05 M	Stability of ozone and no effect of carbonates at pH > 12; mechanism proposed and a new step: reaction $HO\bullet + OH^- \rightarrow O^- + H_2O$; free radicals such as $O_2^{-\bullet}$ and $O_3^{-\bullet}$; process simulation	66 (1989)
100-mL glass syringes; spectrophotometric analysis; oxygen measurements by GC; 0–46°C, pH 0–4, $HClO_4$	Formation of oxygen is proposed and confirmed by isotopic interchange; primary step: dissociation of oxygen at acid pH; ozone first-order kinetics; activation energy: 79.5 kJmol^{-1}	67 (1991)
Batch–semibatch photoreactors; 5.3 W low pressure Hg 1 amp; colorimetric analysis, 20°C, pH 2–10, NaOH, and H_2SO_4 continuous feeding	Kinetics: no UV: $-r_{O3} = k_1 10^{0.395(\text{pH 14})} C_{O3}^{1.5}$ With UV: $-r_{O3} = k_2 10^{0.064(\text{pH 14})} C_{O3}^{1.5} I^{0.9} + k_1 10^{0.395(\text{pH 14})} C_{O3}^{1.5}$ Chloride and nitrate: weak scavengers of $HO\bullet$	68 (1996)
Pulse radiolysis experiments; 20°C, pH > 11	Kinetic modeling with ozone decomposition mechanisms of Reference 21 and Reference 24 with some minor modifications; inorganic carbon reactions included; mechanism of Reference 24 better at high pH	69 (1992)

TABLE 2.7 (continued)
Works on Aqueous Ozone Decomposition Kinetics

Reacting System and Operating Conditions	Main Observations	Reference No. (Year)
Batch reactor; natural organic matter (NOM): 0–0.25 mM as organic carbon; spectrophotometric analysis with NOM-free water; colorimetric analysis: with NOM in water; pH 7.5 phosphate buffer, t-butanol, and p-chlorobenzoate in some runs	Kinetic modeling of ozone decomposition following mechanisms in Reference 21 and Reference 24; predictions of free radical and molecular ozone concentrations with time; proposition of some HO + NOM reactions to fit experimental results	25 (1997)
100-mL glass syringe reactors; 22–45°C, pH 0–4, HClO$_4$; presence of H$_2$O$_2$ produced from γ-ray irradiation of oxygen; spectrophotometric analysis	Mechanism proposed and tested Formation of H$_2$O$_2$ checked; initiation and termination reactions are surface-catalyzed	70 (1998)
Semibatch bubble photoreactor; spectrophotometric analysis; low pressure Hg lamp; 20°C, pH 7	Kinetic modeling of ozone decomposition by UV radiation; determination of the ozone quantum yield: 0.64	71 (1996)
1 cm stirred spectrophotometric quartz-glass cells; spectrophotometric analysis; batch flask, colorimetric analysis; activated carbon or black carbon; different organics: methanol, acetate, p-chlorobenzoate, NOM, etc.	Catalytic effect of activated carbon and a carbon black; effects of promoters and inhibitors of ozone decomposition; ozone first-order kinetics Catalytic reaction generates hydroxyl radicals	27 (1998)
500-mL graduated glass cylinder; 20°C, pH 7.5 phosphate buffer; NOM from 18 surface waters; p-chlorobenzoate addition to determine HO concentration[27]	Promoting and inhibiting effects of NOM; ozone decomposition in two steps: $t < 1$ min and $t > 1$ min: ozone first-order kinetics; rate constant correlated to physical parameters of surface waters	44 (1999)
Stopped flow spectrophotometer; spectrophotometric analysis; 25°C, pH 10.4–13.2 NaOH, 0.5 M ionic strength with HClO$_4$	Use of mechanism in Reference 24 with some modifications; small amounts of H$_2$O$_2$ adsorbed on the spectrophotometer cell can alter the kinetic results; prediction of ozone and other free radicals half-life	72 (2000)
38.3-L batch reactor; liquid ozone UV photometry analyzer, 18–30°C, 2.6 and 7, phosphate buffer; ionic strength: 0.01 M	Mechanism and rate equation as in Reference 45 Rate constant and activation energies determined	73 (2002)

References

1. Riebel, A.H. et al., Ozonation of carbon-nitrogen bonds. I. Nucleophilic attack of ozone, *J. Am. Chem. Soc.*, 82, 1801–1807, 1960.
2. Buxton, G.V. et al., Critical review of data constants for reactions of hydrated electrons, hydrogen atoms, and hydroxyl radicals (•OH/•O$^-$) in aqueous solution, *J. Phys. Chem. Ref. Data*, 17, 513–886, 1988.
3. Ayres, G.H., *Análisis Químico Cuantitativo*, Ediciones del Castillo, Madrid, 1970.

4. Doré, M., *Chimie des Oxydants et Traitement des Eaux,* Technique et Documentation-Lavoisier, Paris, 1989.
5. Lin, S.H. and Yeh, K.L., Looking to treat wastewater? Try ozone, *Chem. Eng.*, May, 112–116, 1993.
6. Hoigné, J., Chemistry of aqueous ozone and transformation of pollutants by ozonation and advanced oxidation processes, in *The Handbook of Environmental Chemistry, Vol. 5, Part C, Quality and Treatment of Drinking Water II,* in J. Hrubec, Ed., Springer-Verlag, Heidelberg, 1998, 83–141.
7. Morrison, R.T. and Boyd, R.N., *Organic Chemistry,* Allyn & Bacon, Newton, MA, 1983.
8. Kuczkowski, R.L., Ozone and carbonyl oxides, in *1,3 Dipolar Cycloaddition Chemistry.* John Wiley & Sons, New York, 1984, 197–276.
9. Bailey, P.S., The reactions of ozone with organic compounds, *Chem. Rev.*, 58, 925–1010, 1958.
10. Erikson, R.E. et al., Mechanism of ozonation reactions. IV. Carbon-nitrogen double bonds, *J. Org. Chem.*, 34, 2961–2966, 1969.
11. White, H.M. and Bailey, P.S., Ozonation of aromatic aldehydes, *J. Org. Chem.*, 30, 3037–3041, 1965.
12. Hoigné, J. and Bader, H., Rate constants of the reactions of ozone with organic and inorganic compounds. I. Non dissociating organic compounds, *Water Res.*, 17, 173–183, 1983.
13. Hoigné, J. and Bader, H., Rate constants of the reactions of ozone with organic and inorganic compounds. II. Dissociating organic compounds, *Water Res.*, 17, 185–194, 1983.
14. Mehta, Y.M., George, C.E., and Kuo, C.H., Mass transfer and selectivity of ozone reactions, *Can. J. Chem. Eng.*, 67, 118–126, 1989.
15. Legube, B., Contribution á l'etude de l'ozonation de composés aromatiques en solution aqueuese. Ph.D. thesis, Université de Poitiers, France, 1983.
16. Eisenhauer, H.R., The ozonation of phenolic wastes, *J. Water Pollut. Control Fed.*, 40, 1887–1899, 1968.
17. Eisenhauer, H.R., Increased rate and efficiency of phenolic waste ozonation, *J. Water Pollut. Control Fed.*, 43, 200–208, 1971.
18. Glaze, W.H., Kang, J.W., and Chapin, D.H., The chemistry of water treatment processes involving ozone, hydrogen peroxide and ultraviolet radiation, *Ozone Sci. Eng.*, 9, 335–342, 1987.
19. Weiss, J., Investigations on the radical HO_2 in solution, *Trans. Faraday Soc.*, 31, 668–681, 1935.
20. Staehelin, S. and Hoigné, J., Decomposition of ozone in water: rate of initiation by hydroxyde ions and hydrogen peroxide, *Environ. Sci. Technol.*, 16, 666–681, 1982.
21. Staehelin, S. and Hoigné, J., Decomposition of ozone in water in the presence of organic solutes acting as promoters and inhibitors of radical chain reactions, *Environ. Sci. Technol.*, 19, 1206–1212, 1985.
22. Buhler, R.E., Staehelin, J., and Hoigné, J., Ozone decomposition in water studied by pulse radiolysis. 1. HO_2/O_2^- and HO_3/O_3^- as intermediates, *J. Phys. Chem.*, 88, 2560–2564, 1984.
23. Staehelin, J., Buhler, R.E., and Hoigné, J., Ozone decomposition in water studied by pulse radiolysis. 2. OH and HO_4 as chain intermediates, *J. Phys. Chem.*, 88, 5999–6004, 1984.
24. Tomiyasu, H., Fukutomi, H., and Gordon, G., Kinetics and mechanism of ozone decomposition in basic aqueous solution, *Inorg. Chem.*, 24, 2962–2966, 1985.

25. Westerhoff, P. et al., Applications of ozone decomposition models, *Ozone Sci. Eng.*, 19, 55–73, 1997.
26. Peyton, G.R. and Glaze, W.H., Destruction of pollutants in water by ozone in combination with ultraviolet radiation. 3. Photolysis of aqueous ozone, *Environ. Sci. Technol.*, 22, 761–767, 1988.
27. Jans, U. and Hoigné, J., Activated carbon and carbon black catalyzed transformation of aqueous ozone into OH-radicals, *Ozone Sci. Eng.*, 20, 67–89, 1998.
28. Christensen, H.S., Sehensted, H., and Corfitzan, H., Reactions of hydroxyl radicals with hydrogen peroxide at ambient and elevated temperatures, *J. Phys. Chem.*, 86, 55–68, 1982.
29. Eriksen, T.E., Lind, J., and Merenyi, G., On the acid-base equilibrium of the carbonate radical, *Radiat. Phys. Chem.*, 26, 197–199, 1997.
30. Draganic, Z.D. et al., Radiolysis of aqueous solutions of ammonium bicarbonate over a large dose range, *Radiat. Phys. Chem.*, 38, 317–321, 1991.
31. Neta, P., Huie, R.E., and Ross, A.B., Rate constants for reactions of inorganic radicals in aqueous solution, *J. Phys. Chem. Ref. Data*, 17, 1027–1284, 1988.
32. Chen, S. and Hoffman, M.Z., Reactivity of the carbonate radical in aqueous solution. Tryptophan and its derivatives, *J. Phys. Chem.*, 78, 2099–2102, 1974.
33. Chen, S., Hoffman, M.Z., and Parsons, G.H., Jr., Reactivity of the carbonate radical toward aromatic compounds in aqueous solution, *J. Phys. Chem.*, 79, 1911–1912, 1975.
34. von Gunten, U. and Hoigné, J., Bromate formation during ozonation of bromide containing waters: interaction of ozone and hydroxyl radicals, *Environ. Sci. Technol.*, 28, 1234–1242, 1994.
35. von Gunten, U., Bruchet, A., and Costentin, E., Bromate formation in advanced oxidation processes, *J. Am. Water Works Assoc.*, 88, 53–65, 1996.
36. Westerhoff, P. et al., Numerical kinetic models for bromide oxidation to bromine and bromate, *Water Res.*, 32, 1687–1699, 1998.
37. Haag, W.R. and Hoigné, J., Ozonation of bromide containing waters: kinetics of formation of hypobromous acid and bromate, *Env. Sci. Technol.*, 17, 261–267, 1983.
38. Taube, H., Reactions of solutions containing O_3, H_2O_2, H^- and Br^-. The specific rate of reaction O_3+Br^-, *J. Am. Chem. Soc.*, 64, 2468–2474, 1942.
39. Zehavi, D. and Rabini, J., Oxidation of aqueous bromide ions by hydroxyl radicals. Pulse radiolytic investigation, *J. Phys. Chem.*, 76, 312–319, 1972.
40. Shukairy, H.M., Miltner, R.J., and Summers, R.S., Bromide's effect on DBP formation, speciation and control. Part 1. Ozonation, *J. Am. Water Works Assoc.*, 86, 72–79, 1994.
41. Schwarz, H.A. and Bielski, B.H., Reactions of $HO_2\bullet$ and $O_2^-\bullet$ with iodine and bromine and the I_2^- and I atom reduction potentials, *J. Phys. Chem.*, 90, 1445–1448, 1986.
42. von Gunten, U. and Hoigné, J., Advanced oxidation of bromide containing waters: bromate formation mechanisms, *Environ. Sci. Technol.*, 32, 63–70, 1998.
43. Tanaka, J. and Matsumura, M., Kinetic studies of removal of ammonia from seawater by ozonation, *J. Chem. Technol. Biotechnol.*, 77, 649–656, 2002.
44. Westerhoff, P. et al., Relationship between the structure of natural organic matter and its reactivity towards molecular ozone and hydroxyl radicals, *Water Res.*, 33, 2265–2276, 1999.
45. Acero, J.L. and von Gunten, U., Influence of carbonate on the ozone/hydrogen peroxide based advanced oxidation process for drinking water treatment, *Ozone Sci. Eng.*, 22, 305–328, 2000.

46. Sotelo, J.L. et al., Ozone decomposition in water: kinetic study, *Ind. Eng. Chem. Res.*, 26, 39–43, 1987.
47. Rothmund, V. and Burgstaller, A., Velocity of decomposition of ozone in aqueous solution, *Monatsheft*, 34, 665–692, 1913.
48. Sennewald, K., The decomposition of ozone in aqueous solution, *Z. Physik. Chem. (Leipzig)*, A164, 305–317, 1933.
49. Alder, M.G. and Hill, G.R., The kinetics and mechanism of hydroxide ion catalyzed ozone decomposition in aqueous solution, *J. Am. Chem. Soc.*, 72, 1884–1886, 1950.
50. Stumm, W., Der Zerfall von Ozon in wässriger Lösung, *Helvetica Chem. Acta*, 37, 773–778, 1954.
51. Kilpatrick, M.L., Herrick, C.C., and Kilpatrick, M., The decomposition of ozone in aqueous solution, *J. Am. Chem. Soc.*, 78, 1784–1789, 1956.
52. Hewes, C.G. and Davison, R.R., Kinetics of ozone decomposition with organics in water, *AIChE J.*, 17, 141–147, 1971.
53. Merkulova, B.P., Louchikov, B.S., and Ivanovskii, M.D., Kinetics of ozone decomposition in sulfuric acid solutions, *Izv. Vyssh. Ucheb. Zaved. Khim. Technol.*, 14, 818–823, 1971.
54. Hoigné, J. and Bader, H., The role of hydroxyl radical reactions in ozonation processes in aqueous solutions, *Water Res.*, 10, 377–386, 1976.
55. Rizzuti, L., Augugliaro, V., and Marucci, G., Ozone absorption in alkaline solutions, *Chem. Eng. Sci.*, 31, 877–880, 1976.
56. Sullivan, D.E. and Roth, J.A., Kinetics of ozone self-decomposition in aqueous solution, *AIChE Symp. Series*, 76, 142–149, 1979.
57. Teramoto, M. et al., Kinetics of the self-decomposition of ozone and the ozonation of cyanide ion and dyes in aqueous solutions, *J. Chem. Eng. Japan*, 14, 383–388, 1981.
58. Haruta, K. and Takayama, T., Kinetics of oxidation of aqueous bromide ion by ozone, *J. Phys. Chem.*, 85, 2383–2388, 1981.
59. Sheffer, S. and Esterson, G.L., Mass transfer and reaction kinetics in the ozone/tap water system, *Water Res.*, 16, 383–389, 1982.
60. Forni, L., Bahnemann, D., and Hart, E.J., Mechanism of the hydroxide ion initiated decomposition of ozone in aqueous solution, *J. Phys. Chem.*, 86, 255–259, 1982.
61. Gurol, M.D. and Singer, P.C., Kinetics of ozone decomposition: a dynamic approach, *Env. Sci. Technol.*, 16, 377–383, 1982.
62. Roth, J.A. and Sullivan, D.E., Kinetics of ozone decomposition in water, *Ozone Sci. Eng.*, 5, 37–49, 1983.
63. Sehested, K. et al., Ozone decomposition in aqueous acetate solutions, *J. Phys. Chem.*, 91, 2359–2361, 1987.
64. Nadezhdin, A.D., Mechanism of ozone decomposition in water: the role of termination, *Ind. Eng. Chem. Res.*, 27, 548–550, 1988.
65. Sotelo, J.L. et al., Effect of high salt concentrations on ozone decomposition in water, *J. Env. Sci. Health*, A24, 823–842, 1989.
66. Nakareseisoon, S. and Gordon, G., The very slow decomposition of ozone in highly basic solutions, *Ozone Sci. Eng.*, 11, 49–58, 1989.
67. Sehested, K. et al., The primary reaction in the decomposition of ozone in acidic solutions, *Env. Sci. Technol.*, 25, 1589–1596, 1991.
68. Ku, Y., Su, W., and Shen, Y., Decomposition kinetics of ozone in aqueous solution, *Ind. Eng. Chem. Res.*, 35, 3369–3374, 1996.
69. Chelkowska, K. et al., Numerical simulations of aqueous ozone decomposition, *Ozone Sci. Eng.*, 14, 33, 1992.

70. Sehested, K. et al., On the mechanism of the decomposition of acidic O_3 solutions, thermally or H_2O_2-initiated, *J. Phys. Chem.*, 102, 2667–2672, 1998.
71. Gurol, M.D. and Akata, A., Kinetics of ozone photolysis in aqueous solution, *AIChE J.*, 42, 3283–3292, 1996.
72. Nemes, A., Fábián, I., and Gordon, G., Experimental aspects of mechanistic studies on aqueous ozone decomposition in alkaline solution, *Ozone Sci. Eng.*, 22, 287–304, 2000.
73. Hsu, Y. et al., Ozone transfer into water in a gas-inducing reactor, *Ind. Eng. Chem. Res.*, 41, 120–127, 2002.
74. von Gunten, U. and Laplanche, A., Oxidation and disinfection with ozone, an overview, *Proc. International Specialized Symposium IOA*, pp. 39–73, Toulouse, France, 2000.

3 Kinetics of the Direct Ozone Reactions

This chapter discusses the kinetics of direct ozone reactions. It is evident that the direct ozonation of molecular organic compounds is a treatment that can be used extensively in the direct ozonation of inorganic compounds.

Ozone reactions in water and wastewater are heterogeneous parallel-series gas liquid reactions in which a gas component (ozone) transfers from the gas phase (oxygen or air) to the water phase where it simultaneously reacts with other substances (pollutants) while diffusing. The main aim of the kinetic study is to determine the rate constant of the reactions and mass transfer coefficients. This is achieved by establishing the corresponding kinetic law.[1] In contrast to chemical equilibrium, kinetic laws are empirical and must be determined from experiments. According to the type of experiments, the ozonation kinetic study can follow one of two different approaches. The first approach is based on experimental results of homogeneous ozonation reactions. This is the case where ozone and any compound are dissolved in water and then mixed and their concentrations with time are observed. The kinetic law, in this case, relates the chemical reaction rate to the concentration of reactants (and products, in the case of reversible reactions). Thus, for any general irreversible ozone direct reaction with a compound B,

$$z_{O3}O_3 + z_B B \longrightarrow z_P P \tag{3.1}$$

and z_{O3}, z_B, and z_P are the stoichiometric coefficients of ozone, B and P, respectively. The kinetic law corresponding to the ozone or B chemical reaction rates are

$$r_{O_3} = z_{O_3} k C_{O3}^n C_B^m \tag{3.2}$$

and

$$r_B = z_B k C_{O3}^n C_B^m \tag{3.3}$$

where k, n, and m are the reaction rate constant and reaction orders for ozone and B, respectively. Note that z_{O3} and z_B have negative values due to convention. Both equations are related by the stoichiometric coefficients

$$r_{O3} = \frac{z_{O3}}{z_B} r_B = \frac{1}{z} r_B \tag{3.4}$$

In most of the studies, however, the ratio between coefficients, $z_B/z_{O3} = z$, is considered, so that Reaction (3.1) becomes

$$O_3 + zB \longrightarrow z'P \qquad (3.5)$$

where z has a negative value when used in the rate law.

The second possibility is the study of ozonation kinetics as a heterogeneous process — that is, as it develops in practice. In this second case, the absorption rate of ozone or ozonation rate, N_{O3}, represents the kinetic law of the heterogeneous process. The stoichiometric equation is now

$$O_3(g) \xrightarrow{k_L a} O_3(l) \qquad (3.6)$$

and Reaction (3.1). Step (3.6) represents the mass transfer of ozone from the gas to the water phase, and $k_L a$ is the volumetric mass transfer coefficient (see Chapter 4).

The kinetic law equation is, in many cases, a complex expression (see Chapter 4) that is deduced from transport phenomena studies, and it depends not only on the concentrations, chemical rate constants, and reaction orders as in the homogeneous case, but also on physical properties (diffusivities), equilibrium data (ozone solubility), and mass transfer coefficients.[2,3]

Both the homogeneous and heterogeneous approaches have advantages and drawbacks. For example, the homogeneous approach does not have the problem of mass transfer, and rate constants can be obtained straightforwardly from experimental data of concentration time. Unfortunately, this approach does not allow a comparison between mass transfer and chemical reaction rates, and it is not suitable for very fast ozone reactions unless expensive apparatus, such as the stopped flow spectrophotometer, are available. The "heterogeneous approach" presents the problem of mass transfer that must be considered simultaneously with the chemical reaction, but any type of ozone reaction kinetics can be studied with simple experimental apparatus. This approach allows the mass transfer coefficients of the ozonation system to be measured and, as a consequence, the relative importance of both physical and chemical steps are established.

Regardless of the approach used, homogeneous or heterogeneous, the kinetic study allows the determination of the rate constant, reaction orders (and mass transfer coefficients, in this case). Kinetic law equations, r_{O3} or N_{O3}, come from the application of mass balances of reacting species that depend on the type of flow the phases (gas and water) have through the reactor (see Appendix A1). Mass-balance equations in a reactor are also called the reactor design equations. Also, the type of reactor operation is fundamental. For example, the reactor can be a tank where aqueous solutions of ozone and the target compounds are charged. This is the homogeneous discontinuous or batch reactor case. The mass balance is a differential equation with at least one independent variable, the reaction time. In a different situation, two nonmiscible phases, ozone gas and water containing ozone, and the target compound, respectively, can be simultaneously fed to the reactor (tank, column, etc.), and the unreacted ozone gas and treated water continuously leave the

Kinetics of the Direct Ozone Reactions

reactor. This is the case for heterogeneous continuous reactors where operation is usually carried out at stationary conditions. Reaction time is not a variable and the mass balance is an algebraic or differential equation depending on the flow type of the gas and water phases.

Reactors are also usually classified according to the way reactants are fed and on the type of flow of phases. Regarding the way reactants are fed, the reactors can be:

- Discontinuous or batch type
- Continuous type

Regarding the type of flow, applied only to the continuous reactor type, reactors are ideal or nonideal. There are two situations for the ideal reactors:

- Perfectly mixed reactors
- Plug flow reactors

In the ideal reactors, the type of flow is based on assumptions that allow the equations of mass balance of species to be established. There are also other phenomena, such as the degree of mixing that influences the performance of reactors. In the cases that will be treated here, perfect mixing or no mixing at all will be considered only for ideal reactors. For a more detailed study of the influence of the degree of mixing the reader should refer to other works.[4]

In nonideal reactors, the flow of phases through the reactor does not follow the hypothesis of ideality, and a nonideal flow study should first be undertaken. A summary of ideal reactor design equations and details about nonideal flow studies with ozone examples are presented in Appendix A1 and Appendix A3, respectively.

Because our purpose in this book is to explain the kinetics of ozonation reactions in this work only the fundamentals of heterogeneous gas–liquid reactions are considered (see Chapter 4). The homogeneous case is a more common situation that is discussed in numerous chemical reaction engineering books.[1,5–7]

3.1 HOMOGENEOUS OZONATION KINETICS

When a homogeneous reaction is studied, the rate law is exclusively a function of the concentration of reactants, rate constant of the reaction, and reaction orders.

The kinetic study of homogeneous ozone reactions could preferentially be carried out in three different ideal reactors (see Appendix A1):

- The perfectly mixed batch reactor
- The continuous, perfectly mixed reactor
- The continuous plug flow reactor

3.1.1 BATCH REACTOR KINETICS

In practice, the reactions are usually carried out in small flasks that act as perfectly mixed batch reactors. In these reactors, the concentration of any species and temperature are constant throughout the reaction volume. This hypothesis allows the material balance of any species, i, present in water to be defined as follows:

$$\frac{dN_i}{dt} = r_i V \qquad (3.7)$$

where Ni and V are the molar amount of compound i charged and the reaction volume, respectively, and r_i is the reaction rate of the i compound. Since ozone reactions are in the liquid phase, there is no volume variation and, hence, Equation (3.7) can be expressed as a function of concentration, once divided by V:

$$\frac{dC_i}{dt} = r_i \qquad (3.8)$$

Another simplification of the ozonation kinetics is due to the isothermal character of these reactions so that the use of the energy balance equation is not needed.

In a general case, ozonation experiments aimed at studying the kinetics of direct ozone reactions are developed in the presence of scavengers of hydroxyl radicals and/or at acid pH so that the ozone decomposition reaction to yield hydroxyl radicals is inhibited (see Chapter 2). This is so because the chemical reaction rate, r_i, presents two contributions due to the direct reaction itself and the hydroxyl radical reaction. Thus, for an ozone-reacting compound B, the chemical reaction rate is

$$-r_B = zk_D C_{O3} C_B + k_{HOB} C_{HO} C_B \qquad (3.9)$$

where k_{HOB} and C_{HO} are the rate constant of the reaction between B and the hydroxyl radical and its concentration, respectively. Addition to the reacting medium of hydroxyl radical scavengers and/or carrying out the reaction at acid pH yields negligible or no contribution of the free radical reaction [second term on the right side in Equation (3.9)] to the B chemical reaction rate. In this way, the kinetics of B would be due exclusively to the direct reaction with ozone. It should be highlighted, however, that appropriate treatment of the concentration–time data of the scavenger substance, usually of known hydroxyl and ozone reaction kinetics, allows the concentration of hydroxyl radical to be known (see the R_{CT} concept in Chapter 7). Also, the scavenger substance should not react directly with ozone. For example, p-chlorobenzoic acid (pCBA),[8] atrazine,[9] or benzene[10] are appropriate candidates. The rate constants of their direct reactions with ozone are very low (<10 $M^{-1}s^{-1}$) and those corresponding to their reaction with hydroxyl radical are well established. In these cases, the concentration of hydroxyl radical can be determined from the chemical reaction rate of the scavenger, r_S, which in a batch reactor is

$$r_S = -\frac{dC_S}{dt} = k_{HOS} C_{HO} C_S = k'_{HOS} C_S \qquad (3.10)$$

Kinetics of the Direct Ozone Reactions

where k_{HOS} and C_S are the rate constant of the reaction between the hydroxyl radical and the scavenger substance and its concentration, respectively. Then, the concentration of hydroxyl radicals can be determined from the slope of the straight line resulting from a plot of the logarithm of C_S against time. Thus, once C_{HO} is determined, knowing the value of k_{HOB} (see Chapters 7 through 9), the direct rate constant k_D can also be determined after applying Equations (3.8) and (3.9). Note that the concentration of hydroxyl radical can also be calculated from the R_{CT} concept,[8] which also requires a reference or scavenger compound and the measured data of ozone concentration time (see also Chapter 7).

The ozone reaction is carried out with one of the reactants (ozone or B) in excess so that the process behaves as a pseudo-nth order reaction. For example, if it is assumed that compound B is in excess, then its concentration remains constant with time while that of ozone diminishes. Application of the material balance of ozone in a batch reactor [$i = O_3$, Equation (3.8)] once the contribution of the hydroxyl free radical reaction has been neglected leads to

$$\frac{dC_{O3}}{dt} = -k'_D C_{O3}^n \qquad (3.11)$$

where k'_D is the pseudo-nth order rate constant for ozone and the minus sign indicates the negative value of the stoichiometric coefficient of ozone [–1 in Equation (3.5)]:

$$k'_D = k_D C_B^m \qquad (3.12)$$

with k_D and m being the actual rate constant of Reaction (3.5) and reaction order for B, respectively.

Integration of Equation (3.11) leads to

- For $n = 1$:

$$\mathrm{Ln}\frac{C_{O3}}{C_{O3o}} = -k'_D t \qquad (3.13)$$

and

- For $n \neq 1$

$$\frac{C_{O3}^{1-n}}{1-n} = \frac{C_{O3o}^{1-n}}{1-n} - k'_D t \qquad (3.14)$$

where C_{O3o} is the concentration of ozone at $t = 0$. In most ozone reactions, Equation (3.13) has been confirmed from experimental data so that reactions are first order for ozone. If this procedure is applied to experiments where the concentration of B has been changed, different values of k'_D are obtained. Equation (3.12) expressed in logarithmic form becomes

$$Lnk'_D = Lnk_D + mLnC_B \qquad (3.15)$$

According to Equation (3.15), a plot of the left side against the logarithm of the concentration of B should result in a straight line with slope equal to the reaction order for B. The true rate constant for ozone for Reaction (3.5) is obtained from the intercept of this line. In most of the ozone reactions, the value of m is also 1, so that the direct ozonation can be catalogued as a second-order irreversible reaction. Note that this procedure can also be applied to experiments where the ozone concentration is in excess and the concentration does not change with time during the reaction period. In these cases, $k_B = zk_D$, and the rate constant for B is directly determined. Examples of these procedures can be found in the works of Hoigné and coworkers,[11–14] who determined and compiled multiple data on rate constants for inorganic and organic compound–ozone direct reactions. In addition to these works, Table 3.1 presents a list of other research works on homogeneous ozonation kinetics where these procedures were carried out.

The procedure shown above allows the direct determination of the absolute rate constant of the ozone–compound reaction and it can be called the *absolute method*. Another possible approach involves ozone experiments where the aqueous solution initially contains the target compound, B, and another compound called the reference compuond, R, of known ozone kinetics, that is, with known rate constant and stoichiometry. This is called the *competitive kinetics method*. Both mass-balance equations of B and R applied to the batch reacting system [Equation (3.8)-type equations] are divided by each other to yield the following equation:

$$\frac{dC_B}{dC_R} = z_{rel} k_{rel} \frac{C_B}{C_R} \qquad (3.16)$$

where z_{rel} is the ratio of stoichiometric coefficients of the ozone–B and ozone–R reactions and k_{rel}, the ratio of their corresponding reaction rate constants. After variable separation and integration, Equation (3.16) leads to

$$\ln \frac{C_B}{C_{B0}} = z_{rel} k_{rel} \ln \frac{C_R}{C_{R0}} \qquad (3.17)$$

which indicates that a plot of the logarithm of the left side against the logarithm of the ratio between the concentration of R at any time and at the start of ozonation leads to a straight line of slope $z_{rel} k_{rel}$. Knowing z_{rel} and the rate constant of the ozone–R reaction, the target rate constant is obtained from the slope of the plotted straight line. This method has also been applied to different works (see Table 3.1). The method has the advantage that there is no need for the ozone concentration to be known. However, the reference compound should have a reactivity toward ozone

TABLE 3.1
Works on Homogeneous Ozonation Kinetics

Compound	Observations	Reference No. (Year)
Phenol	SF, AKC, 5 to 35°C, pH = 1.5–5.2, $n = m = 1$, $1/z = 2$, $k = 895$ (25°C, pH = 1.5), $k = 29520$ (25°C, pH = 5.2), AE = 5.74 kcalmol^{-1}	15 (1979)
Dyes, CN$^-$ and CNO$^-$	SF, 25°C, for dyes: AKO3, $n = m = 1$, pH 4–7, $k = 2.8 \times 10^4$ (Naphthol yellow), $k = 1.8 \times 10^6$ (Methylene blue); for CN$^-$, AKC, $n = 0.8$, $m = 0.55$ at $9.4 < $ pH $ < 11.6$, $k = 310$	16 (1981)
Bromide	Reactor: 10-cm quartz cell, pH = 1.2–3.6, 5–30°C, $k = 4.9 \times 10^9 \exp(-10^4/RT) + 1.7 \times 10^{12} \exp(-1100/RT) C_{H+}$	17 (1981)
Nondissociating organics	Determination of rate constants of different nondissociating organic compound–ozone reactions; pH = 1.7–7, 20°C, presence of hydroxyl radical scavengers; different methods applied (ozone or B in excess, absolute and competitive methods to determine rate constants); $1/z$ between 1 and 2.5; for k values see Reference 8	11 (1983)
Dissociating organics	Determination of rate constants of different dissociating organic compound-ozone reactions; 20°C, pH varies depending on compound; different methods applied (ozone or B in excess, absolute and competitive methods to determine rate constants); for k values see Reference 9	12 (1983)
Phenanthrene	SF, AKO3, pH = 2–7, 10–35°C, $n = m = 1$, $k = 19400$ (pH = 2.2), $k = 47500$ (pH = 7) at 25°C, AE between 7 (pH = 3) and 12 kcalmol^{-1} at other pH values	18 (1984)
Benzene	SF, AKC, pH = 3 and 7, 5–35°C, $1/z = 1$; reaction orders vary with pH; rate constant at pH 7 likely implies radical reactions, $k = 0.012$ (pH = 3), $k = 12.2$ s^{-1} (pH = 7), AE between 20.9 (pH = 3) and 3.3 kcalmol^{-1} (pH = 7)	19 (1984)
Cyanide	SF, AKC, pH = 2.5–12, 20°C, $n = 1$, $m = 0.63$, $k = 550 \pm 200$ (pH = 7) and other	20 (1985)
Inorganic compounds	Different methods (absolute and competitive, stopped flow, etc.); rate constants at different pH, 20°C, $n = m = 1$; for k values see Reference 10	13 (1985)
Naphthalene	1-L batch reactor, AKC, pH = 5.6, 1°C, $n = m = 1$, $k = 550$ M^{-1}s^{-1}, $1/z = 2$	21 (1986)
Haloalkanes, olefins, pesticides, and other	Determination of rate constants of different organic compound–ozone reactions; 20°C, pH varies depending on compound; different methods applied (ozone or B in excess, absolute and competitive methods to determine rate constants); $1/z$ between 1 and 4; for k values see Reference 11	14 (1991)

TABLE 3.1 (continued)
Works on Homogeneous Ozonation Kinetics

Compound	Observations	Reference No. (Year)
Herbicides	AKC, CK also applied pH = 2 and 7 in the presence of carbonates; 20°C, at pH 2: k(MCPA) = 11.7, k(2,4D) = 5.27; at pH 7: k(MCPA) = 41.9, k(2,4D) = 29.14; for other k values see Reference 19	22 (1992)
Naphthol, Naphthoate	CK, Naphthalene: reference compound; 1-propanol as scavenger, pH = 2.1–3.6, 20°C: k_{NOH} = 8600 and k_{NO^-} = 5 × 10^9	23 (1995)
Ammonia	SF, AKC, pH = 8–10, 25°C, $n = m = 1$; the O_3/H_2O_2 oxidation also investigated, $k = 12.3$ (pH = 8), $k = 27.0$ (pH = 10)	24 (1997)
Benzene	SF, AKC, pH = 5.2–5.4, 25°C, $1/z = 1$, $m = n = 0.5$, $k = 2.67$ s^{-1}	25 (1997)
Pentachlorophenol	SF, AKO3, pH = 2 to 7, 10–40°C, $n = m = 1$, $k = 7.55 \times 10^4$ (pH = 2) and $k = 2.49 \times 10^7$ (pH = 7), AE: 27 kJmol^{-1}	26 (1998)
Ethene (A), Methyl (B), and Chlorine (C) substituted derivative	SF, AKC, $n = m = 1$, 25°C, $k_A = 1.8 \times 10^5$, $k = 14$ (TCE), $k = 8 \times 10^5$ (propene) and other (see Reference 24)	27 (1998)
Nitrobenzene and 2,6-DNT	Batch reactors, AKC, pH = 2, t-butanol as scavenger, 2–20°C, k(NB) = 259exp(–1403/T), k(DNT) = 1.2 × 10^{6}exp(–3604/T)	28 (1998)
Alicyclic amines	2.5-L batch reactor. AKO3, pH 7 phosphate buffer, $n = m = 1$ (assumed), $k = 6.7$ (pyrrolidine), $k = 9.8$ (piperidine), $k = 29000$ (morpholine), $k = 4000$ (piperazine) and other (see Reference 25)	29 (1999)
Aminodinitrotoluene	Batch reactor, CK, resorcinol: reference compound, pH 5, 20°C, $k_{a\text{-ADNT}} = 1.45 \times 10^5$, $k_{4\text{-ADNT}} = 1.8 \times 10^5$	30 (2000)
Atrazine and ozone by-products	Batch reactor, AKC, pH = 2, 20°C, $k_{ATZ} = 6$, $k_{CDIT} = 0.14$, $k_{DEA} = 0.18$, $k_{DIA} = 3.1$ and other (see Reference 27)	31 (2000)
Cyclopentanone (CP) and methyl-butyl-ketone (MBK)	Batch reactor, AKC, 0.05–0.5 M HClO$_4$, k(CP) = 7.95 × 10^{11}exp(–18000/RT), k(MBK) = 5.01 × 10^{10}exp(–16200/RT) AE in calmol^{-1}	32 (2000)
3-Methyl-piperidine	Batch reactor, AKC, pH = 4–6, t-butanol as scavenger, $k = 6.63$	33 (2001)
Microcystin-LR	SF, AKO3, pH 2–7, $k = 3.4 \times 10^4$ (pH 7), $k = 10^5$ (pH 2), AE = 12 kcalmol^{-1}	34 (2001)
MTBE and ozone by-products	Batch rector, AKC, pH 2, 5–20°C, $k_{MTBE} = 0.14$, $k = 0.78$ (t-butylformate) and other	35 (2001)
Metal-diethylenetrimine-pentaacetate (DTPA)	SF, AKC, 25°C, pH = 7.66, $k_{CaDTPA(2-)} = 6200$, $k_{ZnDTPA(3-)} = 3500$, $k_{Fe(III)OHDTPA(3-)} = 240000$ and other (see Reference 31)	36 (2001)
Blue dye 81	SF, AKC, $1/z = 1$, 22°C, only pseudo first-order rate constant is given	37 (2001)

TABLE 3.1 (continued)
Works on Homogeneous Ozonation Kinetics

Compound	Observations	Reference No. (Year)
Six dichlorophenols	SF, AKC, pH 2–6, 5–35°C, $1/z = 2$, $n = m = 1$, $k_{2,4DCP} = 10^7$ (pH = 6) and other, AE ranging from 44.5 to 55.3 kJmol^{-1}	38 (2002)
Naphthalenes and nitrobenzene sulphonic acids	Batch reactor, AKO3, pH 3–9, atrazine to determine hydroxyl radical concentration, k_D at 20°C: 1,5-naphthalene-disulphonic acid: 41; 1-naphthalene-sulphonic acid: 252; 3-nitrobenzene-sulphonic acid: 22	39 (2002)
Methyl-t-butyl-ether	2-L batch reactor for determination of reaction orders: $n = m = 1$; 1-L continuous perfectly mixed reactor for rate constant data determination: pCBA as scavenger to determine hydroxyl radical concentration; $k_D = 1.4 \times 10^{18} \exp(-95.4/RT)$, AE in kJmol^{-1}	10 (2002)

Notes: SF = stopped flow spectrophotometer used. AKC = absolute method to determine k with compound in excess (see Section 3.1). AKO3 = absolute method to determine k with ozone in excess. CK = competitive kinetics to determine k. AE = activation energy. n = ozone reaction order. m = compound reaction order. Units of k in $M^{-1}s^{-1}$ unless otherwise indicated. Other stoichiometric ratio values of ozone direct reactions, determined from homogeneous aqueous solutions, are given in Table 5.6.

similar to that of the target compound, B, and the accuracy of the rate constant determined will depend on that of the rate constant of the ozone–R reaction.

3.1.2 Flow Reactor Kinetics

Kinetic studies of homogeneous direct ozone reactions can also be carried out in flow reactors such as the ideal plug flow and continuous stirred tank reactors (see Appendix A1). In these cases, in addition to the reactor itself, other experimental parts are needed that make the procedure more complex. Parts needed include tank reservoirs to contain the aqueous solutions of ozone and the target and scavenger compounds B and S, and pumps to continuously feed both solutions to the reactor and measurement devices for the flow rates.

These reactors present another drawback with respect to batch reactors related to the concentration–time data obtained. Thus, in batch reactors, the concentration–time data obtained from only one experimental run is sufficient to determine the apparent rate constant and B, or ozone reaction order, depending on the reactant in excess. Because flow reactors usually operate at steady state, only one value of the concentration (at the reactor outlet) is obtained in each experimental run. Thus, determination of the apparent rate constant is less accurate. On the other hand, the procedure for the determination of the direct rate constant in flow reactor experiments is similar to that shown above for the batch reactor. It starts with the application of the corresponding reactor design equation (see Appendix A1), which is the mass balance of ozone or B; application of experimental data to this equation allows the determination of the rate constant and reaction orders. Furthermore, in a plug flow reactor, the procedure is the same as in a batch reactor since the mathematical expressions of the design equation of both reactors coincide. The only difference is that t (the actual reaction time in a batch reactor) is the hydraulic residence time, τ the reactor volume to volumetric flow rate ratio (see Appendix A1), and that application of the integrated Equations (3.13) or (3.14) or (3.17) requires data at steady state from experiments at different hydraulic residence time. In a continuous perfectly mixed reactor, the mathematical procedure is easier because the design equation is now an algebraic expression. For example, in a case where a reaction is carried out in the presence of a hydroxyl radical scavenger with ozone concentration in excess, the design equation or mass balance of B, once the steady state has been reached, is (see also Appendix A1)

$$v_0 C_{Bo} - v C_B = z k_D C_{O3}^n C_B^m = k_D'' C_B^m V \qquad (3.18)$$

where v_0 and v represent the volumetric flow rates of the aqueous solution containing B at the reactor inlet and outlet, respectively, and C_{B0} and C_B are their corresponding concentrations. Equation (3.18) has two unknowns, k_D'' and m, so more than one experiment should be needed to determine these parameters. This drawback, however, can be eliminated since ozone direct reactions are usually first-order reactions with respect to ozone and B (see works quoted in Table 3.1) so that (with $m = 1$) from just one experiment, k_D'' can be determined. Although unusual in the literature, this procedure has been applied by Mitani et al.[10] to determine the rate constant of the direct reaction between ozone and methyl-t-butyl-ether (MTBE); the authors also considered in Equation (3.18) the reaction rate term due to the hydroxyl radical-MTBE. As a consequence, the experiment was carried out in the presence of pCBA to express the concentration of hydroxyl radicals as a function of the R_{CT} parameter[8] (see also Chapter 7). They also measured the concentration of ozone at the reactor outlet and determined directly from the design equation the actual direct rate constant as follows:

$$k_D = \frac{v_0 C_{B0} - v C_B - k_{HOB} C_{HO} C_B V}{z C_{O3} C_B V} \qquad (3.19)$$

The value of k_{HOB} had been previously determined from the competitive oxidation of MTBE and benzene.[10] Data on reaction rate constant for the ozone-MTBE system are also given in Table 3.1.

3.1.3 Influence of pH on Direct Ozone Rate Constants

Data on rate constant values thus far determined (see Table 3.1) show that the reactivity of ozone with some inorganic and organic dissociating compounds extraordinarily

Kinetics of the Direct Ozone Reactions

changes with pH, which constitutes a fundamental aspect of ozone kinetics. Note that indirect reactions of ozone are also pH dependent so that in an ozonation process the increase in ozonation rate with pH can be due not only to indirect reactions but also to direct reactions. This could lead to some errors interpreting the ozonation kinetics because at high pH the direct ozone reaction also can be responsible for the removal of pollutants (i.e., phenols; see Chapter 7). Because of its importance in water pollution, the reaction between ozone and a phenol compound is given here as a typical example. Phenol compounds dissociate according to the equilibrium

$$C_6H_5OH \rightleftarrows C_6H_5O^- + H^+ \quad (3.20)$$

Thus, depending on the pH and pK values, the degree of dissociation is

$$\alpha = \frac{1}{1+10^{pK-pH}} \quad (3.21)$$

This means that the ratio of concentrations of the dissociating and nondissociating forms of the phenol compounds varies with pH and with the reactivity with ozone. At low pH value the phenol compound is present only in its nondissociating form ($\alpha = 0$) while at a pH higher than the pK only the dissociating form is in water ($\alpha = 1$). As was shown in Chapter 2 for the activation of electrophilic aromatic substitution reactions, the activating character of substituting groups strongly influences the reaction rate. Thus, the $-O^-$ group is a stronger activating group than the $-OH$ group and, thus, the reactivity of phenols with ozone increases with pH, as observed experimentally. In fact, the rate constant of the ozone–phenol reaction follows Equation (3.22):

$$k_D = k_{Ndis}(1-\alpha) + k_{dis}\alpha \quad (3.22)$$

where k_{Ndis} and k_{dis} are the rate constants of the reactions of ozone with the nondissociating and dissociating forms of the phenol compound, respectively. Equation (3.22) takes into account the contribution of both direct reactions of ozone. In a general case, the variation of k_D with pH follows a sigmoidal curve as shown in Figure 3.1 for a general ozone–phenol compound reaction case. Figures similar to Figure 3.1 have been reported in the literature.[12,40]

As deduced from literature data,[12] rate constants of phenol compound–ozone reactions at alkaline pH are very high. This means that the kinetics of ozonation is accomplished in a few seconds or even less. Therefore, determination of the concentration at different times has to be followed with special apparatus such as the stopped-flow spectrophotometer. Examples of this approach are listed in Table 3.1. In this kind of study, the reaction is carried out in a spectrophotometer cell where the reaction can be stopped at different microseconds of time, thus allowing the determination of the concentration time profile (see references in Table 3.1). Note that in some cases scavengers (*t*-butanol, carbonates) of hydroxyl radicals were added

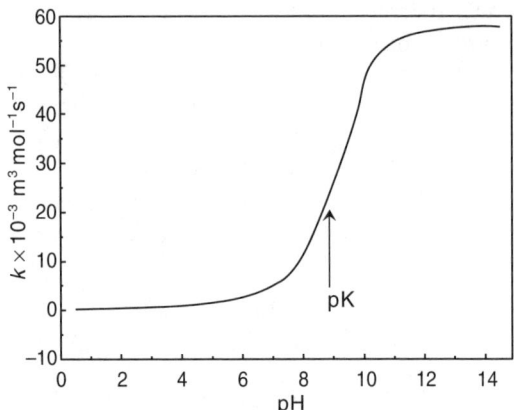

FIGURE 3.1 Influence of pH on the rate constant of the direct reaction of ozone with a phenol compound (x and y axes show arbitrary values).

to the aqueous solutions of ozone and compound B to eliminate any possible competition between the ozone decomposition and hydroxyl radical reactions in water. This is particularly important when the ozone–B reaction has a low reactivity, as will be discussed in Chapter 5.

3.1.4 Determination of the Stoichiometry

As can be deduced from Reaction (3.5), the stoichiometric ratio, z, represents the number of moles of compound B consumed per mole of ozone consumed. This parameter should be calculated so we can know the difference between the disappearance rates of ozone, r_{O3}, and compound B, r_B. In addition, the stoichiometric ratio is a fundamental parameter to establish the kinetic regime of the ozone absorption when ozone undergoes fast reactions in water and, hence, to define the kinetics of the heterogeneous reaction of ozone with compound B, as will be shown in Chapter 5.

In practice, the inverse of z is usually determined because z is in most cases a number lower than unity. The stoichiometric ratio is, as a rule, determined from the mixture of aqueous solutions of different known concentrations of ozone and the target compound B. After mixing, the ratio of the initial concentrations of the resulting solution, C_{Bo}/C_{O3o}, becomes C_B/C_{O3}. Then, the apparent stoichiometric ratio is

$$z_{ap} = \frac{C_{O3o} - C_{O3}}{C_{Bo} - C_B} \tag{3.23}$$

In all cases, when z_{ap} is plotted against C_{Bo}/C_{O3o}, a curve like that shown in Figure 3.2 is obtained. As can be deduced from Figure 3.2, the true stoichiometric ratio of the ozone–B reaction, $1/z$, should be the asymptotic value of the curve corresponding to high values of the ratio between initial concentrations of B and ozone. At low values

Kinetics of the Direct Ozone Reactions

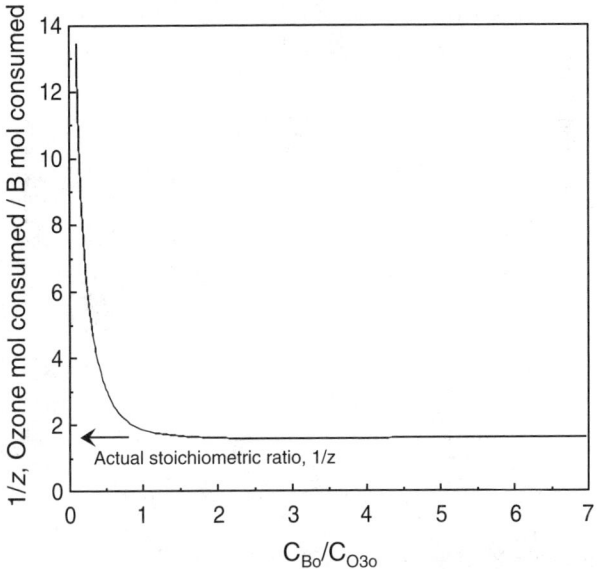

FIGURE 3.2 Variation of the apparent stoichiometric ratio of a general direct reaction between ozone and a given compound using the homogeneous method. Determination of the stoichiometric ratio, $1/z$ (x and y axes show arbitrary values).

of this ratio, ozone is consumed not only by compound B but also in different reactions with intermediate compounds, which result in high values of z_{ap}. When the ratio C_{Bo}/C_{O3o} increases, reactions of intermediates diminish and, as a consequence, so does z_{ap}. When C_{Bo} is in excess, ozone is consumed exclusively by reacting with B so that z_{ap} remains constant regardless of the C_{Bo}/C_{O3o} value. For this situation, z_{ap} represents the inverse of the true stoichiometric ratio, $1/z$. The literature also reports numerous stoichiometric ratio values determined in this way for different ozone reactions (see also Tables 3.1 and 5.6 for examples).

3.2 HETEROGENEOUS KINETICS

When ozone reactions are carried out in practice, that is, by feeding air or oxygen containing ozone into the aqueous solution of the target compound B, the process is heterogeneous and kinetic equations usually express the amount of ozone absorbed per unit time and volume. These equations, in contrast to those of the homogeneous process, not only involve chemical reaction parameters (i.e., the chemical reaction rate constant) but also transport coefficients (i.e., the mass transfer coefficient). Thus, the heterogeneous study allows the relative importance of chemical and physical steps that develop simultaneously to be established. In this sense, ozonation kinetics can be classified as fast, moderate, or slow reactions, and any one of these situations leads to different rate laws. Because of this diversity, it is advisable to introduce the the fundamentals of gas–liquid reactions in Chapter 4 before discussing heterogeneous ozonation kinetics in Chapter 5.

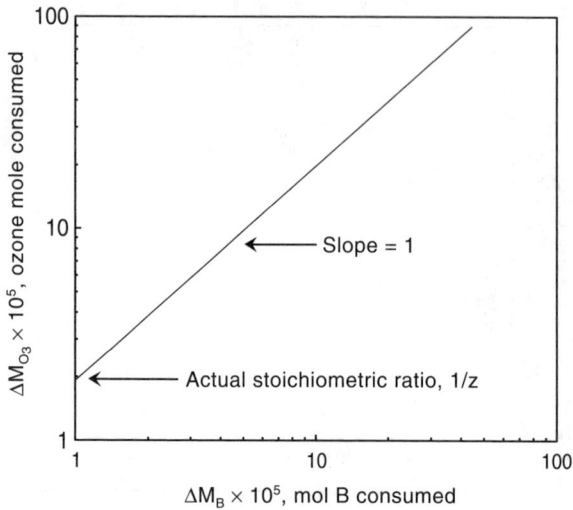

FIGURE 3.3 Variation of molar amounts of ozone and B consumed in a heterogeneous ozonation experiment at different reaction times. Determination of the stoichiometric ratio, $1/z$ (x and y axes show arbitrary values).

3.2.1 DETERMINATION OF THE STOICHIOMETRY

The stoichiometric coefficient of any general ozone direct Reaction (3.5) can also be determined from heterogeneous ozonation experiments where gas containing ozone is continuously fed to an aqueous solution where the target compound B is dissolved at known concentration. As in the homogeneous case, the main difficulty is due to the competition of secondary reactions for the available ozone. In the heterogeneous case, molar amounts of reacted ozone and B are calculated at different reaction periods and plotted together in logarithmic coordinates as shown in Figure 3.3. The ratio of these amounts, that is, the ordinate of the resulting curve or straight line, represents the global stoichiometry of Reaction (3.5), $1/z_G$, which is defined as the moles of ozone consumed in all reactions ozone undergoes in water per mol of B consumed. When the slope of the plotted line is unity, secondary reactions consuming ozone are negligible and z_G is, in fact, z_{app}. At higher reaction periods, the slope likely deviates from unity; that is, $1/z_G$ increases as a consequence of the importance of secondary ozone reactions (see Figure 3.3). The procedure was used by Sotelo et al.[41] to determine the true stoichiometric ratio of the ozone–resorcinol and ozone–pholoroglucinol direct reactions (see also Section 5.3.1).

REFERENCES

1. Levenspiel, O. *Chemical Reaction Engineering*, 3rd ed., John Wiley & Sons, New York, 1999.
2. Danckwerts, P.S.V. *Gas–Liquid Reactions*, McGraw-Hill, New York, 1970.
3. Chaudhari, R.V. and Ramachandran, P.A., Three-phase slurry reactors, *AIChE J.*, 26, 177–201, 1980.

4. Trambouze, P., Van Landeghem, H. and Wauquier, J.P., *Les Reacteurs Chimiques. Conception/Calcul/Mise en Oeuvre*, Editions Technic, Paris, 1984.
5. Moore, J.W. and Pearson, R.G. *Kinetics and Mechanism. A Study of Homogeneous Chemical Reactions*, 3rd ed., John Wiley & Sons, New York, 1981.
6. Smith, J.M. *Chemical Engineering Kinetics. Chemical Engineering Series*, 2nd ed., McGraw-Hill, New York, 1970.
7. Butt, J.B., *Reactor Kinetics and Reactor Design*, Prentice-Hall, Englewood Cliffs, New Jersey, 1980.
8. Elovitz, M.S. and von Gunten, U., Hydroxyl radical/ozone ratios during ozonation processes. I. The R_{CT} concept, *Ozone Sci. Eng.*, 21, 239–260, 1999.
9. von Gunten, U., Hoigné, J. and Bruchet, A., Oxidation in ozonation processes. Application of reaction kinetics in water treatments. In *Proceedings 12th Ozone World Congress*, Vol. 1, pp. 17–25, Lille, France, 1995.
10. Mitani, M.M. et al., Kinetics and products of reactions of MTBE with ozone and ozone/hydrogen peroxide in water, *J. Haz. Mater.*, B89, 197–212, 2002.
11. Hoigné, J. and Bader, H. Rate constants of the reactions of ozone with organic and inorganic compounds. I. Nondissociating organic compounds, *Water Res.*, 17, 173–183, 1983.
12. Hoigné, J. and Bader, H. Rate constants of the reactions of ozone with organic and inorganic compounds. II. Dissociating organic compounds, *Water Res.*, 17, 185–194, 1983.
13. Hoigné, J. et al., Rate constants of the reactions of ozone with organic and inorganic compounds. III. Inorganic compounds and radicals, *Water Res.*, 19, 993–1004, 1983.
14. Yao, C.C.D. and Haag, W.R., Rate constants of direct reactions of ozone with several drinking water contaminants, *Water Res.*, 25, 761–773, 1991.
15. Li, K.Y., Kuo, C.H., and Weeks, J.L., A kinetic study of ozone-phenol reaction in aqueous solutions, *AIChE J.*, 25, 583–591, 1979.
16. Teramoto, M. et al., Kinetics of the self-decomposition of ozone and the ozonation of cyanide ion and dyes in aqueous solutions, *J. Chem. Eng. Japan*, 14, 383–388, 1981.
17. Haruta, K. and Takayama, T., Kinetics of oxidation of aqueous bromide ion by ozone, *J. Phys. Chem.*, 85, 2383–2388, 1981.
18. Cornell, L.P. and Kuo, C.H., A kinetic study of ozonation of phenanthrene in aqueous solutions, *Trans. Air Pollut. Control Assoc.*, 2, 275–286, 1984.
19. Kuo, C.H. and Soong, H.S., Oxidation of benzene by ozone in aqueous solutions, *The Chem. Eng. J.*, 28, 163–171, 1984.
20. Gurol, M.D., Bremen, W.M., and Holden, T.E., Oxidation of cyanides in industrial wastewater by ozone, *Env. Progr.*, 4, 46–51, 1985.
21. Legube, B. et al., Ozonation of naphthalene in aqueous solution. II. Kinetic studies of the initial reaction step, *Water Res.*, 20, 209–214, 1986.
22. Xiong, F. and Graham, N.J.D., Rate constants for herbicide degradation by ozone, *Ozone Sci. Eng.*, 14, 283–301, 1992.
23. Chan, W.F. and Larson, R.A., Ozonolysis of aromatic compounds in aqueous solutions containing nitrite ion, *Ozone Sci. Eng.*, 17, 627–635, 1995.
24. Kuo, C., Yuan, F. and Hill, D.O., Kinetics of oxidation of ammonia in solutions containing ozone with or without hydrogen peroxide, *Ind. Eng. Chem. Res.*, 36, 4108–4113, 1997.
25. Kuo, C.H. et al., Vapor and liquid phase ozonation of benzene, *Ozone Sci. Eng.*, 19, 109–127, 1997.
26. Kuo, C.H. and Huang, C.H., Kinetics of ozonation of pentachlorophenol in aqueous solutions, *Ozone Sci. Eng.*, 20, 163–173, 1998.

27. Dowideit, P. and von Sonntag, C., Reaction of ozone with ethene and its methyl- and chlorine-substituted derivatives in aqueous solution, *Env. Sci. Technol.*, 32, 1112–1119, 1998.
28. Beltrán, F.J., Encinar, J.M., and Alonso, M.A., Nitroaromatic hydrocarbon ozonation in water. 1. Single ozonation, *Ind. Eng. Chem. Res.*, 37, 25–31, 1998.
29. Pietsch, J. et al., Kinetic and mechanistic studies of the ozonation of alicyclic amines, *Ozone Sci. Eng.*, 21, 23–37, 1999.
30. Spanggord, R.J., Yao, D., and Mill, T., Kinetics of aminodinitrotoluene oxidations with ozone and hydroxyl radical, *Env. Sci. Technol.*, 34, 450–454, 2000.
31. Acero, J.L., Stemmler, K., and von Gunten, U., Degradation kinetics of atrazine and its degradation products with ozone and OH radicals: a predictive tool for drinking water treatment, *Env. Sci. Technol.*, 34, 591–597, 2000.
32. Zimin, Y.S. et al., Kinetics of ketone oxidation by ozone in water solutions. *Kinetics Catal.*, 41, 745–748, 2000.
33. Carini, D. et al., Ozonation as pre-treatment step for the biological batch degradation of industrial wastewater containing 3-methyl-piperidine, *Ozone Sci. Eng.*, 23, 189–198, 2001.
34. Shawwa, A.R. and Smith, D.W., Kinetics of microcystin-LR oxidation by ozone, *Ozone Sci. Eng.*, 23, 161–170, 2001.
35. Acero, J.L. et al., MTBE oxidation by conventional ozonation and the combination ozone/hydrogen peroxide: efficiency of the processes and bromate formation, *Env. Sci. Technol.*, 35, 4252–4259, 2001.
36. Stemmler, K., Glod, G., and von Gunten, U., Oxidation of metal-diethylenetriamine-pentaacetate (DTPA)-complexes during drinking water ozonation, *Water Res.*, 35, 1877–1886, 2001.
37. Ledakowicz, S. et al., Ozonation of reactive Blue 81 in the bubble column, *Water Sci. Technol.*, 44, 47–52, 2001.
38. Qiu, Y., Kuo, C., and Zappi, M.E., Ozonation kinetics of six dichlorophenol isomers, *Ozone Sci. Eng.*, 24, 123–131, 2002.
39. Calderara, V., Jekel, M., and Zaror, C., Ozonation of 1-naphthalene, 1,5-naphthalene and 3-nitrobenzene sulphonic acids in aqueous solutions, *Env. Technol.*, 23, 373–380, 2002.
40. Beltrán, F.J. et al., Degradation of o-clorophenol with ozone in water, *Trans. Inst. Chem. Eng., Part B Process Safety Environ. Protection*, 71, 57–65, 1993.
41. Sotelo, J.L., Beltrán, F.J., and González, M., Ozonation of aqueous solutions of resorcinol and pholoroglucinol. 1. Stoichiometric and absorption kinetic regime, *Ind. Eng. Chem. Res.*, 29, 2358–2367, 1990.

4 Fundamentals of Gas–Liquid Reaction Kinetics

The kinetics of heterogeneous reactions is governed by absorption theories of gases in liquids accompanied by chemical reactions. The fundamentals of these theories are necessary to understand the phenomena developing during the ozonation of compounds in water. The necessary steps for studying the kinetics of gas–liquid reactions are described below. Since ozone reactions can be considered irreversible, isothermic, and second-order[1] (for a general ozone–B reaction) or pseudo first-order (the ozone decomposition reaction), the discussion that follows mainly refers to this type of gas–liquid reaction. Nonetheless, some fundamental aspects on series-parallel reactions are also given. Note that ozonation of compounds in water yields a series of by-products that also react with ozone. Therefore, the kinetics of series-parallel gas–liquid reactions constitutes another part of this study. As a first step, the physical absorption of a gas in a liquid is discussed below.

4.1 PHYSICAL ABSORPTION

In a gas–liquid reacting system, diffusion, convection, and chemical reaction proceed simultaneously, and the behavior of the system can be predicted with the use of models that simulate the situation for practical purposes. These models are based on those describing the gas physical absorption phenomena, that is, they are based on gas absorption theories. In a general case, when gas and liquid phases are in contact, the components of one phase can transfer to the other to reach equilibrium. If it is assumed that one component A of a gas phase is transferred to the liquid phase, the rate of mass transfer or absorption rate of A is:

$$N_A = k_G(P_{Ab} - P_i) = k_L(C_A^* - C_{Ab}) \tag{4.1}$$

where the SI units of N_A are in molm^{-2}s^{-1}; k_G and k_L are the individual mass transfer coefficients for the gas and liquid phases, respectively; P_{Ab} and P_i are the partial pressures of A in the bulk gas and at the interface, respectively; and C_A^* and C_{Ab}, are the concentrations of A at the interface and in the bulk of the liquid, respectively (see Figure 4.1). One of the two problems in the rate law is to find some mathematical expression for the mass transfer coefficients. The other one is to determine the interfacial concentrations, P_i or C_A^*. Theoretical expressions for mass transfer coefficients

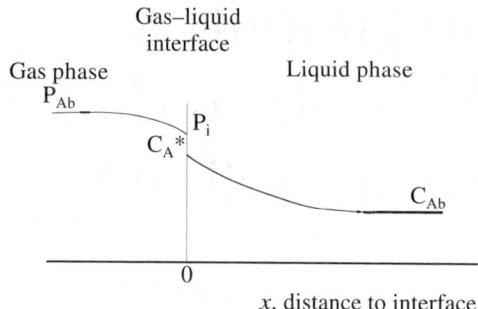

FIGURE 4.1 Concentration profile of a gas component A with the distance to the interface during its physical absorption in a liquid.

can be found from the solution of the microscopic mass-balance equation of the transferred component A that, applied to the liquid phase, is as follows:

$$D_T \nabla^2 C_A = U \nabla C_A + \frac{\partial C_A}{\partial t} \quad (4.2)$$

where the term on the left side represents the molecular and turbulent transport rate of A and the first and second terms on the right side represent the terms of convection and accumulation rates of A, respectively. Equation (4.2) is conveniently simplified according to the hypothesis of the absorption theories. The two most often applied absorption theories are the film and surface renewal theories.

4.1.1 THE FILM THEORY

Lewis and Whitman[2] proposed that when two nonmiscible phases are in contact, the main resistance to mass transfer is located in a stationary layer of thickness δ close to the interface, called the film layer. It is also assumed that mass transfer through the film is only due to diffusion and that the concentration profiles with distance to the interface are reached instantaneously. This is then called the theory of the pseudo stationary state.

In a gas–liquid system there are two films, one for each phase. In most common situations the gas is bubbled into the liquid phase, so the interfacial surface is due to the external surface of bubbles. Concentration profiles of the gas component being absorbed for both the gas and liquid films are shown in Figure 4.2. The film theory assumes a plane interfacial surface when the bubble radius is much lower than the film thickness, δ, a situation that occurs in most of the gas–liquid systems. According to these statements (diffusion, stationary state, and one direction for mass transfer), Equation (4.2) reduces to:

$$D_A \frac{\partial^2 C_A}{\partial x^2} = 0 \quad (4.3)$$

Fundamentals of Gas–Liquid Reaction Kinetics

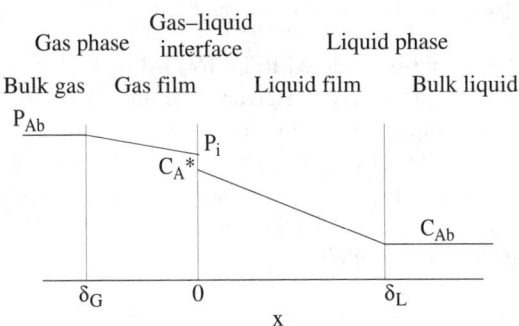

FIGURE 4.2 Concentration profile of a gas component A with the distance to the interface during its physical absorption according to film theory.

Equation (4.3) can be solved with the following conditions:

$$x = 0 \quad C_A = C_A^*$$
$$x = \delta_L \quad C_A = C_{Ab} \tag{4.4}$$

where δ_L and C_{Ab} are the liquid film layer and concentration of A in bulk liquid, respectively. Solution of this mathematical model leads to the concentration profile of A:

$$C_A = C_A^* - \frac{C_A^* - C_{Ab}}{\delta_L} \tag{4.5}$$

By application of Fick's law, the rate of mass transfer or absorption rate is:

$$N_{Ao} = -D_A \left.\frac{dC_A}{dx}\right|_{x=0} = \frac{D_A}{\delta_L}\left(C_A^* - C_{Ab}\right) \tag{4.6}$$

Note that a similar expression could be obtained for the gas phase if Equation (4.3) is applied to the gas film layer. Comparing Equation (4.6) with the general Equation (4.1), the film theory yields the following equation for k_L:

$$k_L = \frac{D_A}{\delta_L} \tag{4.7}$$

Thus, δ_L, the film thickness, is the characteristic parameter of film theory.

4.1.2 SURFACE RENEWAL THEORIES

In these models, the liquid is assumed to be formed by elements of infinite width that are exposed to the interface for a given time and then are replaced by other elements from the bulk liquid. While the liquid elements are at the interface, mass transfer occurs by diffusion in a nonstationary way. The most often used surface renewal theory is that proposed by Danckwerts,[3] which assumes a distribution function of exposition times for the liquid elements. For this surface renewal theory, Equation (4.2) reduces to the following:

$$D_A \frac{\partial^2 C_A}{\partial x^2} = \frac{\partial C_A}{\partial t} \tag{4.8}$$

with the boundary limits:

$$\begin{aligned} t &= 0 & C_A &= C_{Ab} \\ t &> 0 \quad x = 0 & C_A &= C_A^* \\ & x \to \infty & C_A &= C_{Ab} \end{aligned} \tag{4.9}$$

After application of Fick's law,[4] once the concentration profile of A with time and position $[C_A = f(x,t)]$ has been determined from the solution of Equation (4.8), the mean absorption rate of A is as follows:

$$\overline{N}_A = \int_0^\infty N_A(t)\psi(t)dt = \sqrt{D_A s}\left(C_A^* - C_{Ab}\right) \tag{4.10}$$

where $N_A(t)$ is the absorption rate at $x = 0$ and $\psi(t)$ is the distribution function for exposition times of liquid elements, defined as:[3]

$$\psi(t) = s\exp(-st) \tag{4.11}$$

and s is the surface renewal velocity of any element, the parameter that characterizes the Danckwerts theory.

If Equation (4.10) and Equation (4.1) are compared, the mass transfer coefficient k_L is defined as:

$$k_L = \sqrt{D_A s} \tag{4.12}$$

4.2 CHEMICAL ABSORPTION

When the gas component A, being absorbed, simultaneously undergoes a chemical reaction in the liquid, the microscopic mass-balance Equation (4.2) presents an additional term due to the chemical reaction rate law, r_A:

Fundamentals of Gas–Liquid Reaction Kinetics

$$D_T \nabla^2 C_A + r_A = U \nabla C_A + \frac{\partial C_A}{\partial t} \qquad (4.13)$$

This is the case for ozonation reactions. To determine the gas absorption rate (or ozonation rate for ozone processes) we must also follow the steps shown above for the case of physical absorption, that is:

- Solving the microscopic differential mass-balance equation to determine the concentration profile of the gas being absorbed with distance to the interface
- The application of Fick's law to yield the gas absorption rate

Solution of the mass-balance Equation (4.13) depends on the absorption theory applied as explained below for the film and Danckwerts theories. The cases that are treated correspond to irreversible first- (or pseudo first-) and second-order reactions that are usually the types of simple ozonation reactions in water.

4.2.1 Film Theory

4.2.1.1 Irreversible First-Order or Pseudo First-Order Reactions

These reactions have the following stoichiometry:

1. First-order reaction:

$$A \xrightarrow{k_1} P \qquad (4.14)$$

with

$$r_A = -k_1 C_A \qquad (4.15)$$

2. Pseudo first-order reaction:

$$A + zB \xrightarrow{k_1'} P \qquad (4.16)$$

with

$$r_A = -k_1' C_A C_B = k_1 C_A \qquad (4.17)$$

The ozone self-decomposition reaction in water is the typical example that follows this kinetics.

According to film theory, the mass-balance Equation (4.13) becomes:

$$D_A \frac{\partial^2 C_A}{\partial x^2} = k_1 C_A \qquad (4.18)$$

solved with the boundary conditions given in Equation (4.4). Solution of this system leads to the following concentration profile of A with the distance to the interface:

$$C_A = C_A^* \frac{\sinh\left[\left(1-\frac{x}{\delta_L}\right)Ha_1\right]}{\sinh Ha_1} + C_{Ab} \frac{\sinh\left[\left(\frac{x}{\delta_L}\right)Ha_1\right]}{\sinh Ha_1} \quad (4.19)$$

where Ha_1 is the dimensionless Hatta number for an irreversible first-order reaction defined as follows:

$$Ha_1 = \frac{\sqrt{k_1 D_A}}{k_L} \quad (4.20)$$

The square of Ha_1 represents the ratio of the maximum chemical reaction rate through the film layer and the maximum physical absorption rate:

$$Ha_1^2 = \frac{k_1 C_A^* a \delta_L}{k_L a C_A^*} \quad (4.21)$$

where a is the specific interfacial area and $k_L a$ is the volumetric mass transfer coefficient in the liquid phase. Note that the product $a\delta_L$ is the ratio of the liquid film and liquid total volumes. Thus, Ha_1 indicates the relative importance of chemical reaction and mass transfer rates in the gas–liquid system.

Application of Fick's law at the gas–liquid interface leads to the gas absorption rate equation, or to the rate or kinetic law for this type of reaction:

$$N_{Ao} = -D_A \frac{dC_A}{dx}\bigg|_{x=0} = M_1 \frac{Ha_1}{\tanh Ha_1}\left[1 - \frac{C_{Ab}}{C_A^*}\frac{1}{\cosh Ha_1}\right] \quad (4.22)$$

where M_1 is the maximum physical absorption rate and $k_L C_A^*$ is expressed per unit of interfacial surface. Note that in Equation (4.21), the maximum physical absorption rate is expressed per unit of volume.

In Equation (4.22), it is convenient to express C_{Ab} as a function of chemical and mass transfer parameters. Then, the mass transfer rate at the other edge of the film layer, $(x = \delta_L)$, $N_{A\delta}$, and the chemical reaction rate in the bulk of the liquid, R_{b1}, are needed:

$$N_{A\delta} = N_A\big|_{x=\delta_L} = M_1 \frac{Ha_1}{\sinh Ha_1}\left[1 - \frac{C_{Ab}}{C_A^*}\cosh Ha_1\right] \quad (4.23)$$

and

Fundamentals of Gas–Liquid Reaction Kinetics

$$R_{b1} = k_1 C_{Ab} \frac{\beta}{a} \tag{4.24}$$

where units of both rates are given per surface of interfacial area, with β the liquid holdup, defined as the ratio of liquid to total (gas plus liquid) volumes.

By equalizing Equation (4.23) and Equation (4.24), the ratio between concentrations of A in the bulk of the liquid and at the interface, C_{Ab}/C_A^*, can be obtained. Then, after substitution in Equation (4.22), Equation (4.25) is obtained:

$$N_{Ao} = -D_A \left. \frac{dC_A}{dx} \right|_{x=0} = M_1 \frac{Ha_1}{\tanh Ha_1} \left[1 - \frac{M_1 \dfrac{Ha_1}{\sinh Ha_1 \cosh Ha_1}}{R_{b1} + M_1 \dfrac{Ha_1}{\tanh Ha_1}} \right] \tag{4.25}$$

Also, if C_{Ab}/C_A^* is expressed as a function of $\beta/a\delta_L$, a dimensionless number that represents the ratio of the volumes of the total liquid (film plus bulk liquids) and film layer, the following alternative equation is obtained:[5]

$$N_{Ao} = -D_A \left. \frac{dC_A}{dx} \right|_{x=0} = M_1 \frac{Ha_1}{\sinh Ha_1} \left[\cosh Ha_1 - \frac{1}{\cosh Ha_1 + \dfrac{\beta}{a\delta_L} Ha_1 \sinh Ha_1} \right] \tag{4.26}$$

Equation (4.25) or Equation (4.26) constitutes the general kinetic equations for first-order (or pseudo first-order) gas–liquid reactions.

As can be deduced from Equation (4.25), the absorption rate is a function of three maximum rates:

- R_{b1}, the maximum chemical reaction rate in the bulk liquid = $k_1 C_{Ab} \beta/a$
- M_1, the maximum physical absorption rate at the interface = $k_L C_A^*$
- R_F, the maximum chemical reaction rate through the film layer = $k_1 C_A^* a \delta_L$

Also, depending on the values of Ha_1, the absorption rate develops in different kinetic regimes:[6]

- For the fast kinetic regime when $Ha_1 > 3$, then $C_{Ab} = 0$ and:

$$N_{Ao} = M_1 Ha_1 \tag{4.27}$$

- The moderate kinetic regime when $3 < Ha_1 < 0.3$ with the general Equation (4.25) or Equation (4.26)

- The diffusional kinetic regime when $Ha_1 < 0.3$ and $C_{Ab} = 0$ with:

$$N_{Ao} = M_1 \qquad (4.28)$$

- The slow kinetic regime when $Ha_1 < 0.3$ and $C_{Ab} \neq 0$ with the general Equation (4.25) or Equation (4.26)
- The very slow kinetic regime when $Ha_1 \ll 0.01$ with:

$$N_{Ao} = R_{b1} \qquad (4.29)$$

As can be deduced, depending on the relative importance of mass transfer and chemical reaction rates, the kinetics of the gas–liquid reaction will be the chemical reaction rate (the case of very slow kinetic regime) or the physical absorption rate (slow and diffusional regimes). Note that for slow kinetic regimes, the gas–liquid reaction is a two-series process where mass transfer through the film layer first occurs and then the chemical reaction develops in the bulk liquid. Then the absorption rate equation is:

$$N_{Ao} = k_L\left(C_A^* - C_{Ab}\right) = R_{b1} \qquad (4.30)$$

Since the general Equation (4.25) is rather complex, the absorption rate is usually expressed as a function of another dimensionless number called reaction factor E:

$$E = \frac{N_{Ao}}{M_1} \qquad (4.31)$$

As deduced from Equation (4.31), the reaction factor can be defined as the number of times the maximum physical absorption rate increases due to the chemical reaction. Note that this definition has only physical meaning when the kinetic regime is fast or moderate (for $C_{Ab} = 0$). However, according to Equation (4.31), values of E can be lower than unity (the cases of slow kinetic regime or some others with moderate regime), although they have no practical use. In Figure 4.3, a plot of E against Ha_1 is shown with the zones of different kinetic regimes. Note that for a slow kinetic regime ($Ha_1 < 0.3$) Equation (4.26) is used to show the variation of E with the Hatta number. This is because, in the slow kinetic regime, reactions develop in the bulk liquid and the volume ratio parameter $\beta/a\delta_L$ has a great influence on the gas absorption rate.

4.2.1.2 Irreversible Second-Order Reactions

These reactions are typical for most ozone direct reactions in water. The stoichiometric equation is:

$$A + zB \xrightarrow{k_2} P \qquad (4.32)$$

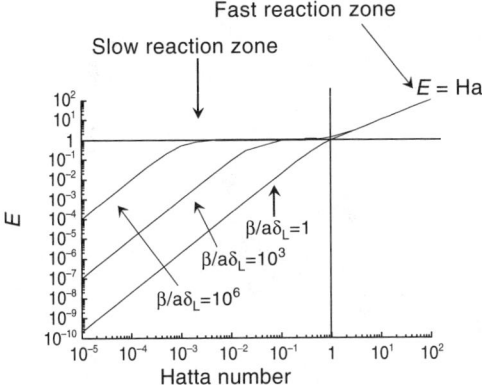

FIGURE 4.3 Variation of the reaction factor with the Hatta number for one irreversible first- or pseudo first-order gas–liquid reaction.

and the chemical reaction rate for the disappearance of A is:

$$r_A = -k_2 C_A C_B \tag{4.33}$$

For this case, both microscopic mass-balance equations of A and B have to be solved simultaneously:

$$D_A \frac{\partial^2 C_A}{\partial x^2} = k_2 C_A C_B \tag{4.34}$$

$$D_B \frac{\partial^2 C_B}{\partial x^2} = z k_2 C_A C_B \tag{4.35}$$

Boundary conditions of this mathematical model are:

$$x = 0 \quad C_A = C_A^* \quad \frac{dC_B}{dx} = 0$$
$$x = \delta_L \quad C_A = C_{Ab} \quad C_B = C_{Bb} \tag{4.36}$$

However, it is not possible to find an analytical solution to this system. Van Krevelen and Hoftijzer[7] found an approximate solution by assuming that the concentration of B through the reaction zone in the film layer is constant, C_{Br}, so that the system transforms to a pseudo first-order one. Figure 4.4 shows the concentration profiles of the actual and assumed situations. In this way, the rate equations deduced in Section 4.2.1.1 could be applied with $k_1 = k_2 C_{Br}$. By applying the proposed approximation,[7] the following equation was found for C_{Br}:

$$C_{Br} = C_{Bb}\left[1 - (E-1)\frac{zC_A^*}{C_{Bb}}\right] \tag{4.37}$$

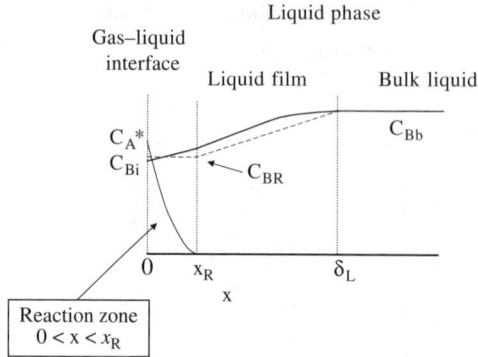

FIGURE 4.4 Concentration profiles of a gas component A and one liquid component B with the distance to the gas–liquid interface during their fast chemical reaction while A is diffusing through the liquid film. Solid line = profiles of B concentration according to film theory. Dotted line = profiles of B concentration according to the simplification of Van Krevelen and Hoftijzer.[7] (From Van Krevelen, D.W. and Hoftijzer, P.J., Kinetics of gas liquid reactions. Part I. General theory, *Rec. Trav. Chim.*, 67, 563–586, 1948. With permission.)

Then Ha_1 [see Equation (4.20)] becomes

$$Ha_1 = \frac{\sqrt{k_2 D_A C_{Br}}}{k_L} = \frac{\sqrt{k_2 D_A C_{Bb}}}{k_L}\sqrt{1-(E-1)\frac{zC_A^*}{C_{Bb}}} \qquad (4.38)$$

Equation (4.38) can be simplified to Equation (4.39):

$$Ha_1 = Ha_2\sqrt{1-(E-1)\frac{zC_A^*}{C_{Bb}}} \qquad (4.39)$$

where Ha_2 represents the Hatta number of the irreversible second-order Reaction (4.32) and has the same physical meaning as Ha_1 in Equation (4.20):

$$Ha_2 = \frac{\sqrt{k_2 D_A C_{Bb}}}{k_L} \qquad (4.40)$$

Finally, the absorption rate of A accompanied by an irreversible second-order chemical reaction with B is given by Equation (4.25) with Ha_1 given by Equation (4.39). The system is solved with Equation (4.31) by a trial-and-error procedure that allows the values of the reaction factor E to be obtained from the values of Ha_2. The solution is usually presented in plots such as Figure 4.5. As deduced from Equation (4.25) and Equation (4.38), the absorption rate is a function of four maximum rates (three of them previously deduced for the first-order reaction kinetics) defined as follows:

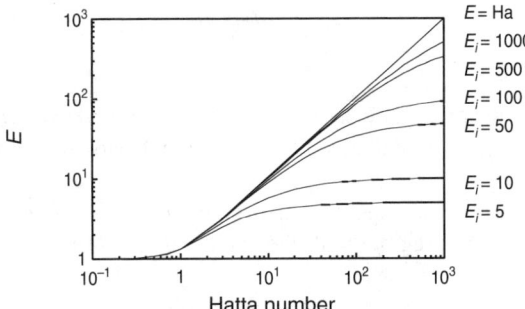

FIGURE 4.5 Variation of the reaction factor with the Hatta number for one irreversible second-order gas–liquid reaction.

- R_{b2}, maximum chemical reaction rate in the bulk liquid:

$$R_{b2} = k_2 C_A^* C_{Bb} \frac{\beta}{a} \quad (4.41)$$

- M_1, maximum physical absorption rate at the interface = $k_L C_A^*$
- R_F, maximum chemical reaction rate through the film layer:

$$R_F = k_2 C_A^* C_{Bb} a \delta_L \quad (4.42)$$

- M_2, maximum physical diffusion of B through the film layer:

$$M_2 = \frac{1}{z} k_L C_{Bb} \quad (4.43)$$

For the case of second-order reactions there are two new kinetic regimes to consider in addition to those listed for first-order reactions:

- Fast kinetic regime, with $C_{Ab} = 0$ and $Ha_2 > 3$:

$$N_{A0} = k_L C_A^* \frac{Ha_1}{\tanh Ha_1} \quad (4.44)$$

where Ha_1 is given by Equation (4.38).

- Instantaneous kinetic regime with $C_{Ab} = 0$ and $Ha_2 > 10 E_i$:

$$N_{A0} = k_L C_A^* E_i \quad (4.45)$$

where E_i is the reaction factor for the instantaneous reaction, defined as follows:

$$E_i = 1 + \frac{D_B C_{Bb}}{zD_A C_A^*} \quad (4.46)$$

where D_B is the diffusivity of compound B in the liquid that can be estimated with the equation of Wilke and Chang[8,9] as indicated in Section 5.1.1.

Note that the fast kinetic regime for first-order reaction is now called the fast pseudo first-order kinetic regime with the following condition to be fulfilled:

$$3 < Ha_2 = \frac{\sqrt{k_2 D_{O3} C_{Bb}}}{k_L} < \frac{E_i}{2} \quad (4.47)$$

The absorption rate law in this case is:

$$N_{A0} = aC_A^* \sqrt{k_2 D_{O3} C_{Bb}} \quad (4.48)$$

Figure 4.6 to Figure 4.12 show the concentration profiles of A and B through the liquid for the different kinetic regimes. Note that for fast or instantaneous kinetic regimes, the chemical reaction develops in a zone or plane of the film layer, respectively, while for slow kinetic regimes, the chemical reaction is in the bulk liquid. Also note that for fast or instantaneous kinetic regime there will not be dissolved gas in the bulk liquid ($C_{Ab} = 0$). It should be highlighted that the rate equations presented are per unit of interfacial surface area (in mols^{-1}m^{-2}). In a practical situation, the problem is that the interfacial surface area is generally unknown. Nonetheless, this problem is overcome by multiplying both sides of the absorption rate equation by the specific interfacial area, a, so that the rate will be given per unit of volume, which is known. Recall that, depending on the relative importance of mass transfer and chemical reaction steps, that is, depending on the kinetic regime, once Equation (4.38) has been accounted for, the general Equation (4.25) simplifies in a way similar to that shown for the case of the first-order reaction. These simplifications will be used in the kinetic study of water ozonation reactions, as shown later.

4.2.1.3 Series-Parallel Reactions

In most cases, ozone reacts not only with the compound initially present in water but also with the intermediate compounds formed in a series-parallel reaction system. Thus, a simplified study of the kinetics (absorption rate law) of these gas–liquid systems is presented here. Let us assume that a gas component A transfers into the liquid where it undergoes the following reactions:

$$A + z_b B \xrightarrow{k_1} z_c C \quad (4.49)$$

$$A + z_c' C \xrightarrow{k_2} z_d D \quad (4.50)$$

Fundamentals of Gas–Liquid Reaction Kinetics

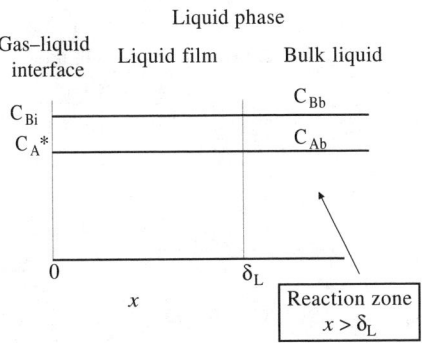

FIGURE 4.6 Film theory: very slow kinetic regime. Concentration profiles for A (gas being dissolved) and B (liquid component) with the distance to the interface.

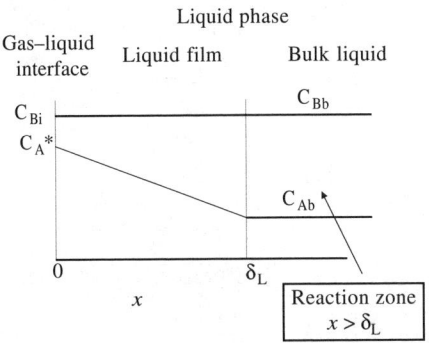

FIGURE 4.7 Film theory: slow kinetic regime. Concentration profiles for A (gas being dissolved) and B (liquid component) with the distance to the interface.

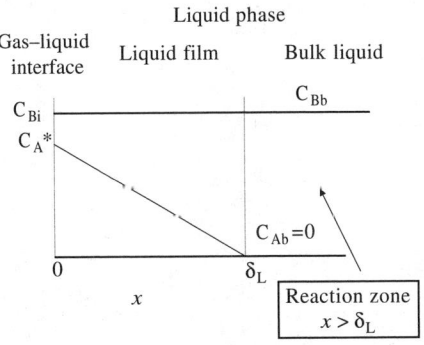

FIGURE 4.8 Film theory: diffusional kinetic regime. Concentration profiles for A (gas being dissolved) and B (liquid component) with the distance to the interface.

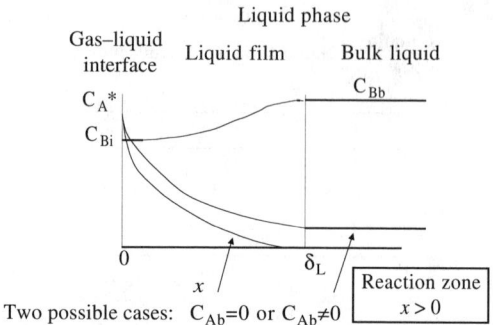

FIGURE 4.9 Film theory: moderate kinetic regime. Concentration profiles for A (gas being dissolved) and B (liquid component) with the distance to the interface.

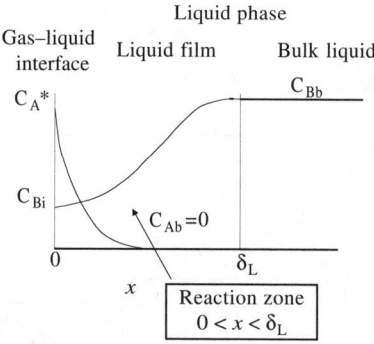

FIGURE 4.10 Film theory: fast (general) kinetic regime. Concentration profiles for A (gas being dissolved) and B (liquid component) with the distance to the interface.

Solution of this system requires the microscopic balance equations of A, B, and C compounds that, in a general form, are represented in Equation (4.51):

$$D_i \frac{d^2 C_i}{dx^2} + r_i = 0 \qquad (4.51)$$

with the following conditions:

$$x = 0 \quad C_A = C_A^* \quad C_B = C_{Bi} \quad C_C = C_{Ci} \quad \frac{dC_B}{dx} = 0 \quad \frac{dC_C}{dx} = 0$$
$$x = \delta_L \quad C_A = C_{Ab} \quad C_B = C_{Bb} \quad C_C = C_{Cb} \qquad (4.52)$$

where C_{Bi} and C_{Ci} are the concentrations of B and C at the interface.

Fundamentals of Gas–Liquid Reaction Kinetics

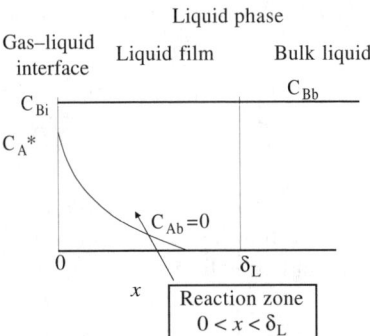

FIGURE 4.11 Film theory: fast pseudo first-order kinetic regime. Concentration profiles for A (gas being dissolved) and B (liquid component) with the distance to the interface.

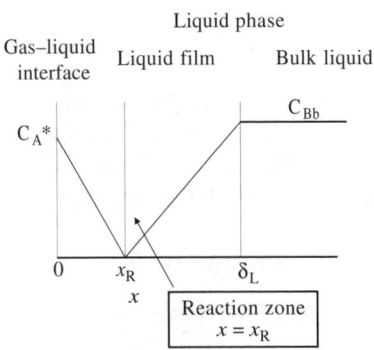

FIGURE 4.12 Film theory: instantaneous kinetic regime. Concentration profiles for A (gas being dissolved) and B (liquid component) with the distance to the interface.

Onda et al.[9] found an expression for the reaction factor, E [Equation (4.53)] after introducing a series of approximations based on that of Van Krevelen and Hoftijzer:[7]

$$E = \frac{Ha_s}{\tanh Ha_s}\left[1 - \frac{C_{Ab}}{C_A^*} \sinh Ha_s\right] \quad (4.53)$$

where

$$Ha_s = Ha_2^2\left[\frac{C_{Bi}}{C_{Bb}} + \frac{k_2 C_{Ci}}{k_1 C_{Cb}}\right] \quad (4.54)$$

and Ha_2 is now the Hatta number of Reaction (4.49). As observed, solution of the mathematical model requires that C_{bi} and C_{ci} be known. Onda et al.[9] applied another equation for the reaction factor, obtained from a total molar balance in differential form:

$$E = 1 - \frac{C_{Ab}}{C_A^*} + \frac{D_B}{z_b D_A}\left(1 + \frac{z_c}{z_c'}\right)\left(1 - \frac{C_{Bi}}{C_{Bb}}\right) - \frac{D_C}{z_c' D_A}\frac{C_{Bb}}{C_A^*} + \frac{C_{Ci}}{C_{Cb}} \quad (4.55)$$

On the other hand, after application of a hypothesis related to the concentration profiles of B and C through the film layer, the following equation that relates the concentration of both compounds was found:[9]

$$\frac{C_{Ci}}{C_{Bi}} = \frac{\dfrac{C_{cb}}{C_{Bb}} + \dfrac{z_c D_B}{z_b D_C}\left(1 - \dfrac{C_{Bi}}{C_{Bb}}\right)}{1 + \dfrac{5 z_c' D_B}{6 z_b D_C}\dfrac{k_2}{k_1}\dfrac{1 - \dfrac{C_{Bi}}{C_{Bb}}}{\dfrac{C_{Bi}}{C_{Bb}}}} \quad (4.56)$$

Finally, a trial-and-error procedure allows the determination of C_{bi} and C_{ci} and the reaction factor E. As it was shown for simple first- or second-order reactions (see Figure 4.3 and Figure 4.5), solution of E at different conditions (that is, at different Ha_2) is usually found in plots with k_2/k_1 as the parameter, as reported in the literature.[9]

4.2.2 Danckwerts Surface Renewal Theory

4.2.2.1 First-Order or Pseudo First-Order Reactions

For the system where Reaction (4.14) develops, the starting microscopic mass-balance equation of A in the liquid is now:

$$D_A \frac{\partial^2 C_A}{\partial x^2} = \frac{\partial C_A}{\partial t} + k_1 C_A \quad (4.57)$$

which should be solved with the boundary conditions (4.9). Danckwerts[10] found an analytical solution for the case of fast reactions ($C_{Ab} = 0$) and $Ha_1 > 1$:

$$N_{A0} = k_L C_A^* \sqrt{1 + Ha_1^2} \quad (4.58)$$

4.2.2.2 Irreversible Second-Order Reactions

For the kinetics of the second-order gas–liquid Reaction (4.32), the case of instantaneous kinetic regime was first solved. For this case, the microscopic mass-balance equations of A and B are:

$$D_A \frac{\partial^2 C_A}{\partial x^2} = \frac{\partial C_A}{\partial t} \quad (4.59)$$

Fundamentals of Gas–Liquid Reaction Kinetics

$$D_B \frac{\partial^2 C_B}{\partial x^2} = \frac{\partial C_B}{\partial t} \tag{4.60}$$

which can be solved with the following boundary conditions:

$$\begin{aligned} x = 0 & \quad C_A = C_A^* \\ x = x_i & \quad C_A = C_B = 0 \\ x \to \infty & \quad C_B = C_{Bb} \end{aligned} \tag{4.61}$$

with x_i being the distance to the interface where A and B react instantaneously (in a reaction plane, see Figure 4.12). This parameter, x_i, has to fulfill Equation (4.62):[10]

$$x_i = 2\chi\sqrt{t} \tag{4.62}$$

where χ is found by trial and error from Equation (4.63):

$$\exp\left(\frac{\chi^2}{D_B}\right) \operatorname{erfc}\left(\frac{\chi}{\sqrt{D_B}}\right) = \frac{C_{Bb}}{zC_A^*}\sqrt{\frac{D_B}{D_A}} \exp\left(\frac{\chi^2}{D_A}\right) \operatorname{erfc}\left(\frac{\chi}{\sqrt{D_A}}\right) \tag{4.63}$$

Thus, solving the mathematical system, the following concentration profiles of A and B can be found:

$$C_A = C_A^* \frac{\operatorname{erf}\left(\frac{x}{2\sqrt{D_A t}}\right) - \operatorname{erfc}\left(\frac{\chi}{\sqrt{D_A}}\right)}{\operatorname{erf}\left(\frac{\chi}{\sqrt{D_A}}\right)} \quad \text{for } x \leq x_i \tag{4.64}$$

$$C_B = C_{Bb} \frac{\operatorname{erf}\left(\frac{x}{2\sqrt{D_B t}}\right) - \operatorname{erfc}\left(\frac{\chi}{\sqrt{D_A}}\right)}{\operatorname{erf}\left(\frac{\chi}{\sqrt{D_A}}\right)} \quad \text{for } x \geq x_i \tag{4.65}$$

and then the absorption rate law is obtained after the application of Fick's law and the exposition time distribution function [see Equation (4.11)]:

$$N_{A0} = k_L C_A^* \frac{1}{\operatorname{erf}\left(\frac{\chi}{\sqrt{D_A}}\right)} \tag{4.66}$$

From the definition of reaction factor or Equation (4.31), the instantaneous reaction factor is defined here as:

$$E_i = \frac{1}{\operatorname{erf}\left(\dfrac{\chi}{\sqrt{D_A}}\right)} \qquad (4.67)$$

In many cases, Equation (4.67) can be simplified to the better-known Equation (4.68):

$$E_i = \sqrt{\frac{D_A}{D_B}}\left[1 + \frac{C_{Bb} D_B}{z C_A^* D_A}\right] \qquad (4.68)$$

which is valid when

$$\frac{C_{Bb}}{z C_A^*} > 25 \qquad (4.69)$$

Also note that E_i from Equation (4.68) coincides with that of the film theory [Equation (4.46)] when $D_A = D_B$.

If the kinetic regime is not instantaneous, the starting microscopic mass-balance equations of A and B with the appropriate boundary conditions are:

$$D_A \frac{\partial^2 C_A}{\partial x^2} = \frac{\partial C_A}{\partial t} + k_2 C_A C_B \qquad (4.70)$$

$$D_A \frac{\partial^2 C_A}{\partial x^2} = \frac{\partial C_B}{\partial t} + z k_2 C_A C_B \qquad (4.71)$$

which can be solved with the following boundary conditions:

$$\begin{aligned} t=0 \quad & C_A = C_{Ab} \quad C_B = C_{Bb} \\ t>0 \quad x=0 \quad & C_A = C_A^* \quad \frac{dC_B}{dx} = 0 \\ x\to\infty \quad & C_A = C_{Ab} \quad C_B = C_{Bb} \end{aligned} \qquad (4.72)$$

An approximate analytical solution to the mathematical model of Equation (4.70) and Equation (4.71) has only been found for the case of the fast kinetic regime. Thus, DeCoursey[11] arrived at the following absorption rate law:

$$N_{A0} = M_1 \sqrt{1 + Ha_2^2 \frac{E_i - E}{E_i - 1}} = M_1 \sqrt{1 + M'} \qquad (4.73)$$

Fundamentals of Gas–Liquid Reaction Kinetics

In contrast to film theory [see Equation (4.25) and Equation (4.38)], Equation (4.73) allows the reaction factor E to be determined in an explicit way:

$$E = \sqrt{1 + \frac{M'^2}{4(E_i - 1)^2} + \frac{E_i Ha_2^2}{E_i - 1} - \frac{M'}{2(E_i - 1)}} \tag{4.74}$$

4.2.2.3 Series-Parallel Reactions

The theory of Danckwerts has also been used to explain the kinetics of series-parallel reaction systems such as that of Reactions (4.49) and (4.50). The approximate and numerical solutions of the microscopic mass-balance equations of the species in solution were also found. Onda et al.[12] have reported the following approximate solution of the reaction factor for second-order Reactions (4.49) and (4.50):

$$E = \sqrt{1 + Ha_S^2 \left[1 - \frac{C_{Ab}}{C_A^*} \frac{1}{1 + Ha_S^2}\right]} \tag{4.75}$$

$$E = 1 - \frac{C_{Ab}}{C_A^*} + \frac{D_B C_{Bb}}{z_B D_A C_A^*}\left(1 + \frac{z_C}{z_C'}\right)\left(1 - \frac{C_{Bi}}{C_{Bb}}\right) + \frac{D_B C_{Bb}}{z_C' D_A C_A^*}\left(\frac{C_{Cb} - C_{Ci}}{C_{Bb}}\right) \tag{4.76}$$

with

$$\frac{C_{Ci}}{C_{Bb}} = \frac{\dfrac{C_{cb}}{C_{Bb}} + \dfrac{z_c D_B}{z_b D_C}\left(1 - \dfrac{C_{Bi}}{C_{Bb}}\right)}{1 + \dfrac{z_c' D_B}{z_b D_C}\dfrac{k_2}{k_1}\dfrac{1 - \dfrac{C_{Bi}}{C_{Bb}}}{\dfrac{C_{Bi}}{C_{Bb}}}} \tag{4.77}$$

Finally, a trial-and-error procedure allows the reaction factor to be known at different values of the Hatta number, Ha_2, of Reaction (4.49). The results are usually expressed as plots of E against Ha_2 as shown in Figure 4.3 or Figure 4.5. Detailed explanations for this reaction system can be found elsewhere.[12]

4.2.3 Influence of Gas Phase Resistance

Thus far the absorption rate law or kinetic equation has been exclusively expressed as a function of the liquid-phase mass-transfer coefficients (k_L or $k_L a$). An additional problem to consider with these equations is the value of the concentration of A at the interface C_A^*, which is related to the partial pressure of A, also at the interface, by Henry's law (see Figure 4.1):

$$P_i = He C_A^* \tag{4.78}$$

where He is the Henry constant. However, in a practical case, P_i is also unknown unless the gas phase resistance to mass transfer is negligible. In this case the partial pressure is constant throughout the gas phase, that is, $P_A = P_i$. This situation holds when the gas being transferred presents a very low solubility in the liquid. For the barely soluble gases, the equations presented thus far can be used directly since the concentration of A at the interface is determined from the experimental partial pressure of A in the gas, which is usually known:

$$P_A = HeC_A^* \qquad (4.79)$$

When the gas phase resistance is not negligible, the mass transfer coefficient through the gas phase has to be considered with the rate equation or absorption rate law. As examples, the cases of slow and fast kinetic regimes are shown below.

4.2.3.1 Slow Kinetic Regime

This is the case where the absorption is a sequential three-step process. For example, in one irreversible first-order reaction, the absorption rate law is given by Equation (4.80):

$$N_{A0} = k_G(P_A - P_i) = k_L(C_A^* - C_{Ab}) = \frac{\beta}{a} k_1 C_{Ab} \qquad (4.80)$$

where the second, third, and fourth terms are the mass transfer rate through the gas and liquid films and the rate of chemical reaction in the bulk liquid, respectively. Combination of these terms allows the concentrations of C_A^* (or P_i) and C_{Ab} to be written as a function of P_A, mass transfer coefficients and chemical reaction rate constant. Then, if this combination is accounted for, the rate of absorption becomes:

$$N_{A0} = \frac{P_A}{\dfrac{1}{k_G} + \dfrac{He}{k_L} + \dfrac{aHe}{\beta k_1}} \qquad (4.81)$$

where the numerator represents the driving force that facilitates the mass transfer, and the three terms in the denominator are the gas, liquid, and chemical reaction resistances to mass transfer, respectively.

4.2.3.2 Fast Kinetic Regime

When the gas phase resistance cannot be neglected, the gas absorption rate is a two-step in-series process: the diffusion of A through the gas film and the diffusion and simultaneous reaction through the liquid film. The absorption rate equation is:

$$N_{A0} = k_G(P_A - P_i) = k_L C_A^* E \qquad (4.82)$$

Fundamentals of Gas–Liquid Reaction Kinetics

which can be conveniently transformed in Equation (4.83) to remove P_i and C_A^*:

$$N_{A0} = \frac{P_A}{\dfrac{1}{k_G} + \dfrac{He}{k_L E}} \tag{4.83}$$

where again the physical meanings of the numerator and the denominator of Equation (4.83) are the driving force and total resistance to mass transfer, respectively. In this case, $1/k_G$ and He/Ek_L are the gas and liquid-phase mass-transfer resistances, respectively.

For other kinetic regimes, rate equations similar to Equation (4.81) and Equation (4.83) are easily deduced. The use of these equations implies the determination of the mass transfer coefficient k_G. For water ozonation reactions, the gas phase resistance to mass transfer is likely to be negligible because ozone is a gas that is barely soluble in water. Then C_A^* or the ozone solubility can be determined through Equation (4.79) directly from the ozone partial pressure in the bulk gas by the application of Henry's law.

4.2.4 Diffusion and Reaction Times

As shown before, the Hatta number indicates the relative importance of mass transfer and chemical reaction steps. This number, although also used in surface renewal theories, comes from the application of the film theory.[7] On the other hand, the concepts of diffusion and reaction times are more related to surface renewal theories.[13] The diffusion time, t_D, is defined as the time between two consecutive renovations of liquid elements. It represents the time one liquid element is at the gas–liquid interface being exposed to mass transfer. In other words, it is the time the dissolved gas A has to diffuse through the liquid element. On the other hand, the reaction time, t_R, is defined as that required for the reaction to proceed at an appreciable rate. The expression for both is:[13]

$$t_D = \frac{D_A}{k_L^2} \tag{4.84}$$

and

$$t_R = \frac{1}{k_1} \tag{4.85}$$

or

$$t_R = \frac{1}{k_1 C_{Bb}} \tag{4.86}$$

for first- and second-order reactions, respectively. Note, however, that the ratio of both times is the square of the Hatta number. By using t_R and t_D, an understanding of the kinetic regimes can also be achieved, as will be shown in Chapter 7. Two situations are presented depending on the relative values of both times:

- When $t_D \ll t_R$ the reactions are classified as slow because before they can develop in a significant way the surface or liquid elements are removed from the gas–water interface. In this case, little or no reaction at all takes place and the absorption behaves as a two-step in-series process: diffusion through the film layer followed by reaction in the bulk liquid.
- When $t_D \gg t_R$ the reactions are fast and the dissolved gas is completely depleted in a liquid element before it is replaced by another gas coming from the bulk of the liquid.

As will be shown in Chapter 7, values of both reaction and diffusion times can also be used to classify the kinetic regimes of ozonation reactions.

References

1. Kuczkowski, R.L., Ozone and carbonyl oxides, in *1,3 Dipolar Cycloaddition Chemistry*, John Wiley & Sons, New York, 1984, 197–276.
2. Lewis, W.K. and Whitman, W.G., Principles of gas absorption, *Ind. Eng. Chem.*, 16, 1215–1220, 1924.
3. Danckwerts, P.V., Significance of liquid film coefficients in gas absorption, *Ind. Eng. Chem.*, 43, 1460–1466, 1951.
4. Bird, R.B., Steward, W.E., and Lightfoot, E.N., *Transport Phenomena*, McGraw-Hill, New York, 1960.
5. Lightfoot, E.N., Steady state absorption of a sparingly soluble gas in an agitated tank with simultaneous irreversible first order reaction, *AIChE J.*, 4, 499–500, 1958.
6. Charpentier, J.C., Mass transfer rates in gas liquid absorbers and reactors, in *Advances in Chemical Engineering*, Vol. 11, Academic Press, New York, 1981, 3–133.
7. Van Krevelen, D.W. and Hoftijzer, P.J., Kinetics of gas liquid reactions. Part I. General theory, *Rec. Trav. Chim.*, 67, 563–586, 1948.
8. Wilke, C.R. and Chang, P., Correlation of diffusion coefficients in dilute solutions, *AIChE J.*, 1, 264–270, 1955.
9. Onda, K., Sada, E., Kobayashi, T., and Fujine, M., Gas absorption accompanied by complex chemical reaction. II. Consecutive chemical reactions, *Chem. Eng. Sci.*, 25, 761–768, 1970.
10. Danckwerts, P.V., *Gas Liquid Reactions*, McGraw-Hill, New York, 1970.
11. DeCoursey, W.J., Absorption with chemical reaction: development of a new relation for Danckwerts model, *Chem. Eng. Sci.*, 29, 1867–1872, 1974.
12. Onda, K., Sada, E., Kobayashi, T., and Fujine, M., Gas absorption accompanied by complex chemical reaction. IV. Unsteady state, *Chem. Eng. Sci.*, 27, 247–255, 1972.
13. Astarita, G., *Mass Transfer with Chemical Reaction*, Elsevier, Amsterdam, 1967.

5 Kinetic Regimes in Direct Ozonation Reactions

This chapter examines in detail the kinetics of direct ozone reactions in water, and considers the different kinetic regimes that ozone reactions present. The main objective of ozonation kinetics focuses on the determination of parameters such as rate constants of reactions and mass-transfer coefficients. First, ways to estimate the ozone properties and solubility or equilibrium constant (Henry's law) are presented since this information is fundamental to any discussion of ozonation kinetics. As already indicated in Chapter 3, the direct reaction between ozone and a given compound B that will be discussed here corresponds to the stoichiometric Equation (3.5), that is, an irreversible second-order reaction (first-order with respect to ozone and B) with z moles of B consumed per mole of ozone consumed. However, with respect to kinetic regimes, the ozone decomposition reaction as first-order kinetics is also discussed.

5.1 DETERMINATION OF OZONE PROPERTIES IN WATER

As observed from the absorption rate law equations deduced in Chapter 4, some properties of both ozone and the reacting compound B should be known to conduct any ozonation kinetic study. These properties are the diffusivity and solubility or equilibrium concentration of ozone in water, C_A^*, which is intimately related to the Henry's law constant, He.

5.1.1 DIFFUSIVITY

Diffusivities of compounds in water can be determined from different empirical correlations. For very dilute solutions, the equation of Wilke and Chang[1] can be used:

$$D_A = 7.4 \times 10^{-12} \frac{(\phi_S MW)^{1/2} T}{\mu_S V_A^{0.6}} \quad (5.1)$$

where D_A is in m^{2}sec^{-1}; T is in K; ϕ_S is an association parameter of the liquid (which is 2.6 for water); MW and μ are the molecular weight and the viscosity of the solvent in *poises*, respectively; and V_A is the molar volume of the diffusing solute in cm^{3}molg^{-1} that can be obtained from additive volume increment methods such as that of Le Bas.[2] The use of Equation (5.1) in aqueous systems leads to an average error of about 10 to 15%. Since Equation (5.1) is not dimensional consistent, the

TABLE 5.1
Calculated Values of Ozone Diffusivity at 20°C

Authors	$D_{O_3} \times 10^9$, m²sec⁻¹	Reference (Year)
Wilke and Chang	1.7[a]	1 (1955)
Nakanishi	2.0	4 (1978)
Matrozov et al.	1.7	5 (1982)
Siddiqi and Lucas	1.6	6 (1986)
Díaz et al.	2.2	7 (1987)
Utter et al.	3–4	8 (1992)
Johnson and Davis	1.4	9 (1996)

[a] The ozone molar volume is 35.5 cm³mol⁻¹ which corresponds to an ozone density of 1.35 gcm⁻³ according to data from the literature.[10]

variable with the specified units must be employed. Other correlations similar to that of Wilke–Chang can also be used to determine the ozone diffusivity. For example, Haynuk and Laudie[2] and Haynuk and Minhas[3] proposed the following correlations, respectively:

$$D_A = 13.26 \times 10^{-9} \left[\frac{1}{(\mu_S \times 10^{-2})^{1.14} V_A^{0.589}} \right] \quad (5.2)$$

and

$$D_A = 1.25 \times 10^{-8} \frac{(V_A^{-0.19} - 0.292)T^{1.52}}{\mu_S^{(9.58/V_A)-1.12}} \quad (5.3)$$

where the terms and units are as in Equation (5.1) (see Table 5.1 for calculated values of ozone diffusivity). It should be noted, however, that of the above three correlations, that of Haynuk and Minhas should be disregarded because of the high deviation observed compared to those of the other two empirical correlations and other values found experimentally.

For the specific case of ozone, some other empirical correlations are available. Thus, the equations of Matrozov et al.:[5]

$$D_A = 4.27 \times 10^{-10} \frac{T}{\mu_S} \quad (5.4)$$

and Johnson and Davis:[9]

$$D_A = 5.9 \times 10^{-10} \frac{T}{\mu_S} \quad (5.5)$$

are usually applied to determine the ozone diffusivity for kinetic studies. Note that the units in Equations (5.4) and (5.5) are the same as in Equation (5.1). Other works in the literature also report on values of ozone diffusivity. Table 5.1 gives a list of these values.

The diffusivity of ozone can also be determined from experimental works of the ozone absorption in aqueous solutions containing ozone fast-reacting compounds. Thus, as shown later, kinetic equations corresponding to instantaneous and fast kinetic regimes of ozone absorption contain the diffusivity of ozone as one of the parameters necessary to know the ozone absorption rate. The procedures are similar to those discussed later for mass-transfer coefficient and rate constant data determination in ozone reactions developing at these kinetic regimes. Thus far, however, to the knowledge of this author, no work on this matter has been reported in the literature. Masschelein[10] has reported a possible procedure based on the ozone uptake by a liquid surface in laminar flow contact conditions. The method, however, implies significant errors in the diffusivity determination. The author, then, suggested that the method could be improved by adding a strong reductor (nitrite, sulfite, etc.) in the water that could enhance the ozone uptake and improve the accuracy of the method.

For compounds B, the diffusivity also needed in some cases (see later in this chapter) is calculated mainly from the Wilke–Chang equation.

5.1.2 Ozone Solubility: The Ozone–Water Equilibrium System

Ozone solubility is a fundamental parameter in the ozonation kinetic studies as it is present in the absorption rate law equations. Ozone–water systems are characterized by a low concentration of dissolved ozone, ambient pressure, and temperature. Henry's law rules the equilibrium of ozone between the air (or oxygen) and water:

$$P_{O3} = He C^*_{O3} \tag{5.6}$$

where He is the Henry's law constant. Equation (5.6) comes from the general criterion of equilibrium of a closed system which, according to thermodynamic rules, postulates that equilibrium is reached when any differential change should be reversible; that is

$$dS = \frac{dQ}{T} = 0 \tag{5.7}$$

where S, Q, and T are the entropy, heat fed to the system, and absolute temperature, respectively. For a closed system at constant pressure and temperature the following specific criterion of equilibrium can be established:[11]

$$dG = 0 \tag{5.8}$$

where G represents the Gibbs free energy. If the closed system is constituted by different phases or subsystems containing n chemical species that transport from one phase to the other, the equilibrium will be reached when these transports stop.

Variation of Gibbs free energy of a given phase depends on pressure, temperature, and concentration changes:

$$dG^{Phase} = \left(\frac{\partial G^{phase}}{\partial T}\right)_{P,N_i} dT + \left(\frac{\partial G^{phase}}{\partial P}\right)_{T,N_i} dP + \sum_{i=1}^{n} \left(\frac{\partial G^{phase}}{\partial N_i}\right)_{T,P,N_{j\neq i}} dN_i \quad (5.9)$$

In Equation (5.9) the last term on the right side represents the contribution of mass transport to the Gibbs free energy variation within one phase where the chemical potential of a given i component (or the partial molar free enthalpy) is:

$$\mu_i^{Phase} = \left(\frac{\partial G}{\partial N_i}\right)_{T,P,N_{j\neq i}} \quad (5.10)$$

For constant pressure and temperature, application of Equation (5.8) to a multiple phase closed system will yield:

$$dG = \sum^{phases} \sum_{i=1}^{n} \mu_i dN_i = 0 \quad (5.11)$$

Since in a closed system there is no variation in the moles of components, Equation (5.11) becomes:[11]

$$\mu_i^{phaseI} = \mu_i^{phaseII} \quad (i=1,2...n) \quad (5.12)$$

which constitutes the specific equilibrium criterion for closed systems of two or more phases at constant pressure and temperature. Then, the chemical potential of a given i component is usually expressed as a function of its fugacity, f_i, according to Equation (5.13):

$$d\mu_i^{phase} = RTd\left(\ln f_i^{phase}\right)_T \quad (5.13)$$

Equation (5.13) finally leads to the equilibrium criterion (5.14) for gas–liquid systems at constant pressure and temperature:

$$f_i^{gas} = f_i^{liquid} \quad (i=1,2...n) \quad (5.14)$$

For the gas phase, the fugacity of any component is defined as a function of its partial pressure p_i or molar fraction y_i times the total pressure P and the fugacity coefficient v_i:

Kinetic Regimes in Direct Ozonation Reactions

$$f_i^{gas} = v_i^{gas} p_i = v_i^{gas} y_i P \qquad (5.15)$$

while in the liquid phase, for very dilute systems the fugacity is a function of the activity coefficient, γ_i, the molar fraction, x_i, and the Henry constant, He:

$$f_i^{liquid} = \gamma_i x_i He_i \qquad (5.16)$$

For gas systems at moderate pressure and well below critical temperature, as in ozone–water systems, the gas phase behaves as an ideal gas and the fugacity coefficients are unity. The activity coefficient depends on the presence of substances (nonelectrolytes, salts, etc.) so that the product of the Henry constant times the activity coefficient can be called the apparent Henry constant, He_{app}.[12] Thus, the equilibrium criterion for the ozone–water system is

$$p_{O3} = y_{O3}P = \gamma_{O3}x_{O3}He = He_{app}x_{O3} \qquad (5.17)$$

In most cases, however, the Henry constant is given as a function of ozone concentration in the water, C_{O3}^*, expressed in moles per liter of solution.

Strictly speaking, the Henry's law constant is a function of temperature, following the relationship:

$$He = He_0 \exp\left(-\frac{H_A}{RT}\right) \qquad (5.18)$$

where T is in K, R is the gas constant, and H_A is the heat of absorption of the gas at the temperature considered. However, when the liquid (in this case, water) contains electrolytes, ionic substances, etc., the Henry's law constant refers to the apparent Henry constant; it is also a function of the ionic strength and some coefficients that depend on the positive or negative charge of the ionic substances present in water. Thus, the effect of salt concentration (salting-out effect) on the Henry constant is considered in the equation of Sechenov:[13]

$$He_{app} = He 10^{K_s c_s} \qquad (5.19)$$

where He is the Henry constant value in salt-free water, c_s is the salt concentration, and K_s is the Sechenov constant that is specific to the gas and salt and that varies slightly with temperature. When no experimental values are available for K_s, the correlation of Van Krevelen and Hoftijzer can be used:[14]

$$He_{app} = He 10^{hI} \qquad (5.20)$$

where I is the ionic strength defined as

$$I = \frac{1}{2}\sum_i C_i z_i^2 \qquad (5.21)$$

where C_i is the concentration of any i ionic species of valency z_i, and h in Equation (5.20) is the sum of the positive and negative ions present in water and in the dissolved gas species. The log-additivity of the salting-out effects in mixed solutions, at low concentrations of salts and, even in the presence of nonelectrolyte substances, led Schumpe[15] to suggest a model considering individual salting-out effects of the ions and the gas:

$$He_{app} = He 10^{\sum(h_i + h_G)C_i} \qquad (5.22)$$

where h_i and h_G are the contribution of a given ion and gas, respectively. Finally, Weisenberger and Schumpe[16] modified the model proposed in Equation (5.22) to be valid for a wider temperature range. To do so, the coefficient related to the gas, h_G, was correlated with temperature to yield

$$h_G = h_{G,0} + h_T(T - 298.15) \qquad (5.23)$$

where $h_{G,0}$ and h_T are parameters specific to the gas being dissolved. Table 5.2 gives a list of parameter values of h_i, $h_{G,0}$, and h_T for different gases, ions, and valid range of temperatures. Although salting-out parameters for different species are tabulated, in an actual case, it is rather difficult to know exactly the ionic species present, their concentration and, therefore, their corresponding h values. This is the reason why most of the experimental works carried out to determine the solubility of ozone or ozone equilibrium in water did not arrive at equations like that of Schumpe et al.[15,16] Thus far, to the knowledge of this author, the only experimental work where this was considered was Rischbieter et al.,[17] as discussed later. Andreozzi et al.[18] also treated the solubility of ozone in water with respect to the salting-out effect, although no correlation was given (see later discussion).

The Henry's law constant is not the only parameter determined in experimental works to establish the ozone–water equilibrium. Other parameters such as the Bunsen coefficient, β, or the solubility ratio, S, have been used. The former is defined as the ratio of the volume of ozone at NPT dissolved per volume of water when the partial pressure of ozone in the gas phase is one atmosphere. The solubility ratio is the quotient between the equilibrium concentrations of ozone in water and in the gas. The equation that relates the three parameters is as follows:

$$He = 8039 \frac{T}{S} = 2268373 \frac{1}{\beta} \qquad (5.24)$$

where He is in PaM^{-1} and S and β are dimensionless parameters. Table 5.3 presents a list of works with the conditions applied and the value of He at 20°C. In some

TABLE 5.2
Parameter Values of Weisenberger and Schumpe Equation (5.23) Corresponding to Gas and Ionic Species and Temperature[a]

h_i for Cations		h_i for Anions		$h_{G,0}$ for Gases and Corresponding h_T for Temperature			
Cation	h_i, m³kmol⁻¹	Anion	h_i, m³kmol⁻¹	Gas	$h_{G,0}$, m³kmol⁻¹	$h_T \times 10^3$, m³kmol⁻¹K⁻¹	Range of Validity, K
H⁺	0	OH⁻	0.0839	H₂	−0.0218	−0.299	273–353
Li⁺	0.0754	HS⁻	0.0851	He	−0.0353	+0.464	278–353
Na⁺	0.1143	Fl⁻	0.0920	Ne	−0.0080	−0.913	288–303
K⁺	0.0922	Cl⁻	0.0318	Ar	0.0057	−0.485	273–353
Rb⁺	0.0839	Br⁻	0.0269	Kr	−0.0071	Not available	298
Cs⁺	0.0759	I⁻	0.0039	Xe	0.0133	−0.329	273–318
NH₄⁺	0.0556	NO₂⁻	0.0795	Rn	0.0477	−0.138	273–301
Mg²⁺	0.1694	NO₃⁻	0.0128	N₂	−0.0010	−0.605	278–345
Ca²⁺	0.1762	ClO₃⁻	0.1348	O₂	0	−0.334	273–353
Sr²⁺	0.1881	BrO₃⁻	0.1116	O₃[b]	0.00396	+0.00179	278–298
Ba²⁺	0.2168	IO₃⁻	0.0913	NO	0.0060	Not available	298
Mn²⁺	0.1463	ClO₄⁻	0.0492	N₂O	−0.0085	−0.479	273–313
Fe²⁺	0.1523	IO₄⁻	0.1464	NH₃	−0.0481	Not available	298
Co²⁺	0.1680	CN⁻	0.0679	CO₂	−0.0172	−0.338	273–313
Ni²⁺	0.1654	SCN⁻	0.0627	CH₄	0.0022	−0.524	273–363
Cu²⁺	0.1675	HCrO₄⁻	0.0401	C₂H₂	−0.0159	Not available	298
Zn²⁺	0.1537	HCO₃⁻	0.0967	C₂H₄	0.0037	Not available	298
Cd²⁺	0.1869	H₂PO₄⁻	0.0906	C₂H₆	0.0120	−0.601	273–348
Al³⁺	0.2174	HSO₃⁻	0.0549	C₃H₈	0.0240	−0.702	286–345
Cr³⁺	0.0648	CO₃²⁻	0.1423	nC₄H₁₀	0.0297	−0.726	273–345
Fe³⁺	0.1161	HPO₄²⁻	0.1499	H₂S	−0.0333	Not available	298
La³⁺	0.2297	SO₃²⁻	0.1270	SO₂	−0.0817	+0.275	283–363
Ce³⁺	0.2406	SO₄²⁻	0.1117	SF₆	0.0100	Not available	298
Th⁴⁺	0.2709	S₂O₃²⁻	0.1149				
		PO₄³⁻	0.2119				
		(Fe(CN)₆)⁴⁻					

[a] *Source*: From Weisenberger, S. and Schumpe, A., Estimation of gas solubilities in salt solutions at temperatures from 273 to 363 K, *AIChE J.*, 42, 298–300, 1996. With permission.

[b] *Source*: From Rischbieter, E., Stein, H., and Schumpe, A., Ozone solubilities in water and aqueous solutions, *J. Chem. Eng. Data*, 45, 338–340, 2000. With permission.

other works, the data thus far reported in the literature have also been converted to the same units for purpose of comparison.[26,32]

The ozone solubility and, consequently, the Henry constant of the ozone–water system is usually determined from experiments of ozone absorption in water. In these experiments ozone is absorbed in water (usually buffered water) at different conditions of pH, temperature, and ionic strength. In many cases, the experimental ozone absorption runs are carried out in small bubble columns or mechanically

TABLE 5.3
Literature Data on Henry's Law Constant for the Ozone Water System

Author and Year		Reference No.
Mailfert, 1894[a]	Pure water, 0–60°C, He = 6384.4 (19°C)	19
	Weak H_2SO_4 solutions, 30–57°C, He = 10506.9 (30°C)	
Kawamura, 1932[a]	Pure water, 5–60°C, He = 8409.2 (20°C)	20
	In H_2SO_4 solutions, from 7.57 N to 0.11 N, 20°C, He = 8701.1 (0.11 N)	
Briner and Perrotet, 1939[b]	Pure water, 3.5 and 19.8°C, He = 7041.1 (19.8°C)	21
	In 35gL^{-1} NaCl solution, 3.5, and 19.8°C, He = 13352.5 (19.8°C)	
Rawson, 1953[a]	Pure water, 9.6–39°C, He = 11619.6 (20.3°C)	22
Kilpatrick et al., 1956	In 0.01 M $HClO_4$ solutions: 15.2–30°C, He = 9092.1 (20°C)	23
Stumm, 1958[a]	0.05 M IS, 5–25°C, He = 7168.8 (20°C)	24
Li, 1977	pH 2.2, 4.1, 6.15, and 7.1, 25°C, He = 17746.7 (pH 7.1)	25
Roth and Sullivan, 1981	Buffered water (phosphates), NaOH or H_2SO_4 to keep pH, 3.5–60°C, pH 0.65–10.2, and He = 10035.1 (20°C, pH 7)[c]	26
Caprio et al., 1982[a]	Pure water, 0.5–41°C, He = 9047.6 (21°C)	27
Gurol and Singer, 1982[a]	pH 3, Na_2SO_4 solution at 20°C, He = 5937.9 (0.1 M IS), and 6955.8 (1.0 M IS)	28
Kosak-Channing and Helz, 1983	In Na_2SO_4 solutions: 5–30°C, pH 3.4, 0–0.6 M IS, and He = 3981 (20°C, 0.1 M)	29
Ouederni et al., 1987	T = 20–50°C, IS = 0.13 M	30
	Sodium sulfate and sulfuric acid for pH 2: He = 7.35 × $10^{12}\exp(-2876/T)$, phosphate buffer for pH 7 He = 1.78 × $10^{12}\exp(-3547/T)$,	
	Correlation for mass-transfer coefficients	
Sotelo et al., 1989	Phosphate buffer solutions: 10^{-3} to 0.5 M IS, 0–20°C, pH 2–8.5, He = 11185.4 (20°C, pH 7, 0.01 M IS)[c]	31
	Phosphate and carbonate buffer solutions: 0.01–0.1 M IS, 0–20°C, pH 7, and He = 8221.7 (20°C, pH 7, 0.01 M IS)[c]	
	In Na_2SO_4 solutions: 20°C, pH 2–7, 0.049–0.49 M IS, He = 11678 (20°C, pH 7, 0.049 M IS)[c]	
	In NaCl solutions: 20°C, pH 6, 0.04–0.49 M IS, and He = 8936.8 (0.04 M IS)[c]	
	In NaCl and phosphates: 20°C, pH 7, 0.05–0.5 M IS, and He = 10699.4 (0.05 M IS)[c]	
Andreozzi et al., 1996	In phosphate buffer solutions: 0–0.48 M IS, 18–42°C, pH 4.75, and He = 5011.8 (0.06 M, 20°C)	18
	In phosphate buffer plus t-butanol solutions: 0–0.48 M IS, 18–42°C, pH 4.71, and He = 5370 (0.06 M, 20°C)	
	In phosphate buffer plus t-butanol solutions: 0.24 and 0.48 M IS, 18–42°C, pH 2–6, and He = 5634.4 (0.24 M, 25°C, pH 6)	
Rischbieter et al., 2000	In different salts: $MgSO_4$, NaCl, KCl, Na_2SO_4, and $Ca(NO_3)_2$, determination of actual He for pure water, He = 9.45 × 10^6 at 20°C	17

Kinetic Regimes in Direct Ozonation Reactions

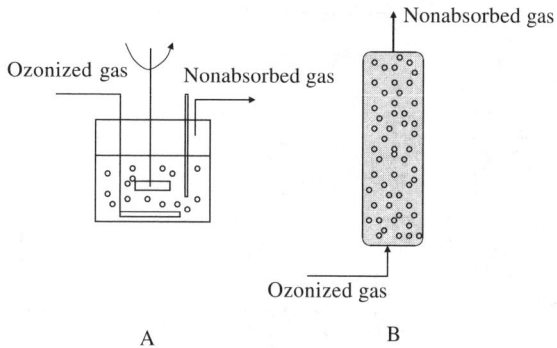

FIGURE 5.1 Experimental contactors in which ozonation kinetic studies are usually carried out: (A) agitated tank; (B) bubble column.

agitated semicontinuous tanks (see Figure 5.1) where a gas mixture (O_2–O_3 or air–O_3) is continuously fed into a volume of buffered water of a given pH which has been previously charged.

In these experiments, in addition, both the gas and water phases are perfectly mixed to facilitate the mathematical treatment of experimental data. According to the hypothesis of perfect mixing, a molar balance of ozone in the water leads to the following equation (see also Appendix A1):

$$\frac{dC_{O3}}{dt} = G_{O3} \tag{5.25}$$

where G_{O3} is the generation rate term of ozone, which varies depending on the kinetic regime of ozone absorption. The absorption of ozone in water is a gas–liquid reaction system because of a fraction of dissolved ozone decomposes in water (see Chapter 4). As a rule, the chemical reaction can be considered an irreversible first- or pseudo first-order reaction, although in some works other reaction orders are also reported (see Chapter 2). The rate constant of this reaction is very low especially when the pH < 7 and the corresponding Hatta number, Ha_1, is <0.01. As a consequence, the kinetic regime of ozone absorption corresponds to a very slow reaction. This means that the ozone absorption rate depends exclusively on the chemical reaction step, and the general Equation (4.25) reduces to that of a homogeneous reaction. Note that at higher pH values a different picture is presented as far as the kinetic regime is concerned. This is treated in the following section. Thus, at pH < 7 and at nonstationary conditions, Equation (5.25) applied to a perfectly mixed reactor becomes:

$$\frac{dC_{O3}}{dt} = k_L a\left(C_{O3}^* - C_{O3}\right) - k_1 C_{O3} \tag{5.26}$$

where $k_L a$ and k_1 are the volumetric mass-transfer coefficient through the water phase and rate constant of the ozone reaction, respectively. Equation (5.26) physically

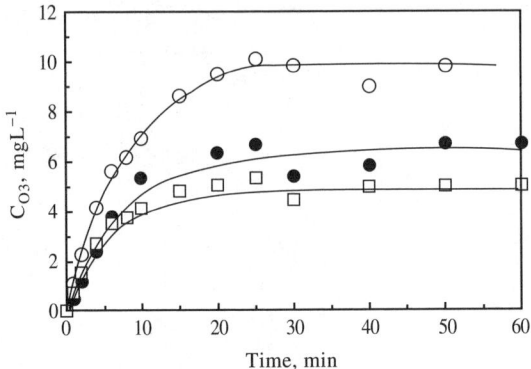

FIGURE 5.2 Typical concentration profiles of ozone against time obtained in ozone absorption in organic-free water at different temperatures T, °C: ○ = 7, ● = 17, □ = 27.

means that the accumulation rate of ozone in water (left side of equation) is the sum of the ozone transfer rate from the gas minus the ozone decomposition rate that results from the chemical reaction that ozone undergoes, that is, the ozone decomposition reaction (right side of equation).

Solving Equation (5.26) determines both C_{O3}^* (the ozone solubility) and the volumetric mass-transfer coefficient, $k_L a$. This procedure has been followed in different works[18,26,29,31] with minor variations. Thus, Sullivan and Roth[33] previously observed that the ozone decomposition reaction followed first-order kinetics and determined the rate constant values at different conditions (see Table 2.7). Figure 5.2 shows a typical profile of the concentration of ozone with time for an absorption experiment in a semibatch well-agitated tank. As observed from Figure 5.2, the concentration of dissolved ozone increases with time until it reaches a stationary value, C_{O3s}. At this time, the accumulation rate term in Equation (5.26) is zero so that

$$k_L a \left(C_{O3}^* - C_{O3s} \right) = k_1 C_{O3s} \quad (5.27)$$

From Equations (5.26) and (5.27) the following is easily obtained:

$$\frac{dC_{O3}}{dt} = \left(k_L a + k_1 \right)\left(C_{O3s} - C_{O3} \right) \quad (5.28)$$

For absorption times lower than that corresponding to the stationary situation and after numerical differentiation of ozone concentration–time data, dC_{O3}/dt is obtained and then plotted against the corresponding $C_{O3s} - C_{O3}$. According to Equation (5.28) this plot should yield a straight line through the origin with slope $k_L a + k_1$. Since k_1 was already known (from homogeneous ozone decomposition experiments), the volumetric mass-transfer coefficient can be determined. From Equation (5.27) the ozone solubility can also be determined as a function of the concentration of ozone at steady state, C_{O3s}. Following this procedure, Roth and Sullivan[26] found the values of C_{O3}^* at different conditions of temperature and pH and arrived at the following equation for He, after applying the Henry's law Equation (5.6):

$$He = 3.84 \times 10^7 C_{OH^-}^{0.035} \exp\left(-\frac{2428}{T}\right) \quad (5.29)$$

where the units of He are atm(molfraction)$^{-1}$.

A similar procedure was used by Sotelo et al.,[31] who studied the ozone absorption in the presence of different salts (carbonates, phosphates, etc.). For the ozone decomposition reaction, they found reaction orders other than 1 (see Table 2.7). They used the general Equation (5.26) with the chemical rate term as kC^n (with n being 1.5 or 2 depending on the buffer type). Note also that in these cases the Hatta number corresponding to this nth order kinetics[14]

$$Ha_n = \frac{\sqrt{kD_{O3}\left(C_{O3}^*\right)^{1-n}}}{k_L} \quad (5.30)$$

was also lower than 0.01, a situation that corresponds to a slow kinetic regime of ozone absorption. In that work,[31] a plot of the sum of the accumulation and reaction rate terms against the ozone concentration, C_{O3}, was used to determine the ozone solubility. According to Equation (5.26), a plot of this type leads to a straight line of slope and origin equal to $-k_L a$ and $k_L a C_{O3}^*$, respectively. Once C_{O3}^* is known, He is obtained from Equation (5.6).

Another typical example of ozone absorption study is Andreozzi et al.[18] These authors also carried out their ozone absorption experiments in a semi-continuous tank where, under the conditions investigated, the volumetric mass-transfer coefficient was much higher than the ozone decomposition rate constant, $k_L a \gg k_1$. According to this conclusion, at steady-state conditions, the ozone concentration C_{O3s} coincides with the concentration of ozone at the gas–water interface; that is, the ozone solubility, $C_{O3s} = C_{O3}^*$. These authors also found values of ozone solubility at different temperature, pH, and ionic strength. They tried to explain their results following Van Krevelen and Hoftijzer type equations [Equation (5.20)]. However, they could not find any general equation of this type because of the absence of data in the literature on salting-out coefficients (h values) that they reported on. In Figure 5.3, values of He obtained from Roth and Sullivan[26] and Sotelo et al.[31] are plotted against pH at different temperature for purpose of comparison.

Finally, the work of Rischbieter et al.[17] considered the model of Weisenberger and Schumpe[16] to determine the ozone solubility in aqueous solutions of different salts. From data on ozone solubility in the absence and presence of salts, these authors determined the h parameter values specific to ozone that were found to be as follows:[17] $h_{G,0} = 3.96 \times 10^{-3}$ m^{3}kmol^{-1} and $h_T = 1.79 \times 10^{-3}$ m^{3}kmol^{-1}K^{-1} for a temperature range between 5 and 25°C (see also Table 5.2 and Table 5.3).

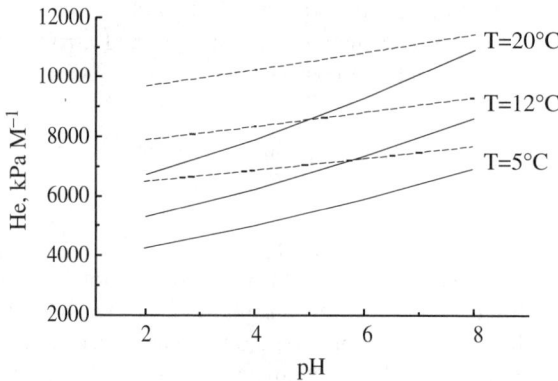

FIGURE 5.3 Variation of the Henry constant for the ozone–water system with pH and temperature. (Solid lines from Roth, J.A. and Sullivan, D.E., Solubility of ozone in water, *Ind. Eng. Chem. Fundam.*, 20, 137–140, 1981. With permission. Dotted lines from Sotelo, J.L. et al., Henry's law constant for the ozone–water system, *Water Res.*, 23, 1239–1246, 1989. With permission.)

5.2 KINETIC REGIMES OF THE OZONE DECOMPOSITION REACTION

The rate of ozone decomposition can be catalogued as a pseudo first-order irreversible reaction. This reaction is, in fact, a nonelementary one constituted by a mechanism of steps that involve free radicals, as explained in Chapter 2. When ozone is absorbed in organic-free water, the system is also a gas–liquid reaction that develops in a given kinetic regime. Knowledge of the kinetic regime of this reaction would help conclude whether the decomposition reaction competes with any other direct ozone reaction for the available ozone (i.e., when a compound B is also present in water). Therefore, in this section, we establish experimental conditions of the different kinetic regimes under which the ozone decomposition reaction can be conducted.

Once the kinetic regime is known from the corresponding Hatta number, Ha_1, the reaction zone (the film or bulk water, see Figures 4.6 to 4.12) can be defined. In this way, a comparison between the importance of the ozone decomposition and ozone direct reactions with any compound B if also present in water can be made. From this comparison, the type of reaction — direct or indirect — through which ozone reacts in water can be determined.

The Hatta number or the reaction and diffusion times are the key parameters to know. The rate constant, individual liquid phase mass-transfer coefficient, and ozone diffusivity are also needed [see Equation (4.20)]. As will be shown later, the kinetic regime will be highly dependent on the pH value. The ozone decomposition at three pH values (2, 7, and 12) will be discussed here.

Application of Equation (4.19), on the other hand, leads to the ozone concentration profile through the film layer. From experiments on ozone decomposition in water carried out at pH 2 and 7, the rate constant of the ozone decomposition reaction

Kinetic Regimes in Direct Ozonation Reactions

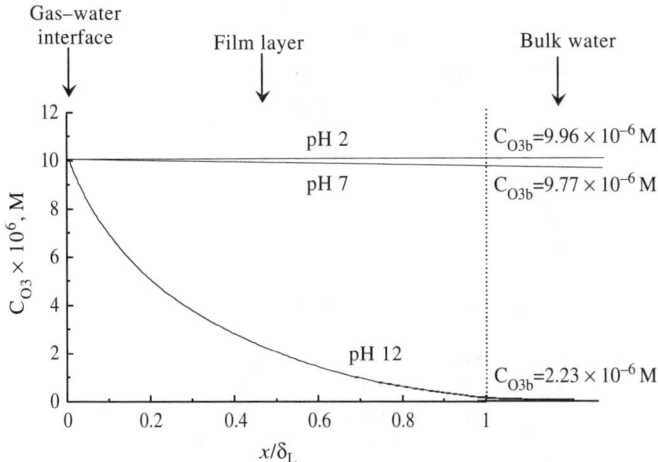

FIGURE 5.4 Variation of the concentration of ozone with the depth of liquid penetration during its absorption in organic-free water at steady state. Conditions: $T = 20°C$, $D_{O3} = 1.3 \times 10^{-9}$ m²sec⁻¹, $k_L = 2 \times 10^{-5}$ msec⁻¹. (From Beltrán, F.J., Theoretical aspects of the kinetics of competitive ozone reactions in water, *Ozone Sci. Eng.*, 17, 163–181, 1995. Copyright 1995, International Ozone Association. With permission.)

was found to be 8.3×10^{-5} sec⁻¹ and 4.8×10^{-4} sec⁻¹, respectively.[34] For a higher pH, say 12, a value of 2.1 sec⁻¹ can be taken, as reported by Staehelin and Hoigné[35] and Forni et al.[36] for the direct reaction between ozone and the hydroxyl ion in organic-free water. When the diffusivity of ozone is taken as 1.3×10^{-9} m²/sec (see 5.1.1), for two values of the liquid phase mass-transfer coefficient of 2×10^{-5} m/sec and 2×10^{-4} m/sec, and Equation (4.19), Beltrán[34] deduced the ozone concentration profile through the film layer corresponding to these two situations at the three pH values studied. Figure 5.4 shows, for example, the results for the case of low mass transfer ($k_L = 2 \times 10^{-5}$ m/sec).

Table 5.4 lists the values of Ha_1. We can see that for pH 2 and 7 the kinetic regime corresponds to a slow reaction, and, hence, the concentration profile of ozone is nearly uniform through the film layer. In this case, the reaction takes place completely in the bulk water. Note that this is in agreement with the kinetic treatment applied in Section 5.1.2 to determine the ozone solubility and the Henry constant. For pH 12, on the contrary, the reaction has passed to a moderate kinetic regime and then there is no available ozone in the bulk water. Here the reaction takes place primarily in the film layer. The treatment applied in Section 5.1.2 does not hold at pH 12.

Beltrán[34] also determined the reaction and diffusion times for the ozone decomposition reaction from data on rate constants at different pH values and the mass-transfer coefficients given above. In Figure 5.5 the reaction time of the ozone decomposition is plotted against pH showing the zones where the kinetic regime is slow or fast. From Figure 5.5 it is deduced that at a pH lower than 12, the ozone decomposition reaction will not interfere with the direct ozone reactions of fast or

TABLE 5.4
Hatta Values of the First-Order Ozone Decomposition Reaction[a]

pH	$k_L = 2 \times 10^{-5}$ msec^{-1}	$k_L = 2 \times 10^{-4}$ msec^{-1}
2	0.016	0.0016
7	0.039	0.0039
12	2.57	0.257

[a] Calculated from Equation (5.30) with $n = 1$ and k at 20°C.

Source: From Beltrán, F.J., Theoretical aspects of the kinetics of competitive ozone reactions in water, *Ozone Sci. Eng.*, 17, 163–181, 1995. With permission.

instantaneous kinetic regime. On the contrary, at a pH higher than 12, the ozone decomposition reaction will be the only way ozone disappears when the ozone direct reactions of compounds present in water, if any, develop in the slow kinetic regime. In other words, depending on the kinetic regimes of the ozone decomposition and ozone–B direct reactions, one of these reactions will be the only way to remove B from water (the ozone decomposition that results from the free radicals that generate). This is of significant importance since the absorption rate law, N_{O3}, will take a different equation (see Chapter 4). Chapter 7 presents a more detailed comparison of the competition between the ozone decomposition reaction and the direct ozone reactions.

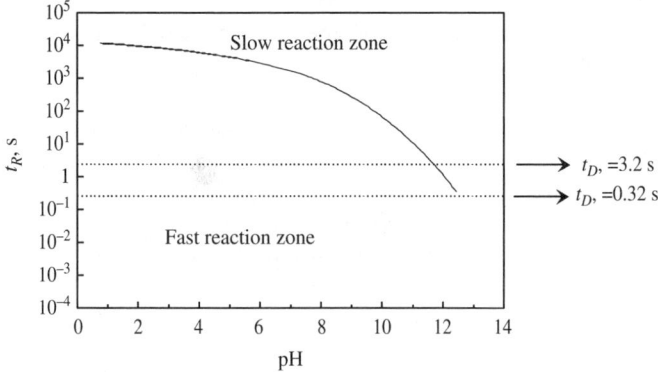

FIGURE 5.5 Reaction time evolution of the ozone decomposition reaction with pH at 20°C. (From Beltrán, F.J., Theoretical aspects of the kinetics of competitive ozone reactions in water, *Ozone Sci. Eng.*, 17, 163–181, 1995. Copyright 1995, International Ozone Association. With permission.)

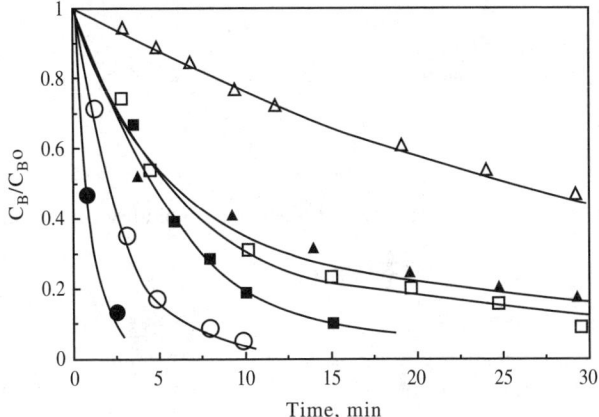

FIGURE 5.6 Effect of hydroxyl radical scavengers on the ozonation of atrazine in water. Conditions: $C_{ATZ0} = 5 \times 10^{-5} = M$; $P_{O3i} = 1050$ Pa. With scavengers: pH: Δ = 2, 0.05 M t-butanol; $\square$ = 7, 0.075 M bicarbonate; $\circ$ = 12, 0.075 M bicarbonate. Without scavengers: $\blacktriangle$ = pH 2; $\blacksquare$ = pH 7; $\bullet$ = pH 12. (From Beltrán, F.J., García-Araya, J.F., and Benito, A., Advanced oxidation of atrazine in water. I. Ozonation, *Water Res.*, 28, 2153–2164, 1994. Copyright 1994, Elsevier Press. With permission.)

5.3 KINETIC REGIMES OF DIRECT OZONATION REACTIONS

A series of steps should be followed before starting the kinetic study of direct ozone reactions. Absorption rate equations for irreversible second-order reactions are not valid when ozone reacts in water not only with the target compound B but also with intermediates formed from the first ozone–compound B direct reaction. For this case, the more complex rate equations for series-parallel reactions hold. Therefore, in the kinetic study of single direct ozonation reactions, appropriate experimental conditions should first be established for the ozone–B reaction to be the only reaction consuming ozone. The competitive effect of the ozone decomposition reaction can be eliminated by considering the kinetic regimes of this reaction and the rate constant of the direct reaction under study (see Section 5.2). At pH lower than 12, if both reactions (ozone decomposition and direct reaction) develop in the same kinetic regime, the former reaction can be stopped by the addition of scavengers of hydroxyl radicals. For example, Figure 5.6 shows the evolution of the concentration of atrazine with time during ozonation experiments in water in the presence and absence of tert-butanol or carbonate, known scavengers or inhibitors of ozone decomposition.[37] As can be seen, the presence of these substances slows the ozonation rate because they trap hydroxyl radicals and inhibit the decomposition of ozone (see the discussion on the ozone mechanism in Chapter 2).

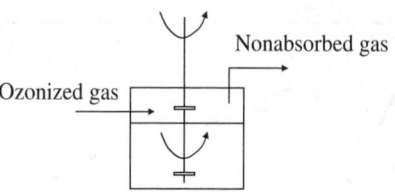

FIGURE 5.7 Experimental agitated cell for kinetic gas–liquid reaction absorption studies.

5.3.1 CHECKING SECONDARY REACTIONS

Once the ozone decomposition reaction has been suppressed, the next step before undertaking the kinetic study of any ozone–B direct reaction is to check the importance of ozone reactions with intermediates. The importance of secondary reactions can be established by calculating the global stoichiometric ratio at different times, as was shown in Section 3.2.1.

In some cases, the effect of secondary reactions is eliminated by reducing the mass-transfer rate of ozone. For example, Beltrán et al.[38] carried out the ozonation of some crotonic acid–derived compounds in an agitated cell and in an agitated tank (see Figure 5.7). The authors observed that the global stoichiometric ratio remained constant (around unity) only when the reactions were carried out in the agitated cell. Thus, in this reactor the ozone–acid reaction was the only one that developed under the conditions investigated. Hence, the agitated cell was the recommended reactor for carrying out the kinetic study for rate constant determination.

5.3.2 SOME COMMON FEATURES OF THE KINETIC STUDIES

Other aspects should be considered in the kinetic study of a gas–liquid reaction such as ozonation. The first one is the establishment of the kinetic regime of ozone absorption due to the variation in the absorption rate law equation, depending on the kinetic regime. For some kinetic regimes the absorption rate law is a simple equation that contains the unknown parameter, mainly the rate constant; for other kinetic regimes the absorption rate law is a complicated equation that will be difficult to deal with (see Chapter 4). Therefore, the appropriate kinetic regime should be not only the regime with the absorption rate law containing the desired parameter but also the regime with the simpler mathematical equation, if possible. Table 5.5 lists the various kinetic regimes that allow the determination of parameters such as the reaction rate constant, volumetric mass-transfer coefficient, etc., together with the corresponding absorption rate equations and conditions to be established. The kinetic regime, as has been shown before, depends on the relative importance of chemical and mass-transfer rate steps. This relationship can be established by calculating the dimensionless Hatta numbers (Ha_2) and the instantaneous reaction factor (E_i), the latter needed only when the reactions are fast or instantaneous. However, *a priori*, the Hatta number is also unknown since parameters such as the reaction rate constant have to be determined [see Equation (4.40) for Ha_2 definition]. Thus, the kinetic study should start from the assumption that at the experimental conditions to be applied the kinetic regime is known and, then, the absorption rate law, N_A (in

TABLE 5.5
Absorption Rate Law Equations for Different Kinetic Regimes of Ozonation[a]

Kinetic Regime	Kinetic Equation	Conditions and Parameter to Determine
Very slow	$N_{O_3} = k_L a(C_{O3}^* - C_{O3}) = \dfrac{dC_{O3}}{dt} + \sum_i r_i$	$Ha_2 < 0.02$, $C_{O3} \neq 0$ Rate constant
Diffusional	$N_{O_3} = k_L a C_{O3}^*$	$0.02 < Ha_2 < 0.3$, $C_{O3} = 0$ Mass-transfer coefficient
Fast	$N_{O_3} = k_L a \dfrac{Ha_2}{\tanh Ha_2}$	$Ha_2 > 3$, $C_{O3} = 0$ Rate constant or mass-transfer coefficients
Fast pseudo first-order	$N_{O_3} = a C_{O3}^* \sqrt{k_D D_{O3} C_M}$	$3 < Ha_2 < E_i/2$, $C_{O3} = 0$ Rate constant or specific interfacial area
Instantaneous	$N_{O_3} = k_L a C_{O3}^* E_i$	$Ha_2 > nE_i$, $C_{O3} = 0$ Mass-transfer coefficient

[a] Equations according to film theory. For stoichiometry, see Reaction (4.32) with A = ozone; N_{O_3} = ozone absorption rate, Msec^{-1}; Ha_2 according to Equation (4.40); E_i according to Equation (4.46); n = function (Ha, E_i); and a = the specific interfacial area.

this case, N_{O3}) (see Chapter 4). This means that some condition referring to the Hatta number has to be confirmed (see also Table 5.5) once the rate constant and/or individual liquid phase mass-transfer coefficient are known. In order to ensure that the hypothesis is solid, some preliminary experiments can be done to classify the kinetic regime as fast or slow. In these experiments, the concentration of dissolved ozone is the key parameter to follow. Thus, the absence of dissolved ozone is definitive proof of fast or instantaneous regime while the opposite situation indicates a slow kinetic regime.

Interpretation of experimental results to study direct ozonation kinetics is accomplished with the use of ozone and B mass-balance equations. The absorption rate law, N_{O3}, is one of the terms of these equations. The mathematical form of the mass-balance equation depends on the reactor type, or, to be more exact, it depends on the type of flow the gas and water phases present through the reactor used. Thus, the second aspect to consider when studying the kinetics of ozonation reactions concerns the type of reactor used for the ozonation experiments. For kinetic studies in the laboratory, experiments are usually carried out in ideal reactors or reactors with ideal flow for water and gas phases (see Appendix A1). Ideal reactors are those that allow the establishment of the mathematical expression of the design equation by applying the same hypothesis, that is, the mass-balance equation of any compound present. In this way, a mathematical expression is readily available to fit the experimental results and determine the kinetic parameters (rate constants, mass-transfer coefficients, etc.).

For a continuous agitated tank (see Figure 5.1) where both the water and gas phase are perfectly mixed and fed continuously to the reactor, the mass-balance equation for, say, the compound B in the ozone–B system, would be:

$$v_0(C_{B0} - C_{Bb}) + \beta r_B V = \beta V \frac{dC_{Bb}}{dt} \tag{5.31}$$

where v_0 and V are the liquid volumetric flow rate and total reaction volume, respectively, and β is the liquid holdup or fraction of liquid in the total volume. Equation (5.31) would reduce to Equation (3.8) when the reactor is semicontinuous, that is, when an aqueous solution of B is initially charged to the reactor.

If the gas phase fed to the ozonation reactor is considered and perfect mixing is also assumed, the ozone mass-balance equation for the gas phase will be:

$$v_g(C_{geb} - C_{gb}) + \beta V G'_{O3} = (1-\beta)V \frac{dC_{gb}}{dt} = Acc \tag{5.32}$$

where v_g is the volumetric gas flow of the gas phase, C_{geb} and C_{gb} are the concentrations of ozone in the bulk gas at the reactor inlet and outlet, respectively, and G'_{O3} takes different forms depending on the kinetic regime of absorption:

- For slow kinetic regime:

$$G'_{O3} = k_L a(C^*_{O3} - C_{O3b}) \tag{5.33}$$

- For fast and instantaneous regime:

$$G'_{O3} = k_L a C^*_{O3} E \tag{5.34}$$

(where E is E_i if the kinetic regime is instantaneous).

The perfect mixing is usually associated with the liquid and gas phases in mechanically agitated tanks and, also, in some cases, with bubble reactors (see Figure 5.1). In this latter device, however, the plug flow is more common for the gas phase flow. Plug flow, on the other hand, is associated with tubular reactors, such as the bubble column. In this case, the concentration of reactants varies along the axial length of the tube with no mixing at all. The ozone mass balance in the gas phase is (see Appendix A1):

$$v_g \frac{dC_{gb}}{dz} - G'_{O3} S\beta = S(1-\beta) \frac{dC_{gb}}{dt} \tag{5.35}$$

In practical situations or even at the laboratory scale, the hypothesis for ideal flows does not hold for the ideal reactor design equations. In these cases, a study of the nonideal flow should be carried out. This study (see Appendix A3) leads to the determination of the residence time distribution function (RTD) and allows the reactor to be modeled as a combination of ideal reactors or as an ideal reactor with some sort of deviation from ideality.[39] In this way, the reactor design equations that hold correspond to those of the ideal reactors that simulate the flow behavior in the

Kinetic Regimes in Direct Ozonation Reactions

real reactor. Also, the RTD function can confirm that the flow through the reactor presents ideal behavior. Some of these models will be discussed in Chapter 11 on kinetic modeling of ozonation processes.

In the next sections, the ozonation kinetic study is conducted by considering the following points:

- Discussion refers to one gas–liquid irreversible second-order reaction between ozone and one compound B present in water with no competition from secondary direct reactions unless otherwise indicated.
- There is no competition of indirect reactions.
- A given kinetic regime will be assumed that, in most cases, will be confirmed once the kinetic parameters have been calculated.
- Unless indicated, the design equations will correspond to a bubble column or mechanically agitated bubble tank where a known volume of the water phase containing compound B is initially charged. Ozone gas is then fed continuously as an oxygen–ozone mixture of known concentration and flow rate. Perfect mixing of both phases, water and gas, will be considered unless otherwise indicated. The reactors are then semibatch ozonation contactors.
- Film theory will be applied unless otherwise indicated.
- The kinetic treatment will go from the instantaneous to the very slow kinetic regime cases.

Table 5.5 shows the parameters usually determined when the kinetic equations for different kinetic regimes are applied.

Other common features of the kinetic study include the use of the ozone solubility and mass-transfer coefficients.

5.3.2.1 The Ozone Solubility

All absorption rate equations contain the ozone solubility term, C_{O3}^* (see Table 5.5). This is also the ozone concentration at the gas–water interface (because equilibrium conditions are assumed to hold constant at the interface). Note that this concentration corresponds to that of ozone at the gas–water interface in equilibrium with the gas leaving the reactor because of perfect mixing conditions (see Appendix A1). Then, application of Henry's law leads to

$$P_{O3s} = HeC_{O3}^* \qquad (5.36)$$

where P_{O3s} is the ozone partial pressure in the gas at the reactor outlet. Since P_{O3s} changes with time, it is more convenient to express C_{O3}^* as a function of the ozone partial pressure at the reactor inlet, P_{O3i}, which remains constant and known. This can be accomplished using the ozone mass balance in the gas phase, which is also perfectly mixed [see Equation (5.32)]. The perfect gas law expresses ozone partial pressures as a function of concentrations:

$$P_{O3s} = C_g RT \qquad (5.37)$$

In many cases, the accumulation rate term in Equation (5.32) can be considered negligible so that the ozone concentration in the gas at the reactor outlet becomes

$$C_{gb} = C_{geb} + \frac{\beta V G'_{O3}}{v_g} \qquad (5.38)$$

Combining Equations (5.36) to (5.38) then allows C^*_{O3} to be expressed as a function of C_{geb}.

5.3.2.2 The Individual Liquid Phase Mass-Transfer Coefficient, k_L

The individual liquid phase mass-transfer coefficient, k_L, is a key parameter to know in order to determine the Hatta number (Ha_2). Although this mass-transfer coefficient can also be determined from chemical methods (see Section 5.3.3), some empirical equations can be used. These equations are applied mainly to very dilute solutions as in most ozonation reactions in drinking water where the aqueous solution contains low concentrations of B. For wastewater ozonation some deviations are found, especially related to the specific interfacial area that affects the volumetric mass-transfer coefficient, $k_L a$, as shown in Chapter 6.

For mechanically stirred reactors the following equation proposed by Van Dierendonck can be used:[40]

$$k_L = 0.42_3 \sqrt{\frac{\mu_L g}{\rho_L}} Sc^{-0.5} \qquad (5.39)$$

where SI units are used and the Schmidt number Sc is defined as

$$Sc = \frac{\mu_L}{\rho_L D_A} \qquad (5.40)$$

with μ_L and ρ_L the viscosity and density of the solution (water in this case).

Calderbank[40] proposed Equation (5.39) for bubble diameters in bubble columns, d_b, greater than 2 mm and Equation (5.41) for $d_b < 2$ mm:

$$k_L = k_{L\ (for\ 2\ mm)} 500\ d_b \qquad (5.41)$$

where the bubble diameter can be calculated from Equation (5.42):

$$\frac{6(1-\beta)}{d_b} = 2\left(\frac{\rho_L g}{\sigma_L}\right)^{0.5} \frac{u_g}{\left(\frac{\sigma_L g}{\rho_L}\right)^{0.25}} \qquad (5.42)$$

where σ_L and u_g are the surface tension of the liquid (water in this case) and the superficial gas velocity, respectively, and the fraction of gas phase, $1 - \beta$, can be calculated from the equation

Kinetic Regimes in Direct Ozonation Reactions

$$1 - \beta = 1.2 \left(\frac{u_g \mu_L}{\sigma_L} \right)^{0.25} \left[\frac{u_g}{\left(\frac{\sigma_L g}{\rho_L} \right)^{0.25}} \right]^{0.5} \quad (5.43)$$

or simply from experimental data of the height of the liquid in the column, with and without the gas fed, h_T and h, respectively:

$$1 - \beta = \frac{h_T - h}{h_T} \quad (5.44)$$

Values of k_L vary between 10^{-5} and 10^{-4} msec^{-1} for laboratory bubble columns and mechanically stirred reactors. In practice, the range of values is also similar, between 3×10^{-5} and 2×10^{-4} msec^{-1}.[41]

5.3.3 INSTANTANEOUS KINETIC REGIME

In the instantaneous kinetic regime the process rate is exclusively controlled by the diffusion rate of reactants, ozone and B, through the liquid film close to the gas–water interface. For this kinetic regime the reaction develops in a plane inside the film layer (see Figure 4.12 for concentration profiles through the film layer). According to film theory, the diffusion rates of ozone and B are the same, once the stoichiometric ratio is accounted for:

$$z \frac{D_{O3}}{x_R} C^*_{O3} = \frac{D_B}{\delta_L - x_R} C_{Bb} \quad (5.45)$$

Equation (5.45) allows x_R, the distance to the interface where the reaction plane is found (see Figure 4.12), to be calculated. x_R is related to the reaction factor with Equation (5.46):

$$x_R = \frac{D_{O3}}{k_L E} \quad (5.46)$$

The absorption rate law given by Equations (4.45) and (4.46) applied to the ozone–B reaction becomes as follows:

$$N_{O3} = k_L C^*_{O3} E_i = k_L C^*_{O3} \left[1 + \frac{D_B C_{Bb}}{z D_{O3} C^*_{O3}} \right] \quad (5.47)$$

with C^*_{O3} calculated from Equations (5.36) to (5.38). Equation (5.47) holds if $Ha_2 > 10 E_i$ (see also Table 5.5). Reactions of ozone with phenols at alkaline conditions (i.e., at pH > pK of the phenol) or reactions of ozone with some dyes are catalogued as instantaneous reactions.[42,43] These reactions present very high rate constant values that make the kinetic regime instantaneous. In this case, the condition of the instantaneous regime can first be checked to establish the experimental conditions to apply, that is, the ozone concentration in the gas, B concentration, etc.

If the stoichiometric ratio, z, is accounted for, the absorption rate law also expresses the chemical disappearance rate of compound B. Then the mass balance of B in water in a semibatch well-agitated reactor becomes:

$$-\frac{dC_{Bb}}{dt} = zN_{O3}a \tag{5.48}$$

Integration of Equation (5.48) taking into account Equations (5.47) and (5.36) to (5.38) yields

$$\ln \vartheta = -\frac{k_L a D_B}{D_{O3}} t \tag{5.49}$$

where

$$\vartheta = \frac{C_{Bb} + \dfrac{zD_{O3}C^*_{O3}}{D_B}}{C_{Bo} + \dfrac{zD_{O3}C^*_{O3}}{D_B}} \tag{5.50}$$

According to this method, a plot on the left side of Equation (5.49) against time should lead to a straight line. From the slope of this line the volumetric mass-transfer coefficient $k_L a$ is obtained. This procedure was used in a previous work[42] where the ozonation of p-nitrophenol was studied. As an example, Figure 5.8 shows the plot mentioned prepared from experimental data of the ozonation of p-nitrophenol at pH 8.5. The instantaneous kinetic regime is confirmed from the values of Ha_2 and E_i. Given the fact that the rate constant of this reaction is about $14 \times 10^6 \ M^{-1}sec^{-1}$ [42] the resultant Hatta number was much higher than E_i and the condition of instantaneous regime is fulfilled. This procedure has also been applied in other works, where the ozonation of resorcinol, phloroglucinol, and 1,3 cyclohexanedione, considered precursors of trihalomethane compounds in water, was studied.[44,45] In these cases, however, the value of $k_L a$ obtained can be taken as a lower limit for this coefficient. This is so because C^*_{O3} was directly calculated by application of Henry's law to the gas at the reactor inlet and not from Equations (5.36) to (5.38), a situation that does not exactly correspond to the perfect mixing conditions of the water phase. Values of C^*_{O3} used in these calculations were higher than the correct ones that should be obtained from the ozone partial pressure at the reactor outlet as indicated above. In any case, the $k_L a$ values were in the range expected for this type of parameter.[41]

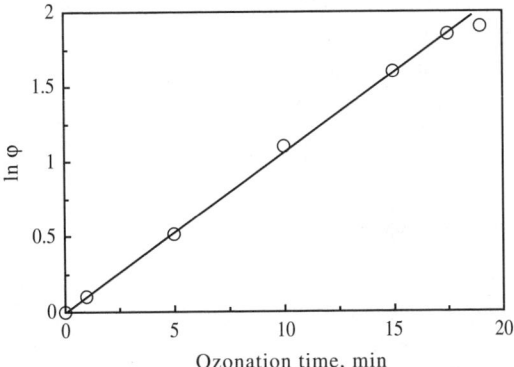

FIGURE 5.8 Determination of the volumetric mass-transfer coefficient from ozonation experiments of *p*-nitrophenol in the instantaneous kinetic regime at pH 8.5. (From Beltrán, F.J., Gómez-Serrano, V., and Durán, A., Degradation kinetics of *p*-nitrophenol ozonation in water, *Wat. Res.*, 26, 9–17, 1992. Copyright 1992, Elsevier Press. With permission.)

A more rigorous treatment, however, made by Ridgway et al.,[43] also determined the volumetric mass-transfer coefficient of a gas–liquid reactor from the results of an instantaneous reaction. In this case, the reaction used was between ozone and the blue dye: indigo disulfonate of potassium. These authors[43] used the Danckwerts theory for instantaneous reactions, that is, Equation (4.68) instead of Equation (4.46) to calculate E_i. They did not neglect the influence of the accumulation rate term, Acc, in Equation (5.32). From Equations (4.68), (5.32), (5.36), and (5.37) they arrived at the following equation for the absorption rate law:

$$N_{O3}a = \frac{C_{Bb} + \dfrac{z\alpha C_{ge} D_{O3}}{D_B} - \dfrac{z\alpha Acc D_{O3}}{v_g D_B}}{\dfrac{z}{k_l a}\sqrt{\dfrac{D_{O3}}{D_B}} + \dfrac{\alpha \beta V D_{O3}}{v_g D_B}} \tag{5.51}$$

where

$$\alpha = \frac{He}{RT} \tag{5.52}$$

Then they introduced the molar balance of B [Equation (5.18)] and, finally, integrated the resulting equation to yield:[43]

$$\ln\left[\frac{C_{Bb} + \dfrac{z\alpha C_{ge} D_{O3}}{D_B} - \dfrac{z\alpha Acc D_{O3}}{v_g D_B}}{C_{Bo} + \dfrac{z D_{O3} C^*_{O3}}{D_B} - \dfrac{z\alpha Acc D_{O3}}{v_g D_B}}\right] = -\frac{1}{\dfrac{1}{k_L a}\sqrt{\dfrac{D_{O3}}{D_B}} + \dfrac{\alpha \beta V D_{O3}}{v_g D_B}} t \tag{5.53}$$

From the slope of a plot similar to that used in Figure 5.8, $k_L a$ is obtained. Note now that the method requires the application of a trial-and-error procedure because the accumulation rate term is unknown. The experiments were carried out at pH 4 and the instantaneous criterion was confirmed (the rate constant for the reaction was about $10^9\ M^{-1}sec^{-1}$). Table 5.6 gives experimental conditions and values of $k_L a$ calculated in these works.

5.3.4 FAST KINETIC REGIME

For the fast kinetic regime the reaction develops in a zone close to the gas–water interface (see Figures 4.10 and 4.11). In comparison to the instantaneous regime, here the reaction is in a zone in the film layer. The condition for this kinetic regime is that $Ha_2 > 3$ (see Table 5.5). The general equation for the absorption rate law is given by Equation (4.44). This equation, however, gives results difficult to use in kinetic determination. A simplification comes from the possibility that the concentration profile of B through the film layer should be constant and the same as in the bulk concentration $C_B = C_{Bb}$ (see Figure 4.11). If this case holds, it is said that the kinetic regime is of fast pseudo first-order. For an ozonation reaction of this type, $Ha_1 = Ha_2$ with $k_1 = k_2 C_{Bb}$. The absorption rate law simplifies to Equation (4.48), which in the case of ozone becomes

$$N_{O3}a = aC_{O3}^* \sqrt{k_2 C_{Bb} D_{O3}} \qquad (5.54)$$

As can be observed, the reaction rate constant, k_2, present in Equation (5.54), results in a simpler mathematical use for its determination.

The procedure to follow for the rate constant determination is similar to that presented above for the volumetric mass-transfer coefficient when the reaction is instantaneous. The first step is the assumption that the fast pseudo first-order kinetic regime holds at the experimental conditions applied. Then the molar balance of B is introduced for a batch system and related to the ozone absorption rate for this kinetic regime with the use of Equation (5.48). Next, Equation (5.55) is obtained:

$$-\frac{dC_{Bb}}{dt} = zaC_{O3}^* \sqrt{k_2 C_{Bb} D_{O3}} \qquad (5.55)$$

Integration of Equation (5.55) between the limits

$$t = 0 \quad C_{Bb} = C_{Bb0}$$
$$t = t \quad C_{Bb} = C_{Bb} \qquad (5.56)$$

leads to the calculation of k_2. Finally, Condition (4.47) has to be checked. This procedure has been applied in some works with some approximations. Sotelo et al.[44,56] determined the rate constants of the reaction between ozone and resorcinol, phloroglucinol, and 1,3 cyclohexanedione by assuming this kinetic regime holds at

TABLE 5.6
Works on Heterogeneous Direct Ozonation Kinetics

Compound	Observations	Reference No. (Year)
Phenol	Wetted wall column, $S = 160$ m^2, 15°C, pH = 1.75–12; moderate kinetic regime with $Ha = 1$, different reaction controlling step according to pH; at alkaline conditions, $n = 0$, $m = 1$. $k_L = 6.7 \times 10^{-5}$ (pH 3 and 4), $k = 1.8 \times 10^9$ (O$_3$-phenolate)	46 (1978)
Phenol	Wetted wall column, $S = 160$ m^2, 9, 15 and 20°C, pH not given; moderate kinetic regime with $Ha = 1$, $AE = 14200$ calmol^{-1}	47 (1978)
Pure water	Semibatch agitated reactor, 12–25°C, PMC, slow kinetic regime; determination of $k_L a$ at different agitation speeds	48 (1980)
Dyes	500-mL washing bottles with plate diffuser, pH 7, phosphate buffer, 20–22°C, $1/z$ between 1.7 (O$_3$-Direct Yellow 12) and 12 (O$_3$-Acid Red 151), PMC, instantaneous kinetic regime	49 (1983)
Phenols	Semibatch reactor, PMC, pH 2.5–3, fast pseudo first-order kinetic regime, CK, determination of relative rate constants	50 (1984)
Maleic acid	0.75-L semibatch stirred bubble reactor, 20°C, pH 2.53 and 2.69, PMC, GCM, moderate–diffusional kinetic regime, $k = 1930$	51
Indigo disulfonate	Semibatch standard agitated tank, diameter = 0.29 m, 25°C, pH < 4, PMC, instantaneous kinetic regime, Danckwerts theory applied, $k_L a = 0.048$ (8 rps)	43 (1989)
o-Cresol	0.75-L semibatch stirred bubble reactor, 20°C, pH 2, phosphate buffer, $1/z = 2$ from homogeneous ozonation, PMC, GCM, fast pseudo first-order kinetic regime, $k = 11955$	52 (1990)
Pure water and resorcinol, and phloroglucinol	0.75-L semibatch stirred bubble reactor, 1–20°C, pH = 2, 7, and 8.5 phosphate buffer, PMC, GCM, slow kinetic regime, $k_L a$ varies depending on gas flow rate and agitation speed: $k_L a = 3.69 \times 10^{-3}$ (pH 7, 20°C, 700 rpm, 30 Lh^{-1}); from homogeneous and heterogeneous ozonation of phenols studied: $1/z = 2$ (O$_3$-resorcinol), $1/z = 1.6$ (O$_3$-phloroglucinol); some intermediates identified	53
2-Hydroxypyridine	1-L semibatch stirred bubble reactor, 20°C, pH 5, 20 mM t-butanol, PMC, slow kinetic regime, three well-mixed reactor models, determination of rate constant of ozone reactions with parent compound and intermediates, and mass-transfer coefficient by fitting experimental results to mass-balance equations	54 (1991)
Malathion	0.5-L semibatch stirred bubble reactor, 10–40°C, pH 2–9, $1/z = 3$ from homogeneous kinetics, PMC, GCM, slow kinetic regime, $n = m = 1$, $k = 98.8$ (20°C, pH 7), $AE = 38.9$ kJmol^{-1}	55 (1991)

TABLE 5.6 (continued)
Works on Heterogeneous Direct Ozonation Kinetics

Compound	Observations	Reference No. (Year)
Resorcinol and phloroglucinol	0.75-L semibatch stirred bubble reactor, 1–20°C, pH = 2, 7, and 8.5 phosphate buffer, PMC, GCM, fast pseudo first-order kinetic regime, $k = 1.01 \times 10^5$ (O_3-Resorcinol, pH 2, 20°C), $k = 1.26 \times 10^6$ (O_3-Resorcinol, pH 7, 20°C); $k = 2.12 \times 10^5$ (O_3-Phloroglucinol, pH 2, 20°C), $k = 9.67 \times 10^5$ (O_3-Phloroglucinol, pH 7, 20°C); kinetic regime goes to instantaneous at pH > 7	56 (1991)
Fenuron	0.5-L semibatch stirred bubble reactor, 10–40°C, pH 2–9, $1/z = 2$ from homogeneous kinetics, PMC, GCM, fast to slow kinetic regime depending on pH, gas constant model, $n = m = 1$, $k = 3.69 \times 10^4$ (20°C, pH 7), $AE = 38.52$ kcalmol^{-1}	57 (1991)
1,3-Cyclohexanedione	0.75-L semibatch stirred bubble reactor, 20°C, pH = 2, 7, and 8.5 phosphate buffer, $1/z = 1$ from both homogeneous ozonation, PMC, GCM, fast pseudo first-order kinetic regime, $k = 1.78 \times 10^5$ (pH 2), $k = 5.43 \times 10^6$ (pH 7), instantaneous kinetic regime, $k_L a = 2.16 \times 10^{-3}$ (pH 8.5)	44 (1991)
Resorcinol and phloroglucinol	As in Reference 35 but at 1–20°C, pH 7 and 8.5 with ozone partial pressure higher than 1780 Pa, PMC, GCM, instantaneous kinetic regime, $k_L a = 1.8 \times 10^{-3}$ (20°C)	45 (1991)
MCPA	0.5-L semibatch stirred bubble reactor, 10–40°C, pH 2–9, $1/z = 2$ from homogeneous kinetics, PMC, GCM, slow kinetic regime, initial rate method, gas constant model, $n = m = 1$ (assumed), $k = 29.65$ (20°C, pH 7), $AE = 13.69$ kcalmol^{-1}	58 (1991)
p-Nitrophenol	0.5-L semibatch stirred bubble reactor, pH 2–12, 20°C, phosphate buffer, ionic strength: 0.1 M, $1/z = 3$ from homogeneous ozonation, PMC, kinetic regimes, diffusional at pH 2, $k_L a = 2.6 \times 10^{-3}$, fast pseudo first-order at pH 6.5, $k = 4.5 \times 10^6$, instantaneous, pH > 7, $k_L a = 2.38 \times 10^{-3}$	42 (1992)
o-Cresol	0.75-L semibatch stirred bubble reactor, 2, 10, and 20°C, pH 7, phosphate buffer, PMC, fast pseudo first-order kinetic regime, $k = 6.99 \times 10^5$ (20°C), kinetic regime changes to fast second order at higher ozone partial pressure	59 (1992)
Gallic acid and epicatachin	0.75-L semibatch stirred bubble reactor, 20°C, pH 4, phosphate buffer, PMC, fast pseudo first-order kinetic regime, CK, phenol, reference compound, $k = 13300$ (O_3-gallic acid), $k = 102700$ (O_3-epicatechin)	60 (1993)
o-Chlorophenol	0.5-L semibatch stirred bubble reactor, 20°C, pH 2–13, phosphate buffer, PMC, fast pseudo first-order to slow kinetic regime, CK, phenol as reference compound, $k = 1640(1 - \alpha) + 5.7 \times 10^7 \alpha$ with α = dissociation degree	61 (1993)

TABLE 5.6 (continued)
Works on Heterogeneous Direct Ozonation Kinetics

Compound	Observations	Reference No. (Year)
Mecoprop	4-L semibatch standard-stirred reactor, pH 2–12, 2–20°C, $1/z = 1$ from homogeneous ozonation, PMC, slow kinetic regime, k at pH < 8, $k = 40$ (pH 2, 20°C), $k = 111.2$ (pH 7, 20°C); O_3/H_2O_2-MCPP was also studied	62 (1994)
Atrazine	1-L semibatch stirred bubble reactor, pH 2, 7, 12, 3–20°C, $1/z = 1$ from homogeneous ozonation, 0.05 M t-butanol at pH 2 to determine k, PMC, very slow kinetic regime, $k = 6.3$ (pH 7, 20°C), $AE = 18.8$ kJmol^{-1}	37 (1994)
Aldicarb	0.25-L agitated cell, pH 7, 7–21°C, phosphate buffer, $1/z = 2$ from homogeneous ozonation, PMC, fast second-order kinetic regime, Danckwerts theory applied, $k = 470300$ (20°C), $AE = 12$ kcalmol^{-1}	63 (1995)
Fluorene (F), Phenanthrene (Ph), Acenaphthene (A)	4-L semibatch standard-stirred reactor, pH 2–12, 4–20°C, $1/z = 2$ (O_3-F and O_3-A) and 1 (O_3-Ph) from homogeneous ozonation; PMC, F: slow kinetic regime, $k = 27$ (pH 2, t-butanol), $AE = 8.8$ kcalmol^{-1}; Ph: slow kinetic regime, CK, naphthalene, reference compound, $k = 2400$ (pH 7, t-butanol); A: fast second-order kinetic regime, $k = 1.08 \times 10^5$ (pH 7)	64 (1995)
Trichloroethylene	1-L semibatch bubble reactor, pH 2, t-butanol, 20°C. $1/z = 1$ from homogeneous ozonation, PMC, very slow kinetic regime, $k = 17.1$	65 (1997)
Cinnamic, crotonic, and hydroxycinnamic acids	0.5-L semibatch stirred reactor or agitated cell. pH 3 and 7, 20°C, $1/z = 1$, PMC, in agitated cell ($a = 19.63$ m^{-1}), fast second-order kinetic regime, Danckwerts theory applied; k between 1.1×10^4 to 1.4×10^4	38 (1998)
2,4,6-Trichlorophenol (TCP) and 2,5-dichloro-hydroquinone (DCHQ)	Wetted sphere absorber, pH 2–7, 15–35°C, fast kinetic regime, Higbie penetration theory, phosphate buffer, for O_3-TCP: $k = 1.44 \times 10^4$ (20°C, pH2), $k = 1.69 \times 10^4$ (20°C, pH 7), for O_3-DCHQ: 1.7×10^5 (25°C, pH 2)	66 (1999)
24 Pesticides	Semibatch stirred bubble reactor, 20°C, pH 7.5, phosphate-carbonate, PMC, slow kinetic regime, $k = 266$ (O_3-BMPC), rest of ozone–pesticide k values from CK, different pesticides as reference compounds, examples: $k = 27600$ (O_3 PCP), $k = 323$ (O_3 MCPA), $k = 1.11$ (O_3-diquat); conditions of kinetic regime did not check	67 (2000)
Phenolic acids	1-L semibatch jacketed bubble (plate diffuser) reactor, pH 2–9, 20°C, phosphate buffer, PMC, fast pseudo first-order, moderate and slow kinetic regimes depending on pH. CK, different reference compounds, k also varied with pH	68 (2000)
Simazine	1-L semibatch jacketed bubble (plate diffuser) reactor, pH 2, 20°C, 0.1 M t-butanol, PMC, very slow kinetic regime, $k = 8.7$	69 (2000)

TABLE 5.6 (continued)
Works on Heterogeneous Direct Ozonation Kinetics

Compound	Observations	Reference No. (Year)
Sodiumdodecylbenzene-sulfonate	2.5-L semibatch jacketed bubble (plate diffuser) column, pH 2, 20°C, $1/z = 1$ from homogeneous ozonation. 0.01 M t-butanol, PMC, very slow kinetic regime, $k = 3.68$	70 (2000)
Alachlor	0.75-L semibatch bubble reactor, pH 2, 20°C, 0.01 M t-butanol, PMC, very slow kinetic regime, $k = 2.8$; other advanced oxidation also studied	71 (2000)
Naphthalene sulfonic acids	5-L semibatch stirred bubble reactor, 20°C, pH 3–9, PMC, slow kinetic regime, determination of rate constant through fitting of experimental results to mass-balance equations, $k = 252$ (O$_3$-1-naphthalene sulfonic acid, pH 3); rate constant of hydroxyl radical compound also determined; maleic acid, oxalic acid, formic acid, and sulfate: by-products	72 (2001)
Resorcinol	1.5-L semibatch stirred bubble reactor, 20°C, pH 2, presence and absence of t-butanol, PMC, slow kinetic regime (conditions not checked), $k = 360$ Maleic acid subproduct	73 (2001)
1,3,6-Naphthalenetri-sulfonic acid	2-L semibatch bubble reactor, 15–35°C, pH 2–9, $1/z = 1$ from homogeneous ozonation; at pH 2 with 0.01 M t-butanol, PMC, slow kinetic regime (conditions not checked), $k = 6.72$ (pH 2, 25°C), $AE = 37.18$ kJmol^{-1}; rate constant of HO compound reaction also determined	74 (2002)

Note: Depending on kinetic regime, k, k_L, and $k_L a$ are determined. PMC = perfect mixing condition for both phases; GCM = gas constant model that assumes ozone gas concentration inside the reactor equals constant ozone gas inlet concentration; CK = competitive kinetics to determine k; AE = activation energy; n = ozone reaction order; m = compound reaction order. Units of k in M^{-1}sec^{-1} unless otherwise indicated. Units of k_L in msec^{-1}. Units of $k_L a$ in sec^{-1}.

their experimental conditions. They considered the term C_{O3}^* as the concentration of ozone in water at equilibrium with the ozone gas entering the reactor so that their k_2 values can be taken as a lower limit of the true ones. With this approximation, once Henry's law is accounted for, Equation (5.55) becomes

$$-\frac{dC_{Bb}}{dt} = \frac{C_{gi}RT}{He} za\sqrt{k_2 C_{Bb} D_{O3}} \tag{5.57}$$

This equation can be integrated to yield

$$\sqrt{C_{Bb}} = \sqrt{C_{Bb0}} - \frac{zaC_{gi}RT}{He}\sqrt{k_2 D_{O3}}\,t = \sqrt{C_{Bb0}} - At \tag{5.58}$$

Kinetic Regimes in Direct Ozonation Reactions

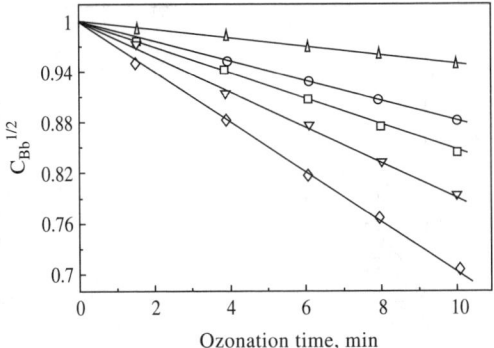

FIGURE 5.9 Verification of Equation (5.58) in the ozonation of resorcinol at different ozone partial pressure. Conditions: pH = 7, 20°C, 30 Lh^{-1} = gas flow rate, 300 rpm = agitation speed; Ozone partial pressure, Pa: △ = 75, ○ = 164, □ = 305, ▽ = 882, ◊ = 13399. (From Sotelo, J.L., Beltrán, F.J., and González, M., Ozonation of aqueous solutions of resorcinol and phloroglucinol. 2. Kinetic study, *Ind. Eng. Chem. Res.*, 31, 222–227, 1991, Copyright 1991, American Chemical Society. With permission.)

Accordingly, a plot of the square root of the concentration of B against the reaction time should yield a straight line (see Figure 5.9 as example). By carrying out experiments at different ozone partial pressures, different values of A are obtained. Then the rate constant can be calculated from a plot of A against C_{gi} or C_{O3}^*, as shown in Figure 5.10 for the case of the ozonation of resorcinol at pH 2 taken from the work of Sotelo et al.[56] As observed, the linearity expected from the relationship between A and C_{O3}^* [Equation (5.58)] only holds for low values of C_{O3}^*. At these conditions, E_i is high enough so that Condition (4.47) is fulfilled. In contrast, Condition (4.47) does not hold at higher values of C_{O3}^*. Deviations from linearity for high values of C_{O3}^* can be attributed to a change in the kinetic regime of ozone absorption and, as a consequence, the assumed ozone absorption rate [Equation (5.54)] also does not hold.

In some cases, however, the reaction of ozone with the compound studied, although in the fast pseudo first-order kinetic regime, cannot be used because of the competition of secondary reactions. In this case, the procedure can also be applied but a differential method must be used. The method is based on the differential initial rate method.[42] In this case, the kinetic regime is applied only to the initial values of the ozone absorption rate and disappearance rate of B since it is evident that at the start of reaction, ozone only reacts with the target compound initially present in water. The starting equation is now:

$$-\frac{dC_{Bb}}{dt}\bigg|_{t=0} = zaC_{O3}^*\sqrt{k_2 C_{Bb0} D_{O3}} \tag{5.59}$$

where C_{O3}^* is expressed as a function of the ozone concentration at the reactor inlet through the use of the Henry and perfect gas laws and the ozone mass-balance

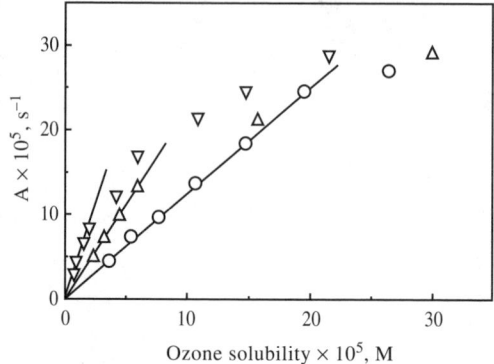

FIGURE 5.10 Variation of A with the ozone solubility in experiments of the ozonation of resorcinol. Conditions: 30 Lh^{-1}, 300 rpm agitation speed, $T = $ °C: ○ = 1, ∆ = 10, ∇ = 20. (From Sotelo, J.L., Beltrán, F.J., and González, M., Ozonation of aqueous solutions of resorcinol and phloroglucinol. 2. Kinetic study, *Ind. Eng. Chem. Res.*, 31, 222–227, 1991. Copyright 1991, American Chemical Society. With permission.)

equation in the gas phase [Equation (5.36) to Equation (5.38)], also applied at the start of the process:

$$C_{O3}^* = \frac{RT}{He}\left[\frac{v_g C_{gi} Vz(-dC_{Bb}/dt)_{t=0}}{v_g}\right] \quad (5.60)$$

According to Equation (5.59), the plot of $(1/z)(-dCB/dt)_{t=0}$ against $C_{O3}^* C_{Bb}^{0.5}$ should lead to a straight line provided the kinetic regime is the fast pseudo first-order type. The rate constant can be obtained from the slope of the line. This procedure has been applied in the kinetic study of the ozonation of *p*-nitrophenol at pH 6.5.[42] As also observed in the work of Sotelo et al.,[56] the linear relationship indicated in Equation (5.59) held only for low values of C_{O3}^*. The possible change of the kinetic regime for high values of C_{O3}^* was confirmed by checking Condition (4.47). Table 5.6 also presents rate constant data of ozone–B reactions calculated in this way.

Note that one of the main difficulties of this method is that the specific interfacial area must be known to follow the procedure. A chemical procedure to determine this parameter is based on the use of experimental data of a fast pseudo first-order ozonation reaction (one with known rate constant) and follows the steps indicated above. In addition to the specific interfacial area, the individual mass-transfer coefficient, k_L, has to be known in order to check the condition of the assumed kinetic regime. Both the specific interfacial area and individual liquid phase mass-transfer coefficient can be calculated from the experimental results of a fast pseudo first-order reaction but with the use of the absorption rate law provided by the Danckwerts theory. This corresponds to Equation (4.58) (with $Ha_1 = Ha_2$), which applied to the mass balance of a compound B that reacts with ozone in the indicated kinetic regime of known kinetics (that is the known rate constant), is as follows:

Kinetic Regimes in Direct Ozonation Reactions

$$-\frac{dC_{Bb}}{dt}\bigg|_{t=0} = zk_laC_{O3}^*\sqrt{1+Ha_2^2} \qquad (5.61)$$

Then, after considering the definition of the Hatta number, Ha_2, the square of Equation (5.61) leads to

$$\left(-\frac{dC_{Bb}}{dt}\bigg|_{t=0}\right)^2 = \left(zk_laC_{O3}^*\right)^2 + \left(zaC_{O3}^*\right)^2 k_2 D_{O3} C_{Bb0} \qquad (5.62)$$

Equation (5.62) represents a straight line called the Danckwerts line.[41] The corresponding plot of the left side against the initial concentration of B corresponding to different experiments allows the simultaneous determination of k_L and a from the ordinate and slope, respectively, of the fitted straight line. Thus far, this procedure has not been applied in ozone reactions although the literature presents examples for other gas–liquid reacting systems.[75,76]

For rate constant determination in the fast pseudo first-order regime, another possibility is the use of an agitated cell to carry out the ozonation experiments. The agitated cell is an agitated tank where the gas is fed just above the liquid surface, as shown in Figure 5.7. The agitation should be kept low enough so as to avoid any disturbance in the plane surface of the liquid but high enough to ensure perfect mixing in the liquid phase. In this reactor the specific interfacial area, a, is known (a = the ratio of the surface of liquid exposed to the gas and the volume of liquid). Since the mass-transfer rate is also very low the amount of ozone absorbed is reduced so that the concentration of ozone in the gas at the reactor outlet is practically coincident with that in the gas at the reactor inlet. Hence, C_{O3}^* can be calculated from the latter by applying the Henry's law constant without the need for the mass balance of ozone in the gas phase. This procedure was also applied in a previous work where the rate constants of the reactions between ozone and some crotonic acid–derived compounds were calculated.[38]

Another possibility to apply the fast pseudo first-order kinetic regime is the use of the competitive kinetic method. With this method, the ozonation of the target compound is carried out in the presence of another compound, or reference compound, R, of known ozonation kinetics (that is, known rate constant and stoichiometric ratio). The reference compound should have a reactivity toward ozone similar to that of the target compound. Both ozone reactions have to develop in the fast pseudo first-order kinetic regime. The method has the advantage that secondary direct reactions have no influence on the rate constant determination since the ozone concentration is removed from the rate equation, as shown below. However, one disadvantage of the method is that the rate constant of the ozone–R reaction must be known with accuracy. The procedure in a semibatch reactor with both the gas and water phases perfectly mixed starts from the application of the mass-balance equations of both the target and reference compounds:

$$-\frac{dC_{Bb}}{dt} = zk_2 C_{Bb} C_{O3} \qquad (5.63)$$

and

$$-\frac{dC_{Rb}}{dt} = z_R k_{2R} C_{Rb} C_{O3} \quad (5.64)$$

where C_{Rb} is the bulk concentration of the reference compound and z_R and k_{2R} are the stoichiometric coefficient and rate constant of its reaction with ozone (that are known), respectively. By dividing both equations and integrating between the limits

$$\begin{array}{llll} t = 0 & C_{Bb} = C_{Bb0} & C_{Rb} = C_{Rb0} \\ t = t & C_{Bb} = C_{Bb} & C_{Rb} = C_{Rb} \end{array} \quad (5.65)$$

Equation (5.66) is obtained:

$$\ln \frac{C_{Bb}}{C_{Bb0}} = \frac{zk}{z_R k_{2R}} \ln \frac{C_{Rb}}{C_{Rb0}} \quad (5.66)$$

Although the reactions develop in a zone in the film layer, the concentrations of B and R must be constant and equal to those they have in the bulk water. If not, the analytical integration indicated above would not be valid. Note that this procedure can also be applied to other different kinetic regimes, provided that the concentrations of B and R are constant within the film layer (see the slow and moderate regime sections). From Equation (5.66) a plot of the $\ln(C_{Bb}/C_{Bb0})$ against $\ln(C_{Rb}/C_{Rb0})$ leads to a straight line of slope equal to $zk_2/z_R k_{2R}$. From the slope, the rate constant of the ozone–B reaction can be finally obtained. Then Condition (4.47) for the fast pseudo first-order kinetic regime must be confirmed. Results of the rate constants obtained from this method are presented in Table 5.6.

Although the fast pseudo first-order kinetic regime is the most recommended one to determine the rate constant of fast gas–liquid reactions, in some cases Condition (4.47) cannot be fulfilled and the use of Equation (4.44) is the only way to determine the rate constant. For these cases, it is convenient to use the absorption rate law given by Danckwerts surface renewal theory [Equation (4.74)] rather than the equation derived from film theory [Equation (4.44)], which is a more difficult mathematical treatment because in the former E is expressed explicitly. Combining Equation (4.31) with the mass balance of B yields, after integration, Equation (5.67):

$$C_{Bb0} - C_{Bb} = zk_l a \int_0^t C_{O3}^* E \, dt \quad (5.67)$$

where E is defined by Equation (4.74). Both E and C_{O3}^* are functions of time, and the determination of the rate constant follows a trial-and-error procedure. First a value of k_2 is assumed, then Ha_2, E_i, and E are determined from Equations (4.40) (for Ha_2), (4.68), and (4.74) at different reaction times. The value of C_{O3}^* is also determined at different times from the mass balance of ozone in the gas [Equation

(5.38)] and Henry's law is applied to the ozone partial pressure at the reactor outlet [Equation (5.36)]. Finally, numerical integration of Equation (5.67) is carried out. The integral value should coincide with the difference between the initial and final concentrations of B, as Equation (5.67) indicates. The last step of the procedure is to check that the condition for the fast reaction is fulfilled. This procedure has been applied to the kinetic study of the ozonation of aldicarb (a hazardous herbicide of water).[63] In this work, the concentration of ozone in the gas at the reactor outlet becomes constant within a few minutes of the start of ozonation, t_i; then Equation (5.67) can be numerically integrated with the boundary conditions:

$$t = t_i \quad C_{Bb} = C_{Bbi}$$
$$t = t \quad C_{Bb} = C_{Bb} \tag{5.68}$$

where C_{Bbi} is the concentration of B at time t_i. The following steps include the trial-and-error procedure that allows the rate constant of the reaction to be determined once the integral of the right side of Equation (5.67) coincides with the difference between concentrations of B at time t_i and at given time t.

The rate constant of a fast ozone reaction can also be calculated from the differential method of initial reaction rates when some secondary direct reactions compete for the available ozone. This is the case with some polynuclear aromatic hydrocarbons, such as acenaphthene.[64] The initial absorption rate is used in the mass balance of B so that

$$-\frac{dC_{Bb}}{dt} = zN_{O3}\Big|_{t=0} = zk_l a C_{O3}^* E\Big|_{t=0} \tag{5.69}$$

where $E_{t=0}$ is given by Equation (4.74). The method does not require any numerical integration, only calculation of the initial disappearance rate of B from the B concentration profile–time data. Using a trial-and-error procedure similar to that of the preceding section the rate constant can be determined. Table 5.6 lists the calculated rate constant data for the ozone–acenaphthene reaction.

5.3.5 MODERATE KINETIC REGIME

When the reaction between ozone and the target compound B is either in the film layer or in the bulk liquid the regime is called moderate (see Figure 4.9). The condition to be fulfilled is that the Hatta number must be between 0.3 and 3 (see also Table 5.5). As shown previously, the absorption rate law equation coincides with the general Equation (4.25) with Ha_1 given by Equation (4.38) and R_{b1} substituted by R_{b2} [Equation (4.41)]. Due to the complexity of the kinetic equation, it is evident that this kinetic regime is not recommended for kinetic studies. Nonetheless, there are two cases where the experimental data obtained at this kinetic regime can be used for this purpose. In one of these cases, there is no dissolved ozone and the rate law equation can be simplified to Equation (4.44), which also applies for the fast kinetic regime (see previous section). In the second case, the concentration of

B can be considered constant through the film layer and the regime can be called moderate pseudo first-order.

5.3.5.1 Case of No Dissolved Ozone

The absorption rate law [Equation (4.25)] simplifies to Equation (4.44). The procedure is similar to that shown for fast reactions. The lack of explicit form of the reaction factor E makes the use of trial-and-error procedures necessary. Thus, the method starts by assuming a value for the rate constant k_2. The Hatta number, Ha_2, is then calculated and, finally, the reaction factor, E, with the help of Equation (4.44). This value should coincide with that obtained from the disappearance rate of B:

$$E = \frac{-\dfrac{dC_{Bb}}{dt}}{zk_l a C^*_{O3}} \tag{5.70}$$

With this procedure, the rate constant of the direct ozonation of maleic acid was determined in a previous work[51] (see Table 5.6 for rate constant data).

5.3.5.2 Case of Pseudo First-Order Reaction with Moderate Kinetic Regime

The absorption rate law is still a complicated equation to be solved. Thus, the best method first applies the competitive kinetic procedure indicated previously for fast pseudo first-order reactions. The starting equations are the mass balances of target and reference compounds, B and R, respectively, in a batch system (recall that batch refers to the water charged to a semibatch reactor). With this method, knowledge of the concentration of ozone is not relevant but concentrations of B and R must be constant through the film layer so that Equations (5.63) and (5.64) hold. If this is the case, the rate constant of the direct ozone–B reaction can be calculated as indicated in Section 5.3.4. However, validation of this procedure requires not only checking the condition for the moderate kinetic regime to be applied (see Table 5.5) but also the uniformity of concentrations of B and R through the film layer, which is the primary reason for choosing the method. Estimations of these concentrations can be made following the procedure developed in Section 4.2.1.3 for series-parallel gas–liquid reactions. Using Equation (4.53) to Equation (4.56), concentrations of B and R at the gas–water interface can be obtained after trial-and-error procedures. The method presented was applied in a work where the competitive ozonation of some phenol compounds was studied.[68] The results are also presented in Table 5.6.

5.3.6 SLOW KINETIC REGIME

The presence of dissolved ozone is the sign of slow reactions in water. In these cases the Hatta number is lower than 0.3 and the reaction factor is lower than unity, if it is defined according to Equation (4.31), that is, defined with respect to the maximum physical absorption rate. Then, the definition of the reaction factor would not have physical meaning. Note, on the contrary, that for the slow kinetic regime the reaction factor defined as the ratio between the actual chemical absorption rate and that of physical absorption is 1:

$$E = \frac{N_{O3}}{k_L a \left(C_{O3}^* - C_{O3b} \right)} = 1 \qquad (5.71)$$

The process presents two steps in series, with the diffusion rate of ozone through the film layer equal to the chemical reaction rate in the bulk liquid. Figure 4.7 shows the general situation for the concentration profiles of ozone (the gas dissolved in our case) and B, the target compound.

In the slow kinetic regime, the influence of the ozone decomposition reaction is the first action to check. In fact, as shown in Section 7.1, at pH < 12, which is the usual situation when the direct ozone reaction studied is in the slow kinetic regime, the ozone decomposition reaction likely competes for the available ozone. In this case, both ozone and free radicals (mainly hydroxyl radicals) react with the target compound B. Thus, some experiments on the ozonation of the target compound should first be carried out in the presence and absence of hydroxyl radical scavengers to avoid the development of indirect reactions. For example, Figure 5.6 shows results of the ozonation of atrazine in the presence and absence of t-butanol and carbonates, known scavengers of hydroxyl radicals. It is seen that, in the absence of scavengers, the remaining concentration of atrazine decreases with time faster than in its presence. It is evident that this effect is due to the additional contribution of free radical reactions (that is, indirect ozone reactions) developed from the ozone decomposition reaction. This is, then, conclusive proof of the competitive effect of ozone decomposition. Reactions for the kinetic study in this case (that is, for the rate constant determination) must be carried out in the presence of scavengers of hydroxyl radicals.

Once indirect reactions are eliminated, the rate constant of the direct reaction ozone–B compound can be obtained directly from the mass balance of B in the batch reactor:

$$r_B = -\frac{dC_{Bb}}{dt} = z k_2 C_{O3b} C_{Bb} \qquad (5.72)$$

Equation (5.72) can be used if concentrations of B and ozone are known with time. This is the typical situation in the slow kinetic regime. The recommended method is the use of an analytical procedure, that is, to integrate Equation (5.72). This can only be done in the absence of secondary reactions that consume ozone since, in that case, concentrations of ozone and B are functions of the conversion of B. However, this is rather difficult in ozonation reactions where ozone is also consumed in unknown direct secondary reactions. In this case, a differential method should be applied. With this method, a plot of $-r_B$ with the product of concentrations of B and ozone would lead to a straight line, according to Equation (5.72). The slope of this line is the rate constant regardless of the presence of secondary reactions. The disappearance rates of compound B, $-r_B$, can be obtained from the experimental curve of the concentration of B with time. The method was used, for example, in the kinetic study of the ozonation of fluorene, a polynuclear aromatic hydrocarbon that reacts slowly with ozone through direct reaction.[64] Thus, the disappearance rates of fluorene were obtained from the first derivative with time of a second-order polynomial equation found, through least squares analysis, to fit the concentration of fluorene with time. Table 5.6 shows values of rate constants obtained with this method.

Another possibility in the kinetic study of slow reactions is the initial method of reaction rates. A drawback of this method is the unknown value of the ozone concentration at the start of reaction. In fact, at the start of the reaction, the concentration of dissolved ozone is theoretically zero; however, the mass balance of ozone in water can be used to avoid this problem. This equation for the case of a semibatch reactor is as follows:

$$k_L a \left(C_{O3}^* - C_{O3} \right) = \sum_i k_{2j} C_{jb} C_{O3b} + k_d C_{OH^-} C_{O3b} + k_T C_{O3} + \frac{dC_{O3b}}{dt} \quad (5.73)$$

where the subindex j represents any compound j present in water that reacts with ozone directly and k_{2j}, k_d, and k_T are the rate constants of the ozone direct, hydroxyl-ion-initiated decomposition, and other initiation reactions, respectively. If Equation (5.73) is applied to the start of ozonation, the ozone accumulation rate term, dC_{O3b}/dt, and intermediate reactions are negligible, and the equation reduces to

$$k_L a \left(C_{O3}^* - C_{O3b} \right)\big|_{t=0} = k_2 C_{Bb0} C_{O3b} \big|_{t=0} = -\frac{r_{B0}}{z} \quad (5.74)$$

From Equation (5.74) the concentration of ozone is

$$C_{O3b}\big|_{t=0} = \frac{k_L a C_{O3}^*}{k_2 C_{Bb0} + k_L a} \quad (5.75)$$

where C_{Bb0} represents the initial concentration of B. Combining Equations (5.74) and (5.75) allows us to determine k_2:

$$k_2 = \frac{-k_L a r_{B0}}{z C_{Bb0} \left[k_l a C_{O3}^* + \frac{r_{B0}}{z} \right]} \quad (5.76)$$

This method was also applied in the kinetic study of the ozonation of fluorene[64] (see Table 5.6).

A third possibility for determining the rate constant, k_2, is the use of the competitive method explained in Section 5.3.4 for the fast pseudo first-order kinetic regime. Note that in slow reactions the concentration of B is constant through the film layer and, as a consequence, the competitive kinetic method can also be applied. In fact, the literature reports some works where the competitive method was used to determine the rate constant of direct ozone reactions[64,50] with the slow kinetic regime of ozone absorption. Table 5.6 shows the reported rate constant values obtained.

5.3.6.1 The Slow Diffusional Kinetic Regime

The slow kinetic regime presents more possibilities, as shown below. One of the subkinetic regimes is diffusional (see Figure 4.8), with the absorption rate equal to the maximum physical absorption rate:

$$N_{O3}a = k_L a C^*_{O3} = M_1 a \qquad (5.77)$$

The reaction factor E is also unity and the combination of the mass balance of B in a semibatch reactor with the absorption rate law [Equation (5.48) and Equation (5.77)], yields:

$$-\frac{dC_{Bb}}{dt} = zN_{O3}a = zk_L a C^*_{O3} \qquad (5.78)$$

Equation (5.78) indicates zero-order kinetics with respect to B, therefore the experimental concentration profile of B with time should be linear. If the concentration of ozone at the interface, C^*_{O3}, is constant during the reaction time, integration of Equation (5.78) with the initial Condition (5.56) leads to a straight line of slope equal to $zk_L a C^*_{O3}$ and the volumetric mass-transfer coefficient can be determined. Note that C^*_{O3} is the concentration in water at the interface in equilibrium with the gas exiting the reactor because of the perfect mixing conditions for both the gas and water phases. Although this concentration usually changes with time, for slow reactions it could remain constant and very close to the value of the ozone concentration at the reactor inlet because of the low values of the ozone absorption rate. C^*_{O3} can then be calculated directly from Henry's law by using the ozone partial pressure at the reactor inlet, which is known and remains constant with time. This procedure was applied in the ozonation of p-nitrophenol at pH 2 with an initial concentration higher than 10^{-4} M.[42] At these conditions, the disappearance rate of p-nitrophenol is constant for the first 15 to 20 min, which indicates zero-order kinetics. The slope of the line is $k_L a C^*_{O3}$. Figure 5.11 shows that a plot of the slope of these lines (the disappearance rate of B) obtained at different ozone partial pressure against the corresponding C^*_{O3} leads to another straight line. The slope of this line was the volumetric mass-transfer coefficient (see also Table 5.6).

5.3.6.2 Very Slow Kinetic Regime

For rate constant determination, the main problem associated with the methods presented above is the need to know the disappearance rate of B, either at any time or exclusively at the start of the process. For this reason, the very slow kinetic regime of absorption is the most appropriated one to determine the rate constant of the ozone–B direct reaction. If this is the case, the ozone absorption rate equals the

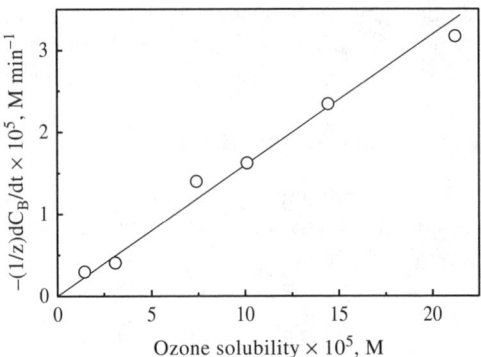

FIGURE 5.11 Determination of the volumetric mass-transfer coefficient from ozonation experiments of *p*-nitrophenol in the diffusional kinetic regime at pH 2. (From Beltrán, F.J., Gómez-Serrano, V., and Durán, A., Degradation kinetics of *p*-nitrophenol ozonation in water, *Wat. Res.*, 26, 9–17, 1992. Copyright 1994, Elsevier Press. With permission.)

maximum chemical reaction rate because the process is chemically controlled. Then, once the stoichiometry ratio, z, is accounted for, the absorption rate is related to the disappearance rate of B, so that, in a batch system (with respect to B) the following is obtained:

$$r_B = -\frac{dC_{Bb}}{dt} = zN_{O3}a = z\beta k_2 C_{O3}^* C_{Bb} \tag{5.79}$$

In laboratory practice, when treating very slow ozone reactions, it is usual to get increased concentrations of dissolved ozone for the first few minutes of reaction and then a steady state concentration of ozone, C_{O3s} which is very close to C_{O3}^*. Thus, integration of Equation (5.79) is carried out with the initial condition

$$t = t_i \quad C_{Bb} = C_{Bbs} \tag{5.80}$$

where t_i is the time needed to reach the steady-state concentration of ozone and C_{Bbs} the corresponding concentration of B to that time. Integration of Equation (5.79) leads to:

$$\ln \frac{C_{Bb}}{C_{Bbs}} = -z\beta k_2 C_{O3s}(t - t_i) \tag{5.81}$$

According to Equation (5.81), a plot on the left side against $(t - t_i)$ should lead to a straight line. The slope of this line is the product of the rate constant, the stoichiometric ratio, liquid holdup, and the steady-state concentration of ozone. The procedure depicted here has also been applied in different works (see Table 5.6). In Figure 5.12, the aforementioned plot is presented for the case of the ozonation of atrazine.[37]

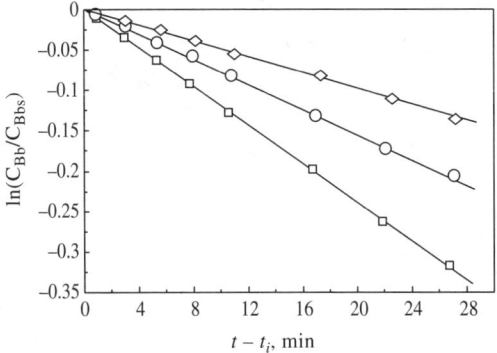

FIGURE 5.12 Determination of the rate constant of the direct reaction between atrazine and ozone from results in a very slow kinetic regime. Conditions: pH = 2, 0.05 M t-butanol, C_{ATZ0} = 4.8 × 10^{-5} M, T °C, P_{O3i}, Pa, respectively: ◊ = 3, 695; ○ = 10, 840; □ = 20, 1100. (From Beltrán, F.J., García-Araya, J.F., and Benito, A., Advanced oxidation of atrazine in water. I. Ozonation, *Water Res.*, 28, 2153–2164, 1994. Copyright 1994, Elsevier Press. With permission.)

5.4 CHANGES OF THE KINETIC REGIMES DURING DIRECT OZONATION REACTIONS

From the previous sections the importance of ozone can be deduced, not only for its practical use in water treatment but also for academic studies with the aim of obtaining kinetic and hydrodynamic parameters. It has been shown that ozone absorption in water develops in different kinetic regimes — instantaneous, fast, or slow — which offer the possibility of determining rate constants, specific interfacial areas, and mass-transfer coefficients. Figure 4.3 and Figure 4.5, which show the changes of E with the Hatta number for first- and second-order irreversible reactions, can also be prepared from experimental results of some ozone direct reactions. In this sense, reactions of ozone with phenols are particularly special because of the dissociating character of phenols. Thus, according to Equation (3.22), the rate constant varies with pH by several orders of magnitude. From the point of view of the kinetic regime of ozone absorption, at constant values of the ozone diffusivity and individual liquid phase mass-transfer coefficient, changes of pH, ozone partial pressure, and phenol concentrations give rise to changes in the kinetic regime of ozonation. From the results reported in preceding papers,[45,53,56] a plot of the experimental reaction factor, E, with the Hatta number, Ha_2, would allow us to observe the different zones where the kinetic regimes hold. This type of plot has been prepared in previous works.[45,39]

5.5 COMPARISON BETWEEN ABSORPTION THEORIES IN OZONATION REACTIONS

With some exceptions, most of the studies concerning the kinetics of the direct reactions of ozone in water have been developed using film theory because of the easier mathematical treatment. However, film theory is based on unrealistic concepts

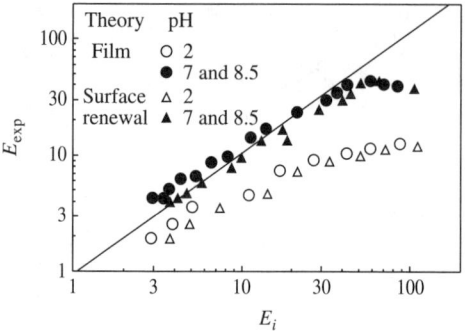

FIGURE 5.13 Comparison between experimental and instantaneous reaction factor determined from different absorption theories and applied to the ozonation of 1,3-cyclohexanedione. (From Sotelo, J.L., Beltrán, F.J., and González, M., The use of surface renewal and eddy diffusivity theories in the ozonation of a THM precursor: 1,3 cyclohexanedione, *Ozone Sci. Eng.*, 13, 421–436, 1991. Copyright 1991, International Ozone Association. With permission.)

and, therefore, it is convenient to compare the kinetic results obtained from this theory with others based on more realistic approaches, such as the surface renewal theory of Danckwerts (see equations in Section 4.2.2). In preceding papers[77,78] this comparison was made for the case of the reaction between ozone and resorcinol, phloroglucinol, and 1,3-cyclohexanedione, compounds that dissociate in water when pH is changed. The experimental reaction factors at different conditions of pH, ozone partial pressure and concentration of B, obtained from the film and Danckwerts theories, were plotted against their corresponding instantaneous factors as shown in Figure 5.13 for the case of 1,3-cyclohexanedione.[77] As can be observed, values of the instantaneous reaction factor from the Danckwerts theory were always higher than those obtained from the film theory, except for the cases of very high ozone partial pressure. This means that the ozone partial pressure range for which the fast kinetic regime of absorption holds is wider when the Danckwerts theory is applied. This is useful information because the rate constant of the reactions of ozone with many organics (phenols, PAHs, etc.) must be determined when the fast, or to be more exact, the fast pseudo first-order kinetic regime holds. From Figure 5.13, it is also observed that the instantaneous kinetic regime ($E_{exp} = E_i$) was reached when some ozonations were at pH 7 and 8.5, regardless of the absorption theory applied. Then, at pH 2, the kinetic regime was never instantaneous. Since differences in the instantaneous reaction factor can give rise to a new distribution of kinetic regime zones, it is convenient to define this new distribution zones with the Danckwerts theory. In fact, in another work[78] other theories such as the eddy diffusivity theory (which are not discussed here) were also applied with results similar to the Danckwerts theory. From the results obtained it was concluded that the ozone partial pressure and pH were the main variables that affect the kinetic regime zones. However, although the range of ozone partial pressures for which kinetic regime zones are established depends on the theory applied, the values of the rate constant and mass-transfer coefficients obtained from the application of both theories were

TABLE 5.7
Rate Constants of Ozone Direct Reactions from Film and Danckwerts Theories in Ozonation Kinetic Studies[a]

Compound and Conditions	Film Theory	Danckwerts Theory
o-Cresol, pH 2, 20°C	1.19×10^4 [52]	1.10×10^4 [52]
Resorcinol, pH 2, 20°C	1.01×10^5 AE = 34.9 [56]	9.68×10^4 AE = 32.9 [78]
Resorcinol, pH 7, 20°C	1.26×10^6 AE = 17.1 [56]	1.23×10^6 AE = 13.6 [78]
Phloroglucinol, pH 2, 20°C	2.12×10^5 AE = 42.5 [56]	2.14×10^5 AE = 40.8 [78]
Phloroglucinol, pH 7, 20°C	9.67×10^5 AE = 53.5 [56]	1.12×10^6 AE = 51.3 [78]
1,3-Cyclohexanedione, pH 2, 20°C	1.78×10^5 [44]	2.03×10^5 [77]
1,3-Cyclohexanedione, pH 7, 20°C	5.43×10^6 [44]	$5,43 \times 10^6$ [77]

[a] Units in $M^{-1}sec^{-1}$, AE = activation energy, in kJmol^{-1}.

similar (see Table 5.7). As a final conclusion it can be said that because of its mathematical simplicity, film theory is the most recommended method for studying the kinetics of direct ozone reactions in water.

References

1. Wilke, C.R. and Chang, P., Correlation of diffusion coefficients in dilute solutions, *AIChE J.*, 1, 264–2270, 1955.
2. Reid, R.C., Prausnitz, J.M., and Sherwood, T.K., *The Properties of Gases and Liquids*, McGraw-Hill, New York, 1977.
3. Poling, B.E., Prausnitz, J.M., and O'Connell, J.P., *The Properties of Gases and Liquids*, 5th ed., McGraw-Hill, New York, 2001, 25.
4. Nakanishi, K., Prediction of diffusion coefficient of non-electrolytes in dilute solution based on generalized Hammond-Stokes plot, *Ind. Eng. Chem. Fundam.*, 17, 253–256, 1978.
5. Matrozov, V. et al., Experimental determination of the molecular diffusion coefficient of ozone in water, *Zh. Prikl. Khim.*, 49, 1070–1073, 1976.
6. Siddiqi, M.A. and Lucas, K., Correlations for prediction of diffusion in liquids, *Can. J. Chem. Eng.*, 64, 839–843, 1986.
7. Díaz, M., Vega, A., and Coca, J., Correlation for the estimation of gas–liquid diffusivity, *Chem. Eng. Commun.*, 52, 271–278, 1987.
8. Utter, R.G., Burkholder, J., and Howard, C.J., Measurement of the mass accommodation coefficient of ozone on aqueous surfaces, *J. Phys. Chem.*, 96, 4973–4979, 1992.
9. Johnson, P.N. and Davis, R.A., Diffusivity of ozone in water, *J. Chem. Eng. Data*, 41, 1485–1487, 1996.
10. Masschelein, W.J., Fundamental properties of ozone in relation to water sanitation and environmental applications, in *Proc. International Specialized Symposium IOA*, 1–30, Toulouse, France, 2000.
11. Costa, E., *Ingeniería Química. 5. Transferencia de materia. 1ª parte*, Alhambra, Madrid, 1988, 101–124.

12. Foussard, J.N. and Debellafontaine, H., Thermodynamic basis for the solubility and diffusivity of ozone in water, in *Proc. International Specialized Symposium IOA*, 35–38, Toulouse, France, 2000.
13. Sechenov, M., Über die Konstitution der Salzlösungen auf Grund ihres Verhaltens zu Kohlensäure, *Z. Phys. Chem.*, 4, 117–125, 1889.
14. Danckwerts, P.V., *Gas Liquid Reactions*, McGraw-Hill, New York, 1970, 32.
15. Schumpe, A., The estimation of gas solubilities in salt solutions, *Chem. Eng. Sci.*, 48, 153–158, 1993.
16. Weisenberger, S. and Schumpe, A., Estimation of gas solubilities in salt solutions at temperatures from 273 to 363 K, *AIChE J.*, 42, 298–300, 1996.
17. Rischbieter, E., Stein, H., and Schumpe, A., Ozone solubilities in water and aqueous solutions, *J. Chem. Eng. Data*, 45, 338–340, 2000.
18. Andreozzi, R. et al., Ozone solubility in phosphate buffer aqueous solutions: effect of temperature, tert-butyl alcohol and pH, *Ind. Eng. Chem. Res.*, 35, 1467–1471, 1996.
19. *International Critical Tables*, 1st ed., Vol. III, McGraw-Hill, New York, 1928, 257.
20. Kawamura, F., Investigation of ozone I. The solubility of ozone in water and in dilute sulfuric acid, *J. Chem. Soc. Japan*, 53, 783–787, 1932.
21. Briner, E., and Perrotet, H., Determination of the solubilities of ozone in water and in aqueous solutions of sodium chloride. Calculation of the solubilities of the atmospheric ozone in waters, *Helv. Chim. Acta*, 22, 397–404, 1939.
22. Rawson, A.E., Ozonization. II. Solubility of ozone in water, *Wat. Wat. Eng.*, 57, 102–111, 1953.
23. Kilpatrick, M.L., Herrick, C.C., and Kilpatrick, M., The decomposition of ozone in aqueous solution, *J. Am. Chem. Soc.*, 78, 1784–1789, 1956.
24. Stumm, W., Ozone for disinfectant for water and sewage, *J. Boston Soc. Civil Eng.*, 45, 68–79, 1958.
25. Li, K.Y., Kinetic and Mass Transfer Studies of Ozone–Phenol Reactions in Liquid–Liquid and Gas–Liquid Systems, Ph.D. thesis, Mississippi State University, Starkville, MS, 1977.
26. Roth, J.A. and Sullivan, D.E., Solubility of ozone in water, *Ind. Eng. Chem. Fundam.*, 20, 137–140, 1981.
27. Caprio, V. et al., A new attempt for the evaluation of the absorption constant of ozone in water, *Chem. Eng. Sci.*, 37, 122–123, 1982.
28. Gurol, M.D. and Singer, P.C., Kinetics of ozone decomposition: a dynamic approach, *Env. Sci. Technol.*, 16, 377–383, 1982.
29. Kosak-Channing, L.F. and Helz, G.R., Solubility of ozone in aqueous solutions of 0–0.6 M ionic strength at 5–30°C, *Environ. Sci. Technol.*, 17, 145–149, 1983.
30. Ouederni, A., Mora, J.C., and Bes, R.S., Ozone absorption in water: mass transfer and solubility, *Ozone Sci. Eng.*, 9, 1–12, 1987.
31. Sotelo, J.L. et al., Henry's law constant for the ozone–water system, *Water Res.*, 23, 1239–1246, 1989.
32. Morris, J.C., The aqueous solubility of ozone in aqueous solution. A review, *Ozone News*, 16, 14–16, 1988.
33. Sullivan, D.E. and Roth, J.A., Kinetics of self-decomposition of ozone in aqueous solution, *AIChE Symp. Series*, 76, 142–149, 1980.
34. Beltrán, F.J., Theoretical aspects of the kinetics of competitive ozone reactions in water, *Ozone Sci. Eng.*, 17, 163–181, 1995.
35. Staehelin, S. and Hoigné, J., Decomposition of ozone in water: rate of initiation by hydroxide ions and hydrogen peroxide, *Environ. Sci. Technol.*, 16, 666–681, 1982.

36. Forni, L., Bahnemann, D., and Hart, J., Mechanism of the hydroxide ion initiated decomposition of ozone in aqueous solution, *J. Phys. Chem.*, 86, 255–259, 1982.
37. Beltrán, F.J., García-Araya, J.F., and Benito, A., Advanced oxidation of atrazine in water. I. Ozonation, *Water Res.*, 28, 2153–2164, 1994.
38. Beltrán, F.J. et al., Comparison of ozonation kinetic data from film and Danckwerts theories, *Ozone Sci. Eng.*, 20, 403–420, 1998.
39. Levenspiel, O., *Chemical Reaction Engineering*, 3rd ed., John Wiley & Sons, New York, 1999.
40. Froment, G.F. and Bischoff, K.B., *Chemical Reactor Analysis and Design*, John Wiley & Sons, New York, 1979.
41. Charpentier, J.C., Mass transfer rates in gas liquid absorbers and reactors, in *Advances in Chemical Engineering*, Vol. 11, 3–133, Academic Press, New York, 1981.
42. Beltrán, F.J., Gómez-Serrano, V., and Durán, A., Degradation kinetics of *p*-nitrophenol ozonation in water, *Wat. Res.*, 26, 9–17, 1992.
43. Ridgway, D., Sharma, R.N., and Hanley, T.R., Determination of mass transfer coefficients in agitated gas–liquid reactors by instantaneous reaction, *Chem. Eng. Sci.*, 44, 2935–2942, 1989.
44. Sotelo, J.L., Beltrán, F.J., and González, M., Kinetic regime changes in the ozonation of 1,3-cyclohexanedione in aqueous solutions, *Ozone Sci. Eng.*, 13, 397–419, 1991.
45. Beltrán, F.J. and González, M., Ozonation of aqueous solutions of resorcinol and phloroglucinol. 3. Instantaneous kinetic regime, *Ind. Eng. Chem. Res.*, 30, 2518–2522, 1991.
46. Augugliaro, V. and Rizzuti, L., The pH dependence of the ozone absorption kinetics in aqueous phenol solutions, *Chem. Eng. Sci.*, 33, 1441–1447, 1978.
47. Augugliaro, V. and Rizzuti, L., Temperature dependence of the ozone absorption kinetics in aqueous phenol solutions, *Chem. Eng. Commun.*, 2, 219–221, 1978.
48. Yocum, F.H., Ozone mass transfer in stirred vessel, *AIChE Symp. Series*, 76, 135–141, 1980.
49. Saunders, F.M., Gould, J.P., and Southerland, C.H., The effect of solute competition on ozonolysis of industrial dyes, *Water Res.*, 17, 1407–1419, 1983.
50. Gurol, M.D. and Nekouinaini, Kinetic behavior of ozone in aqueous solutions of substituted phenols, *Ind. Eng. Chem. Res.*, 23, 54–60, 1984.
51. Sotelo, J.L. et al., Degradación de ácido maleico en disolución acuosa mediante mezclas ozono-oxígeno, *An. Quim.*, 85, 286–291, 1989.
52. Sotelo, J.L. et al., Application of gas absorption theories to *o*-cresol ozonation in water, *Ozone Sci. Eng.*, 12, 341–353, 1990.
53. Sotelo, J.L., Beltrán, F.J., and González, M., Ozonation of aqueous solutions of resorcinol and pholoroglucinol. 1. Stoichiometric and absorption kinetic regime, *Ind. Eng. Chem. Res.*, 29, 2358–2367, 1990.
54. Andreozzi, R. et al., Analysis of complex reaction networks in gas–liquid systems: the ozonation of 2-hydroxypyridine in aqueous solutions, *Ind. Eng. Chem. Res.*, 30, 2098–2104, 1991.
55. Benítez, J., Beltrán-Heredia, J., and González, T., Absorption kinetics of ozone in aqueous solutions of malathion, *Ozone Sci. Eng.*, 13, 487–499, 1991.
56. Sotelo, J.L., Beltrán, F.J., and González, M., Ozonation of aqueous solutions of resorcinol and phloroglucinol. 2. Kinetic study, *Ind. Eng. Chem. Res.*, 31, 222–227, 1991.
57. Benítez, J., Beltrán-Heredia, J., and González, T., Ozonation of aqueous solutions of fenuron, *Ind. Eng. Chem. Res.*, 30, 2390–2395, 1991.

58. Benítez, J., Beltrán-Heredia, J., and González, T., Kinetics of the reaction between ozone and MCPA, *Water Res.*, 25, 1345–1349, 1991.
59. Beltrán, F.J., Encinar, J.M., and García-Araya, J.F., Absorption kinetics of ozone in aqueous *o*-cresol solutions, *Can. J. Chem. Eng.*, 70, 141–147, 1992.
60. Beltrán, F.J., Encinar, J.M., and García-Araya, J.F., Oxidation by ozone and chlorine dioxide of two distillery wastewater contaminants: gallic acid and epicatechin, *Water Res.*, 27, 1023–1032, 1993.
61. Beltrán, F.J. et al., Degradation of *o*-chlorophenol with ozone in water, *Trans. IChemE Process Saf. Env. Pollut.*, 71B, 57–65, 1993.
62. Beltrán, F.J. et al., Oxidation of mecoprop in water with ozone and ozone combined with hydrogen peroxide, *Ind. Eng. Chem. Res.*, 33, 125–136, 1994.
63. Beltrán, F.J. et al., Chemical degradation of aldicarb in water using ozone, *J. Chem. Technol. Biotechnol.*, 62, 272–278, 1995.
64. Beltrán, F.J. et al., Oxidation of polynuclear aromatic hydrocarbons in water. 1. Ozonation, *Ind. Eng. Chem. Res.*, 34, 1596–1606, 1995.
65. Beltrán, F.J. et al., Elimination pathways during water ozonation of volatile organochlorine compounds, *Toxic. Env. Chem*, 63, 107–118, 1997.
66. Rinker, E.B. et al., Kinetics of the aqueous-phase reaction between ozone and 2,4,6-trichlorophenol, *AIChE J.*, 45, 1802–1807, 1999.
67. Hu, J. et al., Evaluation of reactivity of pesticides in water using the energies of frontier molecular orbitals, *Water Res.*, 34, 2215–2222, 2000.
68. Beltrán, F.J. et al., Kinetics of competitive ozonation of some phenolic compounds present in wastewater from food processing industries, *Ozone Sci. Eng.*, 22, 167–183, 2000.
69. Beltrán, F.J. et al., Kinetics of simazine advanced oxidation in water, *J. Env. Sci. Health*, B35, 439–454, 2000.
70. Beltrán, F.J., García-Araya, J.F., and Álvarez, P.M., Sodium dodecylbenzenesulfonate removal from water and wastewater. 1. Kinetics of decomposition by ozonation, *Ind. Eng. Chem. Res.*, 39, 2214–2220, 2000.
71. Beltrán, F.J. et al., Determination of kinetic parameters of ozone during oxidations of alachlor in water, *Water Env. Res.*, 72, 689–697, 2000.
72. Calderara, V., Jekel, M., and Zaror, C., Kinetics of ozone reactions with 1-naphthalene, 1,5-naphthalene and 3-nitrobenzene sulphonic acids in aqueous solutions, *Water Sci. Technol.*, 44, 7–13, 2001.
73. Zaror, C. et al., Kinetics and toxicity of direct reaction between ozone and 1,2-dihydroxybenzene in dilute aqueous solution, *Water Sci. Technol.*, 43, 321–326, 2001.
74. Sánchez-Polo, M., Rivera-Utrilla, J., and Zaror, C.A., Advanced oxidation with ozone of 1,3,6-naphthalenetrisulfonic acid in aqueous solution, *J. Chem. Technol. Biotechnol.*, 77, 148–154, 2002.
75. Jhaveri, A.S. and Sharma, M.M., Absorption of oxygen in aqueous alkaline solutions of sodium dithionite, *Chem. Eng. Sci.*, 23, 1–8, 1968.
76. De Waal, K.J.A. and Okeson, J.C., The oxidation of aqueous sodium sulphite solutions, *Chem. Eng. Sci.*, 21, 559–572, 1966.
77. Sotelo, J.L., Beltrán, F.J., and González, M., The use of surface renewal and eddy diffusivity theories in the ozonation of a THM precursor: 1,3 cyclohexanedione, *Ozone Sci. Eng.*, 13, 421–436, 1991.
78. Beltrán, F.J. and González, M., Study of ozonation of organics in water using unsteady state and turbulent absorption theories, *Environ. Sci. Health Part A Environ. Sci. Eng.*, A27, 1433–1452, 1992.

6 Kinetics of the Ozonation of Wastewaters

Ozone is not confined solely to treatment of natural waters for drinking. Ozone has long been used in wastewater treatment. Although the general objective of ozonation in wastewater treatment is disinfection after the secondary biological treatment[1,2] ozone also plays a variety of other roles, mainly to improve the efficiency of other unit operations such as coagulation–flocculation–sedimentation[3,4] or carbon filtration;[5,6] to remove biologically refractory or toxic compounds to improve biological oxidation units;[7,8] or to reduce the amount of sludge generated in these latter systems.[9,10] Specific literature concerning the application of ozone in the treatment of wastewater (mainly industrial wastewater) dates back to the 1970s, when Rice and Browning[11] published a compendium of cases about ozone application. Thus, both inorganic- and organic-compound-related industries have used ozone for decontamination or disinfection purposes. Rice and Browning[11] divided these industries into 21 categories, as listed in Table 6.1. Wastewater produced in pesticide manufacturing and use, rinsing of wood chips contaminated with pentachlorophenol or other wastes containing 1,4-dioxane, marine aquaria, swine marine slurries from stored livestock wastes, leachates, etc. have all been treated with ozone.[12–17] Among the numerous industrial wastewater types mentioned, ozone is most commonly applied to wastewater containing phenols that are present in numerous industrial processes (coke plants, petroleum refinery, plastics, pulp and paper, textiles, soaps and detergents, food and beverage, etc.). Other wastewaters containing surfactant compounds and dyes have also been treated with ozone.[18,19] Table 6.2 presents a list of papers published since 1995 dealing with the use of ozone in wastewater treatment. Many compounds present in most of these wastewaters directly react with ozone in reactions with very high rate constants. Thus, one expects that ozone is an oxidant truly recommended to reduce or even eliminate the contamination of these wastewaters. However, a rather different result would be obtained if ozonation were the main operation used to decontaminate wastewater. The problem of ozone use in wastewater arises from two facts: the high concentration of fast ozone-reacting compounds (phenols, dyes, some surfactants of benzene sulphonate acid type, etc.) and the presence of other substances (e.g., salts, carbonates, etc.). On the one hand, because of the high concentration of fast ozone-reacting compounds, the ozonation rate is mass transfer controlled (see Table 6.3), and, on the other hand, the presence of ozone decomposition–inhibiting compounds or hydroxyl free radical scavengers halts the ozonation rate when ozone indirect reactions are the main means of pollutant

TABLE 6.1
Wastewater Types Treated with Ozone

Industry	Objectives or Compounds Present
Aquaculture	Shellfish depuration, marine water quality, disease prevention, toxicity
Electric power manufacturing	Biofouling control
Electroplating	Removal of cyanides and cyanates; complex metal cyanides with O_3/UV
Food and kindred products	Sterilization of water for bottle washing, COD reduction of brines, disinfection of processing water
Hospital wastewater	Shower, operating room, kitchen, chemical lab, x-ray lab; target COD for water reuse = 10 mgL^{-1}
Municipal wastewater, inorganics containing water	Removal of Fe and Mn, heavy metals such as Hg, Cr(III), ammonia (among others)
Chloro–alkali production	
Iron, steel, and coke plants	Cyanides, cyanates, phenols, sulfide (among others)
Leather tanneries	Removal of colorants, sulfide
Organic chemical manufacturing plants	Salicylic acid, caprolactam synthesis, alkylamines, organic dyes, chelating agents, etc.
Paints and varnishes	Phenols, methylene chloride
Petroleum refineries	Oils, hydrocarbons, nitroaromatics, phenols, ammonia, mercaptans, etc.
Pharmaceutical industries	Little information available
Photoprocessing	Surfactants, sulfate, phosphates, cyanates, heavy metals
Plastics and resins	Phenols, formaldehyde, synthetic polmers (unsaturated organics, alkylnaphthalene sulfonates), leathers (Zn, phenols, etc.), rubber (olefins, mercaptans, etc.)
Pulp and paper	Bleaching, odor control, mill wastewater treatment, spent sulfite liquor treatment
Soaps and detergents	Alkylbenzene sulfonate surfactants, reduce foaming
Textiles	Organic dyestuff, sizing agents, surfactants, organic and inorganic acids, azoic dyes, azobenzenes

Note: Phenols were also classified as an independent group due to their importance in many wastewater treatment industries, as shown above.

Source: From Rice, R.G. and Browning, M.E., *Ozone Treatment of Industrial Wastewater*, Noyes Data Corporation, Park Ridge, NJ, 1981.

removal (a situation that occurs when the concentration of fast ozone-reacting compounds has been reduced so that the kinetic regime of their ozonation reactions becomes slow). As a result, ozonation is not usually a cost-effective technology if used as the main treatment operation for wastewater because of the large amount of ozone needed. As a consequence, ozone is recommended in wastewater treatment as a complementary agent of other processes, mainly to increase biodegradability, reduce toxicity of recalcitrant compounds, etc.[67]

TABLE 6.2
Recent Literature Concerning Work on Wastewater Ozonation

Wastewater Type	Reacting Features	Reference No. (Year)
Municipal wastewater containing dye and surfactant compounds	Pilot plant system: 3 contact columns, pH 7–7.4, anionic surfactants and nonionic detergents: 3–4, COD: 150–200	19 (1995)
Pulp mill effluents	Batch and semicontinuous ozonation, filtrate bleaching process wastewater COD = 3060, BOD_5 = 540, pH 9.65; aerated stabilization basin wastewater: COD = 1440, BOD_5 = 25, pH 7.1	20 (1995)
Oil shale	Semibatch ozonation, pH 10, COD: 4000, BOD_5/COD: 0.23, phenols: 450, O_3/H_2O_2, and other AOP tested	21 (1995)
Mechanical and chemical pulp mill	Semibatch ozonation, pH 7 (adjusted); different effluents: COD: 1723–615, BOD: 281–708, toxicity reduction	8 (1996)
Printed circuit board rinse water	Batch ozonation, compounds: thiourea, Na gluconate, nitrilotriacetic acid; COD: 45, TOC: 13, O_3/H_2O_2 and catalytic ozonation	22 (1997)
Swine manure wastes	Semibatch ozonation, COD: 54200, BOD_5: 29800, pH 7; compounds present: volatile fatty acids, phenolics, indolics, ammonia, sulfide, phosphates	23 (1997)
Pharmaceutical effluents	AOP treatments: Fenton, O_3/UV, H_2O_2/UV, semibatch ozonation, COD: 670–2700, AOX: 3–5	24 (1997)
Boiling feed water in power plant	Real water treatment plant, pH 6.3–7, COD: 1–10, TOC: 1.5–4, Fe, Mn, chlorides, sulfates, nitrates, O_3, and O_3/H_2O_2	25 (1997)
Dental surgery	Disinfection	26 (1997)
Dyes	Pilot plant, COD: 1071, BOD: 348; ammonia: 21, different dyes: reactive, disperse, sulfur, acid, direct	27 (1998)
Landfill leachates	Semibatch ozonation, COD: 45000–700, pH 5.4–6	16 (1998)
Domestic plus industrial	Coagulation aid, COD: 480, TSS: 110, pH 8.4	28 (1998)
Sewage	Reduce sludge production, TOC: 200, MLSS in aeration tank: 1500–2500; intermittent and continuous ozonation	9 (1998)
Municipal	Pilot plant ozonation, UV, O_3, peracetic acid, disinfection, pH 6.7–8, DOC: 3–24	29 (1999)
Membrane textile effluent	Batch ozonation, COD: 595, BOD_5: 0, pH: 7.95, nonionic surfactants, aldehydes	30 (1999)
Synthetic dyehouse effluent	40-fold diluted dyebath: DOC: 25, pH: 10.94, different dyes and also urea, chloride, carbonate, O_3, O_3/H_2O_2, H_2O_2/UV, semibatch ozonation	13 (1999)
Sludge reduction	Sludge ozonation to solubilization	31 (1999)
Table olive	Semibatch ozonation, different AOP (O_3, O_3/H_2O_2, O_3/UV), COD: 19–25, BOD_5: 2–4.3, pH 13, biodegradability variation	32 (1999)
Textile	Semibatch ozonation, O_3, O_3/H_2O_2, COD: 320, BOD_5: 42–64, pH 8.2	33 (1999)
Domestic	Continuous pilot plant ozonation plus biological aerobic oxidation, COD: 294, BOD_5: 170, pH: 7.2–7.6	34, 35 (1999)

TABLE 6.2 (continued)
Recent Literature Concerning Works on Wastewater Ozonation

Wastewater Type	Reacting Features	Reference No. (Year)
Wine distillery plus domestic sewage	Semibatch ozonation plus biological aerobic oxidation, COD: 21700–300, BOD_5: 13440–187, pH 5.4–10, kinetic study, biodegradability	36, 37 (1999)
Domestic	Batch ozonation, COD: 380, BOD_5: 218, pH 7.6, improve sedimentation	38 (1999)
Domestic	Semibatch pilot plant, pH 7.6–8.6, DOC: 7–16, bromide: 3.48–10.1, total coliforms: 1380–4550, disinfection for reuse in agriculture	39 (2000)
Pharmaceutical	Semibatch ozonation, O_3, O_3/H_2O_2, different compounds: acetylsalicilic, clofibric acid, diclofenac, ibuprofen: 2 μgL^1, pH 7, 10°C	40 (2000)
Pulp mill	Batch and semibatch ozonation, pH 2–10, ethylenediaminetetracetic acid (EDTA): 10–1000	41 (2000)
Sludge from anaerobic degradation	Batch and continuous ozonation, reduce sludge by partial oxidation, TCOD: 7900, TOC: 2900, SS: 9000	42 (2000)
Dyeing and laundering	Dyeing: anionic detergent: 142, COD: 440, chlorides: 8000, pH 7.5, laundering: COD: 1650, anionic detergents: 110, nonionic detergents: 680, pH 10, different AOP treatments	43 (2000)
Agroindustrial-domestic	Continuous pilot ozonation plus aerobic biological oxidation, COD: 2250, BOD_5: 1344, pH 3–7	44 (2000)
Table olive plus domestic sewage	Semibatch ozonation plus aerobic biological oxidation, COD: 1110, BOD_5: 570, nitrites, ammonia n-phenolics, pH 11.1	45 (2000)
Olive oil and table olive plus domestic sewage	Semibatch pH sequential ozonation, olive oil ww: COD 1465, BOD_5: 1240, pH 5.8; table olive ww: COD: 1450, BOD_5: 910, pH 11.3, nitrites, ammonia, phenolics, ozone with pH cycles	46 (2000)
Petrochemical	Ozone plus biological activated carbon; wastewater: benzoic acid and aminobenzoic acid: 500, acrylonitrile: 100, pH 7	7 (2001)
Manufacturing dyes	Ozone as pretreatment of biological oxidation, semibatch ozonation, 3-methyl pyridine: 10^{-3}–10^{-4} M, pH 4–6, kinetic study	47 (2001)
Cherry stillage (2 times diluted)	Semibatch ozonation plus aerobic biological oxidation, COD: 145–180, BOD_5: 100–140, pH 3.8	48 (2001)
Wine distillery plus domestic sewage	Semibatch pH sequential ozonation, domestic ww: COD: 300, BOD_5: 160, pH 7.6; distillery ww: COD: 2500, BOD_5: 1340, pH 3.5, ozone with pH cycles	49 (2001)
Landfill leachate	Semibatch ozonation, pH 8.3, COD: 1400, BOD_5: 170, SS: 270	50 (2001)
Dye production	Semibatch ozonation, different dyes, COD: 18400–2420, pH 0.5–9.3 depending on dye type, O_3/H_2O_2 and Fenton	51 (2001)
Textile	Different AOP treatments, Improve biodegradability, COD: 2154, BOD_5: 1050, TOC: 932, antraquinone, anionic detergent, alkylnaphthalensulfonate, chlorides	52 (2001)

TABLE 6.2 (continued)
Recent Literature Concerning Works on Wastewater Ozonation

Wastewater Type	Reacting Features	Reference No. (Year)
Domestic plus dyestuff	Pilot plant ozonation, COD: 234–38, BOD_5: 5.6–27, SS: 69–12, pH 7.1–7.7, different dyestuff	53 (2001)
Mechanical pulp production	Semibatch ozonation, wet air oxidation, COD: 1600–16500, TOC: 6100–6700, pH 4.8–6.5, tannin+lignin acids, fatty acids, sterols, triglycerides among others	54 (2001)
Fruit cannery effluent	Semibatch ozonation, COD: 12000–45000, pH 9.8–13.5: O_3, O_3/H_2O_2, activated carbon	55 (2001)
Kraft pulp mill effluent	Pilot plant impinging jet bubble ozone column, COD: 750–681, BOD_5: 21.5–18.8, pH: 7.6, aromatic halogen and color-causing compounds	56 (2001)
Secondary and tertiary domestic effluents	Pilot plant ozonation, COD: 30–71, TOC: 0 < 10–26, pH: 7–7.5, different fecal microorganisms, disinfection for reuse	57 (2002)
Paper pulp effluents	Semibatch ozonation, 10 different AOP applied, COD: 1384, TOC: 441, pH 10, comparison and cost estimation	58 (2002)
Reactive dyebath effluent	Semibatch ozonation, comparison of AOPs (O_3, UV/H_2O_2, UV/TiO_2), 15-fold dilution, TOC: 46.8, AOX: 0.102, carbonates: 490.6, pH 10.9, different dyebath	18 (2002)
Textile effluent	Packed bed (Raschig ring) ozone continuous flow column, COD: 1512, BOD: 90.6, pH 10.9, reductions of COD, pH, phytotoxicity reduction	59 (2002)
Domestic sludge	Cylindrical bubble column, MLSS: 10100 with 73% VSS, ozone dosage: 0.01–2 g/gMLSS, significant mineralization at high ozone dosage and solubilization at low ozone dosage	60 (2002)
Fruit cannery (FC) and winery (W) effluents anaerobically treated	After anaerobic oxidation: FC: COD: 525–750, W: COD: 148–370, ozone and ozone/hydrogen peroxide treatment in continuous flow bubble column plus GAC adsorption in fixed bed column, COD and color reductions followed	61 (2002)
Industrial landfill leachates	Treatments: ozone, ozone/hydrogen peroxide, hydrogen peroxide; semibatch bubble column; biological oxidation postreatment; BOD/COD = 0.05, COD: 390–560; increases of biodegradability and up to 50% COD reduction	62 (2002)
Pharmaceutical effluent	Semibatch bubble column, values of biologically treated wastewater: COD: 8034, BOD: 3810, pH 8.7, significant UV absorbance reductions	63 (2002)
Log yard run-off	Pre- and post-ozonation of biological oxidation, magnetically semibatch tank reactor, BOD: N: P: 100: 5: 1, MLSS: 2500: ozone reduces COD (22%) and increases BOD (38%)	64 (2002)
Domestic effluent	Activated sludge ozonation, sludge periodically treated with ozone in a semibatch tank, 75% reduction with 0.05 $gO_3/gVSS$, biological reactor: residence time: 10 d, 2 gL^{-1} SS, slight increase in COD	65 (2002)

TABLE 6.2 (continued)
Recent Literature Concerning Works on Wastewater Ozonation

Wastewater Type	Reacting Features	Reference No. (Year)
Pharmaceutical effluent	Synthetic wastewater prepared from antibiotics, COD: 900, 1.5-L semibatch bubble column; effects of pH and addition of hydrogen peroxide, increases in BOD/COD	66 (2003)

Note: Units in mgL^{-1}

6.1 REACTIVITY OF OZONE IN WASTEWATER

In ozonation processes, the nature of compounds present in water will determine the degree of reactivity with ozone. Compounds with specific functional groups (aromatic rings, unsaturated hydrocarbons, etc.) are prone to ozone attack while other compounds (saturated hydrocarbons, alcohols, aldehydes, etc.) can be considered resistant to ozone attack. In these cases the second type of ozone reaction (indirect reactions) can play an important role, although this will also depend on the concentration of fast ozone-reacting compounds (kinetic regime) and hydroxyl radicals, the way they are generated, inhibiting substances, and pH of water. Based on these observations, when ozone is applied to a real wastewater there will likely be numerous series-parallel ozone reactions, depending on the wastewater complexity. If the presence of initiators, promoters, and inhibitors is of great importance in the treatment of natural water, the unknown nature and concentration of these compounds and others that directly react with ozone constitute the main problem in studying not only the kinetics of wastewater but also predicting ozonation efficiency. Knowledge of the composition of the wastewater results is fundamental to predicting ozone reactivity and potential application. In addition, pH and concentration of the compounds present in the wastewater are other key factors for future kinetic studies.

The chemical composition of the wastewater determines its potential reactivity with ozone. Table 6.3 gives values of the Hatta number for some direct ozone reactions with compounds that could be present in wastewater and the kinetic regime of these ozonation processes. The information given about the recommended ozone system should be used to improve as much as possible the pollutant removal rate. As can be deduced from Table 6.3, pH, concentration, and nature of pollutants are major factors affecting the choice of recommended action. Some of these compounds dissociate in water when pH is increased, enhancing the ozonation rate (see Chapter 2). In these cases, mass transfer limitation constitutes the major problem, and ozone feeding devices are key factors affecting the performance of the ozonation rate. Other compounds such as pesticides are usually present in low concentration (ppm or ppb level) due to solubility limitations. In these cases, chemical ozone reactions control

TABLE 6.3
Reactivity and Kinetic Regimes of Industrial Wastewater Ozonation Related to the Presence of Some Specific Contaminants

Wastewater Type	Specific Contaminant	Concentration, pH, and Rate Data	Hatta Number, Kinetic Regime, and Action to Take
Ash dump[21]	Phenolics	Hundreds of mgL^{-1}, pH = 12, $k = 1.8 \times 10^7$ [68]	$Ha > 10$, Instantaneous, DW, AOP NR
Swine manure wastes[23]	Odor compounds: p-cresol, sulfides	Few to tens mgL^{-1}, pH 7, $k = 7.5 \times 10^5$ (of O$_3$–o–cresol reaction)[69]	$Ha < 10$, Fast to moderate regime, DW, AOP NR
		Tens of mgL^{-1}, pH 7, $k = 3 \times 10^9$ [70]	$Ha > 10$, Fast to instantaneous regime, DW, AOP NR
Pharmaceutical[24]	AOXs: chlorophenol, heptachlor	Few mgL^{-1}, pH 7, $k = 10^8$ [68]	$3 < Ha < 10$, Fast pseudo first-order regime, DW, AOP NR
		Hundreds µgL^{-1}, pH 7, $k = 90$ [71]	$Ha < 0.1$, Slow regime, IW, AOP R
Pulp mill[41]	EDTA	Hundreds mgL^{-1}, pH 8, $k = 20000$ (O$_3$-dimethylamine reaction)[68]	$Ha < 0.5$, Moderate regime, Mainly IW, AOP R
Textile[18,43,52]	Azoic dyes	Few to tens mgL^{-1}, pH 10, $k = 10^8$ [72]	$3 < Ha < 10$, Fast regime, DW, AOP NR
Table olive[45]	Phenolics	Hundreds to thousands of mgL^{-1}, pH 12.9, $k = 1.8 \times 10^7$ (O$_3$-phenol reaction)[68]	$3 < Ha < 20$, Likely fast regime, DW, AOP NR
Olive oil[46]	Phenolics	Thousands of mgL^{-1}, pH 4.9, $k = 5 \times 10^4$ (O$_3$-phenol reaction)[68]	$1 < Ha < 5$, Moderate to fast regime, DW, AOP NR
Petrochemical[7]	Benzoic acid	Hundreds of mgL^{-1}, pH 7, $k < 0.15$ (p-chlorobenzoic-O$_3$ reaction)[68]	$Ha < 0.01$, Very slow regime, IW, AOP R
Herbicide manufacturing	Atrazine and others	Tens to thousands µgL^{-1}, pH 7, $k < 10$ [73]	$Ha < 0.01$, Very slow regime, IW, AOP R
Electroplating, photoprocessing[11]	Cyanides	Tens of mgL^{-1}, pH 10, $k = 10^5$ [70]	$Ha < 3$, Moderate regime, DW, AOP only for complex cyanides
Petrochemical	PAHs; phenanthrene	Tens to thousands µgL^{-1}, pH 7, $k = 3000$ [74]	$Ha < 0.01$, Slow regime, IW, AOP R
Municipal	Ammonia	Tens to hundreds mgL^{-1}, pH 7, $k < 1$ [70]	$Ha < 0.001$, Very slow regime, IW, AOP R
	Detergents: NaDBS	Few mgL^{-1}, pH = 7, $k < 5$ [75]	$Ha < 0.001$, Very slow regime, IW, AOP R
Explosives	Nitrotoluenes	Few mgL^{-1}, pH 7, $k < 10$ [76]	$Ha < 0.01$, Slow regime, IW, AOP R

TABLE 6.3 (continued)
Reactivity and Kinetic Regimes of Industrial Wastewater Ozonation Related to the Presence of Some Specific Contaminants

Wastewater Type	Specific Contaminant	Concentration, pH, and Rate Data	Hatta Number, Kinetic Regime, and Action to Take
Gasoline tank leaking Petroleum industry	BTEX: Benzene, toluene, ethylbenzene, xylene	Few µgL^{-1}, pH 7, $k < 100$ [77]	$Ha < 0.001$, Very slow regime, IW, AOP R
Chemical processes	1,4-dioxane	Hundreds µgL^{-1}, pH 7, $k = 0.32$ [77]	$Ha < 0.001$, Very slow regime, IW, AOP R
Chemical industries Groundwater	Low molecular weight organohalogens: TCE, PCE, DCE	Few to hundreds of µgL^{-1}, pH 7, $k < 100$ [77]	$Ha < 0.001$, Very slow regime, IW, AOP R

Note: Units of k in $M^{-1}s^{-1}$, Ha = Hatta number ($k_L = 5 \times 10^{-4}$ms^{-1} and $D_{O3} = 10^{-9}$ m^{2}s^{-1} to determine Ha). DW = process through direct ozone reaction; IW = process through indirect ozone reaction of ozone; AOP NR = advanced oxidation process not recommended (see Chapter 7); AOP R = advanced oxidation process recommended (see Chapter 7).

---

the process rate, and advanced oxidation processes are recommended (e.g., O_3/H_2O_2). As will be shown in Chapter 7, when ozone reactions develop in the slow kinetic regime (chemical control) the indirect ozone reactions usually predominate. However, the presence of hydroxyl radical scavengers needs to be considered as a limitation. Also, the case of volatile compounds (benzene, toluene, trichloroethylene, etc.) is particularly important since volatility could constitute an important means of pollutant removal. For example, in some work[78] volatility constituted the main means of trichloroethane removal in an ozonation process. In these cases care should also be taken regarding the possible waste of ozone.

Although ozone reactivity with single compounds present in wastewater (Table 6.3) can be predicted, classification of all wastewater in terms of its reactivity with ozone is a rather difficult, if not unrealistic, task. However, as a general rule, high concentration of pollutants would suggest high reactivity with ozone (which is an indication of fast kinetic regime and ozone direct reactions) and low concentration usually means low ozone reactivity and, hence, a factor that favors the development of ozone indirect reactions.

6.2 CRITICAL CONCENTRATION OF WASTEWATER

Because of the changing nature of compounds present in wastewater undergoing ozonation (e.g., phenols becomes unsaturated carboxylic acids and then aldehydes, saturated carboxylic acids, ketones, etc.), the reactivity in terms of kinetic regime of ozonation usually changes from fast to slow. Knowledge of the critical concentration

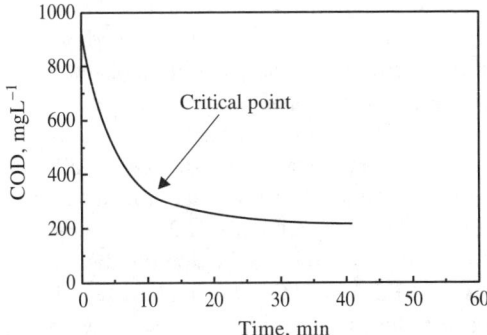

FIGURE 6.1 Typical profiles of COD with time in ozonation experiments of industrial wastewaters showing the critical concentration point (values of COD and time in x and y axes show arbitrary values).

value of any wastewater to change from one degree of ozone reactivity to the other depends on the nature of the wastewater and can be determined from laboratory experimental results. When ozone is applied to some wastewater in a semibatch well-agitated tank, the pollution concentration (measured as chemical oxygen demand, COD) vs. time data usually takes the trend plotted in Figure 6.1. In most cases, two reaction periods will be noted: the first initial period of high ozonation rate where the pollution concentration rapidly falls, and a second period where the ozonation rate continuously decreases with time until the ozonation rate stops and the pollution concentration reaches a plateau value. The critical pollution concentration is that corresponding to the time when both periods coincide (about 10 min in Figure 6.1). In most cases, the pollution of wastewater during the first period is removed by direct ozone reactions that usually develop in the fast kinetic regimes of ozonation. In these cases, the absence of dissolved ozone is a clear indication that a fast or instantaneous kinetic regime of ozonation develops (see Chapter 4). In the second period, ozone likely decomposes into hydroxyl radicals and pollution is mainly removed by indirect ozone reactions. In this second period, ozonation reactions develop in the slow kinetic regime and removal of COD is carried out at a lower rate because carbonate/bicarbonate ions have been formed as a result of partial mineralization during the initial fast reaction period. It should be mentioned, however, that in some cases only one reaction period seems to develop, depending on the nature of wastewater, as will be shown in Section 6.4. In any case, and as a general rule, it can be said that high polluted wastewater ozonation is accomplished through fast kinetic regime ozone direct reactions, while low polluted wastewater ozonation develops through slow kinetic regime ozone indirect reactions.

6.3 CHARACTERIZATION OF WASTEWATER

The nature of the reactions that ozone undergoes in wastewater can be established by characterization of wastewater. As shown above, ozone reactivity depends on the concentration (and also the nature) of pollutants present in wastewater. However, in real wastewater the actual pollution concentration is unknown and surrogate parameters (chemical oxygen demand, COD, total organic carbon, TOC, etc.) are used to express the pollution concentration. The magnitude of these parameters, especially COD, gives an estimate of the potential ozone reactivity.

In addition to COD and TOC (this latter more commonly used in natural water), other parameters are employed to measure the degree of pollution. These parameters include biological oxygen demand (BOD) and the measurement of wastewater absorptivity in the UV-C region, specifically at 254 nm wavelength (A_{254}). Another parameter that can be used is the mean oxidation number of carbon (MOC), which combines the values of COD and TOC to yield more reliable data on pollution concentration (especially during oxidation processes), avoiding the difficulties that some compounds resistant to COD determination present. Methods to measure any of these parameters can be followed elsewhere with the aid of detailed protocols issued by APHA, DIN, etc.[79,80] Here we provide a brief explanation of the importance of these parameters and their applications in water and/or wastewater.

6.3.1 THE CHEMICAL OXYGEN DEMAND

COD is undoubtedly the most general parameter for following the pollution concentration of water in a given physical, chemical, or even biological process treatment. COD also gives a quantitative measurement of the depth of any chemical or biological oxidation step in the treatment of wastewater. This parameter, therefore, has been continuously applied in kinetic studies of water and wastewater treatment (such as ozonation) because, in contrast to other parameters such as TOC (see later), COD supplies information on the magnitude of oxidation steps. COD represents the amount of oxygen needed for complete mineralization of the matter present in water through chemical oxidation. It is also used as a general parameter to express changes in pollution concentration in physico-chemical processes such as flocculation–coagulation–sedimentation, filtration, etc. Thus, pollution concentration is measured in terms of mg of oxygen units per liter of water.

The proportionality between pollution concentration and COD is obtained once the theoretical oxygen demand, ThOD, is accounted for. Thus, this latter parameter represents the amount of oxygen needed to remove 1 mg of pollution. Pollution concentration in mg/L is thus simply the ratio between COD and ThOD:

$$C(mg/L) = \frac{COD(mgO_2/L)}{ThOD(mgO_2/mg)} \quad (6.1)$$

COD, however, has some limitations derived from the presence in water or wastewater of compounds totally or partially resistant to chemical oxidation with dichromate, the chemical oxidant generally used in the analytical method, or volatile compounds that, during COD analysis, stay in the gas phase (COD analysis implies reflux methods). Examples of these compounds are cyclohexane, tetrachloroethylene, pyridine, potassium cyanide, nitrate, etc.[81] Another problem stems from the opposite situation: the presence of compounds that consume dichromate but should not be considered as a fraction of the pollution concentration. These compounds include hydrogen peroxide and/or chloride ions. Hydrogen peroxide is generated in water when ozonation is applied or may be added to the water when the combination of ozone and hydrogen peroxide is used. Chloride ions are common in wastewater. These problems can usually be overcome by using complementary agents, such as mercuric salts, which are added before the COD analysis to precipitate chlorides.[81] The problem of hydrogen peroxide can be solved by first determining the amount of COD due to different concentrations of hydrogen peroxide. This COD must be subtracted from the COD of the wastewater sample.[82]

6.3.2 THE BIOLOGICAL OXYGEN DEMAND

Like COD, BOD measures the pollution in a given wastewater, but it gives the amount of biodegradable matter. It gives the amount of oxygen needed for microorganisms that may be added to the water sample to biodegrade the matter in water. It is, then, a parameter applied mainly to biological systems but it is also measured after other water treatment units (e.g., ozonation) that are used before the biological or secondary treatment. The measure of BOD is generally made after 5 days of digestion (see Reference 79 for an example of the detailed analytical method). Shorter times do not warrant 100% biodegradation and longer times could involve the development of other phenomena that also consume oxygen, such as nitrification. In any case, however, BOD is not an absolute measurement of the biodegradability of water because it depends on the capacity of microorganisms added or already present in the water sample to aerobically digest the matter. In this respect, it is noted that there can be two measurements of BOD: the total biological chemical demand where the presence of particulate matter is accounted for and the BOD that refers to the dissolved matter. The particulate matter refers to the matter retained in 0.45-µm-pore diameter filters.

As far as biodegradability is concerned, the ratio BOD/COD is a more convenient parameter because it takes into account the total amount of pollution the water contains measured as COD. Numerous works express the biodegradability of a water sample in terms of the combined use of BOD and COD,[32,35,42] particularly to indicate changes in biodegradability due to the application of a given treatment (see Section 6.5).

6.3.3 TOTAL ORGANIC CARBON

This is another frequently used general parameter that represents the total amount of organically bound carbon present in dissolved and particulate matter in water. The analytical method involves the transformation (through UV radiation, chemical oxidation, or combined methods) of organic carbon in carbon dioxide, which is measured directly by a nondispersive infrared analyzer.

In many cases the particulate matter (retained in a 0.45-µm-pore diameter filter) is removed and the measurement corresponds to the dissolved organic carbon, DOC. This is the usual TOC value in laboratory-prepared water, where dissolved model compounds are the only species present in the aqueous sample. The particulate or suspended organic carbon is called SOC. In addition to TOC and DOC, another measurement corresponds to the inorganic carbon, IC, which results from carbonate and bicarbonate ions and dissolved carbon dioxide. If the sample contains volatile organic substances, their corresponding carbon measurement represents the purgeable organic carbon, POC, which is also a fraction of TOC. In summary, carbon content of water involves the following parameters: TOC, DOC, IC, SOC, POC, and NPOC (nonpurgeable organic carbon). Detailed protocols to measure the different forms of carbon can be found elsewhere.[79]

Although TOC or DOC yields the quantity of organic matter transformed in CO_2, it is not a recommended parameter to follow any oxidation kinetics, such as ozonation kinetics, because it does not give a quantitative value for the oxidation evolution. This is very often observed when studying ozonation processes. In ozonation, TOC barely decreases with time in many cases, but COD usually does. For example, COD is able to measure the change that occurs when phenol is oxidized to maleic acid and other compounds (COD measurements before and after oxidation give the oxygen needed for this change), but the corresponding TOC values likely remain the same. Then, according to TOC measurements no significant changes would occur, but the actual situation is that phenol has really become maleic acid and other compounds. TOC gives a measure of the mineralization achieved in the ozonation process.

6.3.4 ABSORPTIVITY AT 254 NM (A254)

This parameter represents a partial measurement of the pollution concentration of the water/wastewater. It specifically gives a measure of the amount of aromatic and unsaturated compounds in water. This parameter is often used in natural water to measure the concentration of compounds that are assumed precursors of trihalomethanes and other organochlorine compounds (i.e., chloroacetic acids, among others) when water is chlorinated.[83] These precursors, usually called humic substances, are formed by macromolecules containing aromatic structures that absorb 254 nm UV radiation. The A254 parameter is also useful for wastewater containing phenol compounds.[34,45]

6.3.5 MEAN OXIDATION NUMBER OF CARBON

This parameter also allows the depth of oxidation to be followed by measuring the oxidation state of carbon atoms in any molecule considered. First proposed by Stumm and Morgan[84] and later modified by Mantzavinos et al.,[85] who called it the mean oxidation *state* of carbon, it was renamed by Vogel et al.[81] as the mean oxidation *number* of carbon (MOC). MOC is based on the change of oxidation number of carbon atoms in a molecule when subjected to oxidation. For a given organic molecule, MOC is defined as follows:[81]

$$\mathrm{MOC} = \frac{\sum_{i=1}^{n} OC_i}{n} \tag{6.2}$$

where OC_i is the oxidation number of the ith carbon atom and n is the number of carbon atoms in the molecule. In a solution containing j different molecules, the mean oxidation number is:[81]

$$\text{MOC}_m = \frac{\sum_{j=1}^{m} C_j \text{MOC}_j n_j}{\sum_{j=1}^{m} C_j n_j} \tag{6.3}$$

where the subindex j represents any molecule present in solution and C_j, MOC_j, and n_j are their corresponding concentration, mean oxidation number, and number of carbon atoms, respectively. It is evident that both in drinking water and, especially, wastewater, the concentration of many compounds present is unknown so MOC_m is a rather impractical parameter. Thus, it is defined as the mean oxidation number of carbon of the water content, MOC_w:[81]

$$\text{MOC}_w = 4 - \frac{4M_C}{M_{O_2}} \frac{\text{COD}_{org}}{\text{TOC}} = 4 - 1.5 \frac{\text{COD}_{org}}{\text{TOC}} \tag{6.4}$$

where M_C and M_{O_2} are the atomic mass of carbon and molecular mass of oxygen, respectively, and COD_{org} is the chemical oxygen demand of organic compounds. As can be deduced from Equation (6.4), MOC_w also presents some drawbacks resulting from the presence of inorganic substances that can be oxidized (i.e., nitrites to nitrates) or from the presence of N, S, P heteroatoms bonded to carbon atoms in the organic compound molecules. Thus, Equation (6.4) is deduced by considering that carbon atoms are exclusively bonded to H and O atoms because it is assumed that only carbon atoms are oxidized. However, the presence of N, S, or P atoms bonded to carbons could also consume oxidants as is the case for the oxidation of nitrobenzene, where the nitrogen atom goes from the nitro group to the nitrate ion group, that is, the oxidation number varies from +3 to +5. Chloro substituting groups present this same problem: the chlorine atom also consumes oxygen, so using Equation (6.4) with the experimentally measured COD to determine the MOC_w value will yield values lower than the true one. Applications of MOC_w in water and wastewater treatment processes are described in detailed elsewhere.[81]

6.4 IMPORTANCE OF pH IN WASTEWATER OZONATION

In ozonation systems pH usually exerts a positive effect on the COD removal rate. This effect is due to two factors: the presence of dissociating compounds that react fast with ozone (phenol or aromatic amine compounds) and, in the absence of high concentrations of these compounds, the increase of ozone decomposition to generate hydroxyl radicals. From these observations we can deduce that any increase of COD removal during wastewater ozonation at increasing pH can be due to both the direct or indirect reactions of ozone. This contradictory behavior, however, can easily be explained as follows. Direct reactions can be responsible for the increase of COD removal because of the increase in the rate constant of the reactions between ozone and dissociating compounds in wastewater. To give an example, the case of a wastewater containing phenol can be considered. At pH 4 the rate constant is about 10000 $M^{-1}s^{-1}$ but at pH 9 the rate constant increases up to approximately 10^9 $M^{-1}s^{-1}$ [68] (see also Chapter 3). In the absence of ozone fast-reacting compounds, the increase in pH gives rise to the appearance of hydroxyl radicals because ozone preferentially decomposes in wastewater (there are no compounds to react directly with ozone). In these cases the use of ozone combined oxidations (AOPs) can be recommended. Chapter 7 discusses conditions for establishing the relative importance of direct and indirect reactions. In any case, another general rule of ozonation, relative to the pH value, is that at pH lower than 12 (see Section 7.1) ozone will be consumed only through direct reactions in very concentrated wastewater when ozone fast-reacting compounds are present in high concentration.

In some cases, however, the pH effect is not as evident as might be expected. The existence of both reaction periods, as indicated in Section 6.2, is also somewhat misleading. For example, let us take the case of the ozonation of a domestic wastewater. This wastewater, as many others, usually contains important amounts of carbonates that inhibit the indirect ozone reactions. In Figure 6.2 the evolution of COD with time during the ozonation of such a type of wastewater is shown at different pH values (wastewater was buffered). By looking at the experimental results, two observations can be made: first, the position of the critical point that represents the initial fast COD decrease with time (first fast-reacting period) compared to the slower second one is not as evident as might be expected in light of the above observations. Second, pH has no influence. Generally, the increase in pH leads to an increase in the ozonation rate, and hence to an increase in COD removal rate as explained above. It is evident that at pH 4 ozone direct reactions are the only means of COD removal, so that, in accordance with the results from Figure 6.2, the absence of pH effect could mean the absence of ozone indirect reactions, a conclusion that does not support the two-period proposition. However, if wastewater is decarbonated before

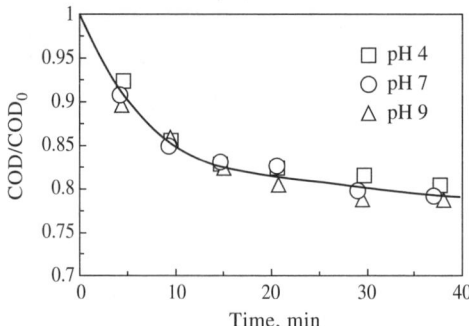

FIGURE 6.2 Effect of pH on the COD variation with time during the ozonation of a domestic buffered wastewater; $COD_0 = 275$ mgL^{-1}.

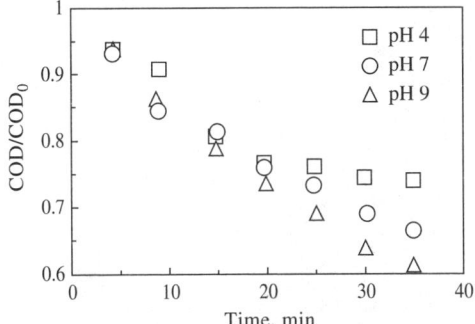

FIGURE 6.3 Effect of pH on the COD variation with time during the ozonation of a domestic decarbonated buffered wastewater; $COD_0 = 275$ mgL^{-1}.

ozonation, and experiments similar to those in Figure 6.2 are carried out, the results are very different, as shown in Figure 6.3. In these cases, when carbonates are not initially present in wastewater, it is observed that pH does have an effect on COD removal rate. At pH 4 both reaction periods are clearly distinguished, the critical COD value being reached at about 15 min. Another observation is that, regardless of pH, values of COD conversion are higher than those observed from Figure 6.2, and the two reaction periods continue to be difficult to distinguish at pH 7 and 9. The explanation for these observations is likely due to the types of ozone reaction. Thus, at pH 4 after the initial reaction period, no fast ozone direct-reacting compounds remain in water, and indirect reactions begin to have an effect. The effect of these reactions, however, is not very important because at pH 4, ozone barely decomposes in water and the concentration of hydroxyl radicals is so low that the COD removal rate approaches zero (the plateau value). At higher pH values the initial starting period should be very short (which is the reason that both periods are not clearly distinguished) and indirect reactions are the main means of ozonation (especially at pH 9). The higher COD removal rate confirms the development of indirect reactions because carbonates are not present in high concentration to inhibit the ozone decomposition in free radicals.

The problem that results from the accumulation of resistant compounds (saturated carboxylic acids, aldehydes, etc.) in the media during ozonation and the subsequent decrease in pH can be partially solved with the help of pH sequential ozonation processes. These processes are carried out at alternating time periods of acid and basic pHs. In this manner, the process efficiency is increased because it benefits from both types of ozone reactions. Thus, depending on the initial pH of wastewater, the ozonation process can start at acid or basic pH to favor direct or indirect reactions. Figures 6.4 and 6.5 show two examples of pH sequential ozonation applied to wastewater from distillery and table olive factories, respectively. Both wastewaters were first diluted with domestic wastewater to establish COD values usually appropriate for secondary treatment in municipal wastewater plants.[46,49] Let us comment first on the pH sequential ozonation of distillery wastewater in Figure 6.4. The pH of this wastewater is about 4, so it is recommended to start ozonation at acid pH.

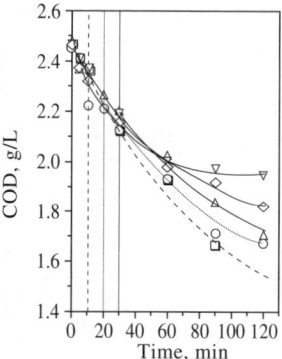

FIGURE 6.4 Single and sequential ozonation of wine distillery processing — synthetic urban wastewater. Evolution of remaining COD concentration with time. Conditions: T = 293 K, gas flow rate = 30 Lh^{-1}, C_{O_3g}(fed) = 20 mgL^{-1}. For acid cycle, pH = 4, for alkaline cycle, pH = 10. Duration of acidic–alkaline cycles, min: ∇ = 120–0, $\Diamond$ = 0–120; $\triangle$ = 10–110; $\circ$ = 20–100; $\square$ = 30–90. (From Beltrán, F.J., García-Araya, J.F., and Álvarez, P., pH Sequential ozonation of domestic and wine distillery wastewater, *Water Res.*, 35, 929–936, 2001. Copyright 2001, Elsevier Press. With permission.)

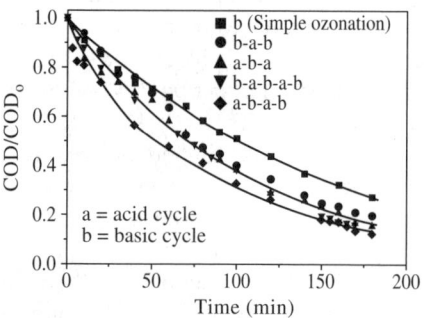

FIGURE 6.5 Single and sequential ozonation of table olive processing — synthetic urban wastewater. Evolution of normalized remaining COD concentration with time. Conditions: T = 293 K, gas flow rate = 20 Lh^{-1}, $k_L a$ = 0.02 s^{-1}, C_{O_3g}(fed) = 45 mgL^{-1} (a = acid cycle, b = alkaline cycle). For acid cycle, pH = 4, for alkaline cycle, pH = 10. (From Rivas, F.J. et al., Two step wastewater treatment: sequential ozonation-aerobic biodegradation, *Ozone Sci. Eng.*, 22, 617–636, 2000. Copyright 2000, International Ozone Association. With permission.)

In Figure 6.4 the evolution of COD with time corresponding to different pH sequential ozonation and conventional ozonation is shown. As can be seen, conventional ozonation at the pH of wastewater leads to a poor degradation rate. When wastewater pH is increased to carry out ozonation at basic pH, the efficiency of ozonation significantly increases and COD reduces from 2.5 to 1.8 gL^{-1}. At acid pH the two reaction periods are clearly seen. COD removal rate is improved when pH sequential ozonation is applied. Thus, as seen in Figure 6.4, two ozonation periods of acid pH (30 min) and basic pH (90 min) lead to the best results as far as COD removal is concerned. In this experiment, COD decreased from 2.5 to about 1.5 g/L. In

Figure 6.5, similar results can be observed for wastewater from a table olive production factory, although the removal efficiency is not as important as in the previous case. In this case, table olive wastewater has a basic pH of about 10 at the start of the process. Again, when ozonation periods of basic and acidic pH are applied, the COD removal rate increases. The objective of pH sequential ozonation is to take advantage of the two kinds of ozone action that are triggered at the right moment. Thus, when the pH is acidic, most of the fast ozone-reacting compounds are removed through direct reactions and resistant compounds are simultaneously generated. To prevent stopping the ozonation process, pH is increased (by adding NaOH), and indirect reactions are favored. The result is an increase in the ozonation rate and COD removal. However, during this period mineralization takes place and carbonate is formed, thus reducing the ozonation rate because of the inhibiting character of these reactions in trapping hydroxyl radicals. When carbonates accumulate in wastewater, pH again changes and becomes acidic, and a new ozonation period starts. In this new period, the objective is the removal of carbonates as carbon dioxide, which is stripped from wastewater. Once this occurs, pH can again be increased to start another period that favors indirect reactions. The number of periods and duration are design aspects that depend on the nature of the wastewater. The optimum combination for the acidic and basic periods is achieved from laboratory experiments. This pH sequential ozonation could be a recommended option in some cases but, as in any process technology, its application will depend on cost.

6.5 CHEMICAL BIOLOGICAL PROCESSES

Numerous works on the biological treatment of wastewater deal with the combined operation of chemical and biological oxidations.[86] In these works the beneficial effects of chemical oxidation as a pretreatment or post-treatment step in biological oxidation have been confirmed. Ozone plays a major role because of the different mechanisms of reaction associated with its use. Thus, in many wastewaters, the application of ozone at appropriate levels usually improves the biodegradability of the wastewater and, in some cases, the rate of sedimentation of the activated sludge and its production.[87] However, ozonation alone is not a recommended technology for the treatment of wastewater. Due to the high levels of organic matter, in many cases, high consumption of ozone is observed with small percentage reductions of COD, although this always depends on the nature of the wastewater treated as stated above. Therefore, before studying the kinetics of the wastewater, preliminary ozonation experiments should be conducted to establish the reactivity of ozone and the beneficial effects that an ozonation stage could add to the whole treatment. Typical experiments include the use of ozone alone or combined with other oxidants such as hydrogen peroxide, UV radiation followed by biological treatments, and measurements of COD, TOC, BOD, etc. The results are usually compared to those obtained in the absence of ozone. For example, Figure 6.6 shows the changes observed in the COD of a domestic wastewater with time in the process of biological oxidation with activated sludge, both previously treated and untreated with ozone.[88] As can be observed, if the wastewater is preozonated, the biological oxidation step allows a COD reduction of about 83% at 35°C while the individual processes lead

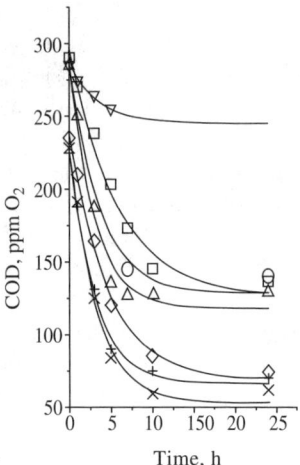

FIGURE 6.6 Variation of COD with time during activated sludge biological oxidation of municipal wastewater with and without preozonation. Ozonation conditions: pH 7.5, 20°C, ozone dose: 100 mgL^{-1}, COD$_0$ = 280–300 mgL^{-1}. Biological oxidation conditions: pH 7.2–7.7, VSS$_0$ = 1100–1200 mgL^{-1}, DO = 3–4 mgL^{-1}, T = 20°C: no ozone: □ = 5, ○ = 20, △ = 35, ▽ = 60. With preozonation: ◊ = 5, + = 20, × = 35. (From Beltrán, F.J., García-Araya, J.F., and Álvarez, P., Impact of chemical oxidation on biological treatment of a primary municipal wastewater. 2. Effects of ozonation on the kinetics of biological oxidation, *Ozone Sci. Eng.*, 19, 513–526, 1997. Copyright 1997, International Ozone Association. With permission.)

to COD reductions of 20% (only ozonation), not shown, and 55% (only biological oxidation). The beneficial effect of preozonation is clear, but ozone alone is not a recommended option.

6.5.1 Biodegradability

Another important advantage of the ozone application is the improvement of wastewater biodegradability. The biological oxygen demand, BOD, is the parameter that measures the biodegradability of a wastewater but the literature also reports the ratio BOD/COD as a more realistic parameter because it also considers the magnitude of pollution (that is, the magnitude of COD). BOD is usually determined after 5 days but in ozonated samples more time is allowed to facilitate the acclimation of microorganisms of the BOD test to the ozonated wastewater. Since after 10 days consumption of oxygen is also due to nitrification processes, BOD at 10 days is a recommended value to calculate the BOD/COD ratio. For example, Figure 6.7 shows the effect of ozone dose on the BOD/COD ratio for an ozonated distillery wastewater.[36] It is observed that biodegradability measured as BOD/COD ratio is deeply affected by the ozone dose applied in the preozonation stage. The improved biodegradability is associated with the partial oxidation of organic matter to give low molecular weight oxygenated compounds rather than complete oxidation to carbon dioxide. In Figure 6.7 the existence of an optimum ozone dose is also observed.

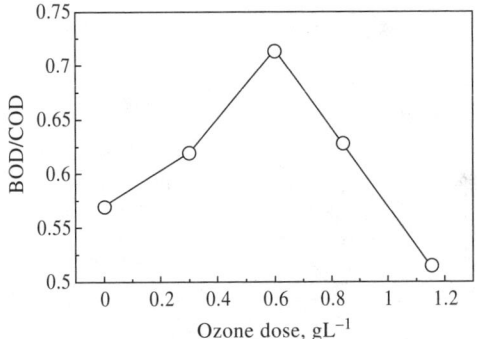

FIGURE 6.7 Influence of ozone dose on the biodegradability of a distillery wastewater induced by ozonation. Conditions: 20°C, pH = 5.4, 30 Lh^{-1} gas flow rate, domestic sewage to vinasse volume ratio = 10. (From Beltrán, F.J., García-Araya, J.F., and Álvarez, P., Wine distillery wastewater degradation. 1. Oxidative treatment using ozone and its effect on the wastewater biodegradability, *J. Agric. Food Chem.*, 47, 3911–3918, 1999. Copyright 1999, American Chemical Society. With permission.)

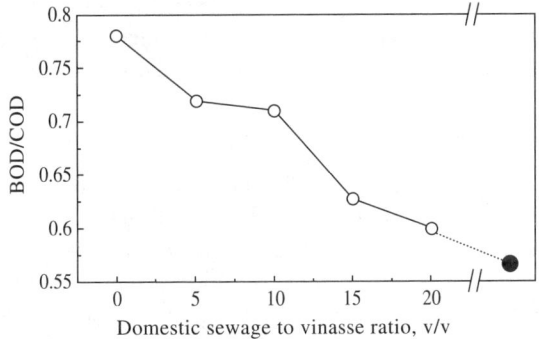

FIGURE 6.8 Influence of domestic sewage to vinasse volume ratio on the biodegradability induced by ozonation. Conditions: 20°C, pH = 5.4, 30 Lh^{-1} gas flow rate, C_{O3gi} = 20 mgL^{-1}, domestic sewage to vinasse volume ratio = 0–20. Black circle corresponds to domestic sewage without vinasses. (From Beltrán, F.J., García-Araya, J.F., and Álvarez, P., Wine distillery wastewater degradation. 1. Oxidative treatment using ozone and its effect on the wastewater biodegradability, *J. Agric. Food Chem.*, 47, 3911–3918, 1999. Copyright 1999, American Chemical Society. With permission.)

Lower values of the ozone dose are not sufficient to accomplish conversions of all refractory organics and yield other compounds more amenable for microorganisms. Conversely, higher ozone doses likely lead to removal of biodegradable compounds formed during the chemical oxidation process. It is also evident that these effects are highly dependent on the nature of the wastewater. For example, Figure 6.8 shows results on the BOD/COD ratio obtained in the ozone–biological oxidation process of a domestic plus distillery wastewater.[36] It can be seen that the percentage composition of the wastewater (domestic to distillery contribution ratio) affects the

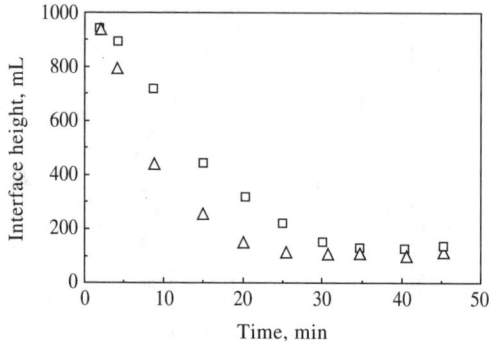

FIGURE 6.9 Variation of wastewater–solid interface height with time during sludge sedimentation of domestic wastewater biologically treated with and without a preozonation step. Conditions: 20°C, pH 7, $COD_0 = 265$ mgL^{-1}; wastewater treatment type: □ = aerobic biological oxidation, △ = ozonation plus aerobic biological oxidation.

BOD/COD. Figure 6.8 shows that the increase of the domestic sewage in the composition of the total wastewater is detrimental to improved biodegradability. It is then deduced that the distillery wastewater composition initially contained organic compounds resistant to biological oxidation that ozonation transformed into other more biodegradable forms while the domestic sewage was not affected.

6.5.2 Sludge Settling

Another important effect of preozonation that concerns the biological process is the improvement of the sedimentation rate. As the literature reports, ozone addition can lead to particle destabilization through different mechanisms.[3] This has also been observed in the activated sludge treatment resulting from an integrated ozone–biological oxidation process, as shown in Figure 6.9 for the case of the sedimentation rate of activated sludge in preozonated and non-preozonated domestic wastewater. The sedimentation rate is measured as the decrease observed in the sludge–clear water interface with time in a 1-L column. As can be observed from Figure 6.9, the preozonated samples showed a faster sedimentation rate. If the design of the sedimentation unit is required, the beneficial effects of preozonation can reduce the surface area of the sedimentation unit, as has also been shown in other work.[38]

6.5.3 Sludge Production

Sludge generated in wastewater biological oxidation processes is becoming an important problem because of its restricted use in landfilling and agriculture. Methods are needed to disintegrate the sludge from wastewater treatment plants. Ozonation has also been reported as a possible technology to reduce the amount of sludge by reacting with solid particles and increasing the biodegradability.[60,65] This is particularly useful in anaerobically produced sludges where biodegradability is very low — one of the problems of the anaerobic system. The beneficial effect of ozone has been

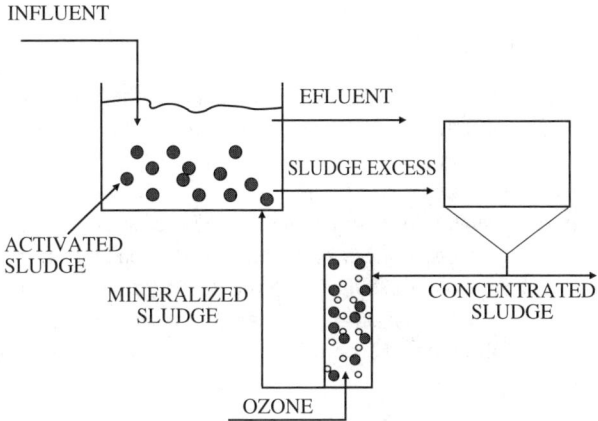

FIGURE 6.10 Use of ozone to reduce the sludge volume in wastewater biological oxidation.

reported to reach the zero sludge production in some cases[10] at an ozone dose of 0.136 gO_3/gSS. In this type of process, ozonation is applied in the returning sludge line for some period of time. The ozonation can also be applied in a tank where the returning sludge line finishes (see Figure 6.10). Ozone is able to destroy microorganisms and produce more organic compounds, partially mineralizing the sludge. Once in the aerobic tank another part of the ozonated sludge is then mineralized, yielding lower amounts of sludge. In this type of process, improvements in the quality of other parameters are also noted. For example, the sludge volumetric index can be kept as low as 100 mLg^{-1}, compared to a value of 800 mLg^{-1} in a nonozonated run for the same period of time.[9]

6.6 KINETIC STUDY OF THE OZONATION OF WASTEWATERS

Thus far the kinetic studies of ozone reactions have been limited to model compounds dissolved in highly purified water. In these studies, the concentration of the target compound B is usually high so that the reactions are fast in most of the cases. In practice, however, there are two possibilities for drinking water or wastewater treatment. The first one is for cases where the concentration of reactants is much lower than at the laboratory scale so that the conditions of slow kinetic regime usually hold. The second one concerns wastewater with multiple compounds with different unknown concentrations. Global parameters are used to follow the degree of pollution, such as the chemical oxygen demand, COD, total organic carbon, TOC, etc. The first situation is more common in drinking water treatment, but in that case ozone indirect reactions are generally the main means of pollutant removal. This case will be discussed later (see Chapter 7). The second case is the treatment of wastewater. In the sections that follow, the kinetic studies of wastewater ozonation are described.

6.6.1 ESTABLISHMENT OF THE KINETIC REGIME OF OZONE ABSORPTION

The first step in modeling the process for the study of the kinetics of wastewater ozonation is the determination of the rate constant of the ozone reactions. It is evident, however, that this is not a viable task because of the number of compounds that constitute the wastewater. Strictly speaking, the ozonation of wastewaters is a multiple series-parallel system of ozone reactions. Therefore, the kinetics is followed using global or surrogate parameters representing the concentration of the wastewater. As mentioned previously, COD is the recommended parameter and, for purposes of simplification, it can be assumed that ozone would react with the matter in water through the following irreversible second-order reaction:

$$O_3 + zCOD \to P \tag{6.5}$$

When the kinetic regime is slow, it is also assumed that ozone decomposes in free radicals that react with the organic matter through the following reaction:

$$HO\bullet + COD \to P' \tag{6.6}$$

In this section, it is assumed that only the kinetics of the direct ozonation Reaction (6.5) proceeds.

The steps for studying the kinetics of the direct wastewater ozonation are similar to those shown previously for single compounds. The first step is to establish the kinetic regime of ozone absorption because this will allow the ozone absorption rate law to be fixed (see Table 5.5). This can be done by determining the experimental reaction factor, E [see Equation (4.31)], with the absorption rate of ozone, N_{O3}, calculated from the difference between the ozone molar rates at the reactor inlet and outlet. This leads to two possible situations. For reaction factors higher than unity, the kinetic regime can be considered fast or instantaneous, while for E values approximately equal to or lower than unity, the kinetic regime is considered slow. In addition, the presence or absence of dissolved ozone confirms one or the other situation. For example, Figure 6.11 shows the changes in COD and dissolved ozone concentration with time corresponding to the ozonation of two different wastewaters of high (2000 mg/L^{-1}) and low (160 mg/L^{-1}) COD.[89] The results show that dissolved ozone was found only after approximately 15 min from the start of the ozonation of the lowest concentrated wastewater (tomato wastewater). Hence, fast or instantaneous reactions developed at any time in the ozonation of the more concentrated wastewater and for the first 15 min of the ozonation of the low concentrated wastewater. The experimental reaction factors also confirmed this conclusion.[89] Next, it was necessary to establish the experimental conditions for the fast pseudo first-order kinetic regime, the kinetic regime recommended to determine the rate constant (see Chapter 4).

Kinetics of the Ozonation of Wastewaters

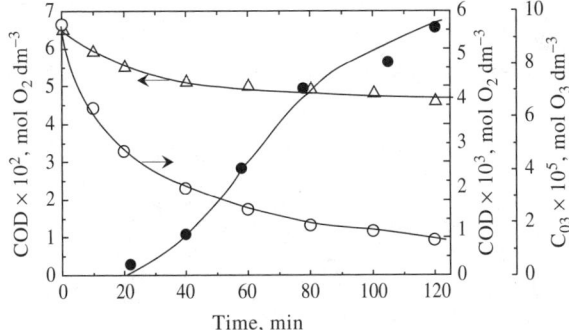

FIGURE 6.11 Variation of COD and dissolved ozone concentration with time during the ozonation of wastewaters. Conditions: distillery wastewater: Δ = COD, tomato wastewater: ○ = COD; ● = dissolved ozone concentration. (From Beltrán, F.J. et al., Kinetic study of the ozonation of some industrial wastewaters, *Ozone Sci. Eng.*, 14, 303–327, 1992. Copyright 1992, International Ozone Association. With permission.)

For the fast pseudo first-order kinetic regime, Condition (4.47) must be fulfilled. Due to the number of reactions present, for the rate constant of ozone direct reactions, k_2, a reaction rate coefficient ε_{O3}, representative of the surrogate Reaction (6.5), is considered. In fact, this coefficient should not be taken as a real rate constant but one that is also affected by the physical properties of the medium.[90,91] As observed in Equation (4.47), the Hatta number depends on the product of the reaction rate constant and the diffusivity of ozone. Values of the ozone diffusivity are known only in very dilute organic-free water. For wastewater, the following equation can be considered:

$$\varepsilon_{O3} = k_2 \frac{D'_{O3}}{D_{O3}} \tag{6.7}$$

where D'_{O3} is the diffusivity of ozone in the wastewater, which is unknown. According to Condition (4.47), values of Ei should be as high as possible for the kinetic regime to be fast pseudo first-order. However, only approximate values of Ei can be obtained since this parameter depends on the diffusivities and ozone solubility. In the case of wastewater ozonation studies, the diffusivity of the organic matter is usually taken as $5 \times 10^{-10} m^2 sec^{-1}$, which can be an average value for the diffusivities of compounds in water.[89] For ozone, the diffusivity in dilute organic-free water is also taken. It is evident that this represents the highest limiting value of this parameter because in the presence of organic matter, ozone diffuses more slowly than in organic-free water. On the other hand, the ozone solubility also depends on the organic matter present in the wastewater (see Chapter 5). However, as shown below (Section 6.6.2), values of the Henry law constant determined by absorbing ozone in a wastewater do not differ very much from those in organic-free water.[92,93] Then, for calculation purposes, the ozone solubility in organic-free or dilute organic water

TABLE 6.4
Rate Coefficient Data Corresponding to the Ozonation of Some Wastewater

Wastewater Type	System Properties	Rate Coefficient, $M^{-1}s^{-1}$	Reference
Wine distillery	Semibatch tank, COD = 2080 mgL^{-1}, 50 Lh^{-1}, inlet P_{O3} = 2229 Pa, pH = 4.8, z = 0.4^a	6240	89
Tomato processing	Semibatch tank, COD = 160 mgL^{-1}, 50 Lh^{-1}, inlet P_{O3} = 425 Pa, pH = 8.5, z = 1.47^b	3.89 × 10^{-4}	89
Domestic	Semibatch bubble column, COD = 65 mgL^{-1}, 30 Lh^{-1}, inlet P_{O3} = 507 Pa, pH = 7.5, z = 1.18	2 × 10^{-5}	90

z values in gCOD/g O_3

^{a} After 60 min reaction; ^{b} after 15 min reaction.

---

can be considered for determining Ei. Thus, actual values of Ei are likely higher than those calculated from Equation (4.46) with these assumptions. Once Ei is known, if $1 < E < Ei$ the fast pseudo first-order kinetic regime can be assumed to develop. Application of this procedure was followed in a previous work.[89]

If ozonation is carried out in a semibatch agitated reactor where gas and wastewater phases can be considered perfectly mixed, the steps that follow for the rate coefficient determination are the same as those shown in Section 5.3.4 for the case of the ozonation of model compounds.[89] Table 6.4 shows results of kinetic studies corresponding to some wastewater ozonation.

6.6.2 DETERMINATION OF OZONE PROPERTIES FOR THE OZONATION KINETICS OF WASTEWATER

In addition to the rate coefficient, ε_{O3}, parameters such as the Henry constant and the volumetric mass transfer coefficient, $k_L a$, are also needed to solve the ozone absorption rate equations. Values of the Henry constant and mass transfer coefficients can be found in the literature or determined as shown in preceding sections for organic-free water or very dilute organic aqueous solutions. With wastewater, the problem is that the presence of different substances may affect the values of both parameters. Therefore, attempts should be made to estimate He and $k_L a$ in more polluted water. The usual procedure could be that shown in Section 5.1.2 for organic-free water where the mass balance of ozone in water is used. In organic-free water, the rate constant of the decomposition reaction of ozone is known, but in a practical case (with wastewater), the rate coefficient is also unknown and the use of the mass balance of ozone in water is not appropriate. However, a similar method can be applied using the mass balance of ozone in the gas phase, provided the kinetic regime of ozone absorption corresponds to slow reactions. The method is based on the fact that, for slow reactions, the ozone absorption rate can be expressed as in Equation (4.30), that is, as a function of the ozone driving force ($C_{O3}^* - C_{O3b}$) and the procedure

Kinetics of the Ozonation of Wastewaters

does not depend on the rate coefficient value. The mass balance of ozone in the gas phase in a semibatch reactor where the gas and wastewater phases are perfectly mixed is given by Equation (5.32). If the accumulation term is assumed negligible (as has been observed in a previous work[91]) from Equations (4.30) and Equation (5.17), once the Henry's and perfect gas laws have been accounted for [Equations (5.36) and Equation (5.37)], $C_{O3}^* = C_g RT/He$), the concentration of ozone in the gas at the reactor outlet becomes as follows:

$$C_{gb} = \frac{C_{geb}}{1 + \dfrac{k_L aRT}{v_g He}\beta V} + \frac{C_{O3b}}{\dfrac{v_g}{k_L a} + \dfrac{RT}{He}\beta V} \qquad (6.8)$$

According to Equation (6.8) a plot of C_{gb} against C_{O3b} corresponding to different times of one ozonation experiment in the real wastewater should lead to a straight line. From the slope and ordinate of this line the Henry constant and mass transfer coefficient can be determined. Note that in the mass balance of ozone [Equation (5.32)] there is no need to neglect the accumulation rate term but the procedure would be more complicated because a trial-and-error method similar to that used by Ridgway et al.[72] (see Section 5.3.3) should be applied.

There is another way to obtain He and $k_L a$ for the ozonation of a wastewater. Again, the kinetic regime has to be slow so there must be dissolved ozone. This procedure applies when the concentrations of ozone in the gas and wastewater phases remain constant with time. This situation usually occurs after some reaction time has elapsed since the start of ozonation. At these conditions (see Reference 92 as an example) the variation of COD with time is also constant ($dCOD/dt$ = constant) and the ozone absorption rate is:

$$-\frac{dCOD}{dt} = zN_{O3s} = k_L a\left(C_{O3s}^* - C_{O3bs}\right) \qquad (6.9)$$

where subindex s denotes steady-state conditions for ozone concentrations. Since the ozone absorption rate, at steady-state conditions, can also be expressed as the difference between the experimental ozone molar rates at the reactor inlet and outlet:

$$N_{O3s} = \frac{m_i - m_o}{\beta V} \qquad (6.10)$$

after considering Henry's law, we obtain:

$$C_{gs} = \frac{He}{RT}\left(\frac{m_i - m_o}{\beta V k_L a}\right) + \frac{He}{RT}C_{O3s} \qquad (6.11)$$

Since at these conditions Equation (6.8) also holds, solving Equations (6.8) and Equation (6.11) will allow both He and $k_L a$ to be determined experimentally. The procedures presented above have been checked in a bubble column where the gas phase was considered in plug flow although the wastewater phase stayed in perfect mixing.[91] For plug flow conditions, the ozone mass balance is applied to a differential reactor volume of height dz and it becomes a first-order differential equation [see Equation (5.35)]. The following steps, as in the case of the semibatch agitated reactor, lead to Equation (6.12):[91]

$$\frac{-dC_{gb}}{C_{gb} - C_{O3b}\dfrac{He}{RT}} = \frac{k_L aRT}{v_g He} Sdh \qquad (6.12)$$

Equation (6.12) can be integrated between the column limits:

$$\begin{aligned} h = 0 & \quad C_{gb} = C_{geb} \\ h = h_T & \quad C_{gb} = C_{gob} \end{aligned} \qquad (6.13)$$

The result of integration is Equation (6.14), which relates the ozone concentration in the gas at the reactor outlet with the ozone concentration in water:

$$C_{gob} = \frac{C_{geb}}{\exp\left[\dfrac{k_L aRTSh_T}{v_g He}\right]} + \left[1 - \frac{1}{\exp\left[\dfrac{k_L aRTSh_T}{v_g He}\right]}\right]\frac{He}{RT} C_{O3b} \qquad (6.14)$$

From Equation (6.14) a plot of C_{gob} against C_{O3b} should lead to a straight line so He and $k_L a$ can be determined from the slope and ordinate. This procedure was followed in a previous work where the ozonation of a food-related wastewater was studied.[92]

This second procedure was also applied to the results obtained in a bubble column. In this case, due to the fact that the ozone partial pressure (or the ozone concentration in the gas) varies along the height of the column, an average value of the ozone concentration in the gas for the whole column at steady-state conditions, C_{gsav}, is estimated using Equation (6.11). Thus:

$$C_{gsav} = \frac{He}{RT}\left(\frac{m_i - m_o}{\beta V k_L a}\right) + \frac{He}{RT} C_{O3s} \qquad (6.15)$$

C_{gsav} can also be estimated as follows:

$$C_{gsav} = \frac{1}{h_T}\int_0^{h_T} C_{gs} \qquad (6.16)$$

Kinetics of the Ozonation of Wastewaters

TABLE 6.5
Data on Henry Constant and Volumetric Mass Transfer Coefficient Corresponding to Ozone-Buffered Water and Ozone–Tomato Wastewater Systems[a]

Ozone System	System Characteristics	$He \times 10^{-6}$	$k_L a \times 10^2$	Reference
Ozone-buffered water	pH 7, 12°C, I = 0.01 M, Phosphate buffer	9.14		93
Ozone-buffered water	pH 7, 22°C, I = 0.01 M, Phosphate buffer	11.75		93
Ozone-buffered water	pH 7, 22°C, I = 0.01 M, 40 Lh^{-1} Phosphate–carbonate buffer	8.49	1.05	93
Ozone–wastewater	Tomato wastewater, COD = 300 mgL^{-1}, pH = 6.5–7.8, 17–18°C, 20 Lh^{-1}	7.66×10^6 [b]	5.91×10^6	92
Ozone–wastewater	Tomato wastewater, COD = 500 mgL^{-1}, pH = 6.5–7.8, 17–18°C, 35 Lh^{-1}	7.05[b]	6.99×10^{-2}	92
Ozone–wastewater	Tomato wastewater, COD = 500 mgL^{-1}, pH = 6.5–7.8, 17–18°C, 35 Lh^{-1}	9.95[c]		92
Ozone–wastewater	Domestic wastewater, COD = 300 mgL^{-1}, pH 7.5, 20°C	7.60[c]		90

[a] He = Henry's law constant in PaM^{-1}; $k_L a$ is in s^{-1}. [b] From Equation (6.14). [c] From Equation (6.15) and Equation (6.18).

where C_{gs}, the ozone gas concentration at any position within the column, can be obtained from Equation (6.12), once integrated with the boundary limits:

$$h = 0 \quad C_{gs} = C_{geb}$$
$$h = h_T \quad C_{gs} = C_{gso} \quad (6.17)$$

Thus, once Henry's law has been accounted for, the integrated Equation (6.16) becomes

$$C_{gsav} = \left(C_{geb} RT - HeC_{O3s}\right) \frac{Hev_g}{(RT)^2 k_L a Sh_T} \left\{1 - \exp\left[-\frac{k_L a v_g Sh_T}{RTHe}\right]\right\} + HeC_{O3s} \quad (6.18)$$

By solving Equation (6.18) and Equation (6.15) both He and $k_L a$ are determined. Table 6.5 presents data on He y and $k_L a$ obtained from the methods indicated above as reported in previous works. Also, values obtained in organic-free water or very dilute organic water are presented for comparative purposes. As can be seen, values of He are not far from those obtained in buffered water, although experimental results are closer to those obtained in carbonate–phosphate buffered water. This is not surprising since the food wastewater used in the work cited[92] was from a tomato processing plant where carbonates were present. On the contrary, some differences are observed for the $k_L a$. Thus, $k_L a$ values of experimental wastewater ozonation are 2.5 to 5 times higher than those in organic-free or very dilute organic water or determined from empirical equations that can also be applied to pure water systems. The results, however, are a logical consequence of the presence of organic and/or inorganic matter in the wastewater. Thus, the literature[94] reports up to 300% increases in $k_L a$ depending on the content and nature of the matter in water. As reported,[94] these results are due to increases in the specific interfacial area, a, which depends, among other factors, on the surface tension of water, which varies with the presence of organic/inorganic substances. In a previous work, an explanation for these phenomena is given.[92] According to these results, $k_L a$ should be calculated from physical absorption experiments, for example, absorbing oxygen in the wastewater studied, provided oxygen does not remove the COD. Then, after considering the effect of diffusivities, the applied $k_L a$ for ozonation can be:

$$k_L a = (k_L a)_{O2} \left(\frac{D_{O2}}{D_{O3}} \right)^{1/2} \tag{6.19}$$

Note that Equation (6.19) is based on the relationship between diffusivity and mass transfer coefficient as deduced by surface renewal theories [see Equation (4.12)]. The film theory, instead, proposes a direct proportionality [see Equation (4.7)], but due to the more realistic approach of surface renewal theories Equation (6.19) is recommended. Also note that in wastewater, diffusivities of oxygen and ozone will be lower than in organic-free water. However, in Equation (6.19) diffusivities of ozone and oxygen in dilute organic water or organic-free water can be used since the retarding effect of the organic/inorganic matter of the water on both diffusivities is likely to be similar.

The main conclusions drawn from the above discussion about He and $k_L a$ are

- Experimentally, He can be determined for the ozone–wastewater systems only where ozonation develops in the slow kinetic regime.
- The mass transfer coefficient, $k_L a$, must be obtained experimentally since in this case the content of the wastewater may significantly affect its value. For this case, however, oxygen absorption experiments could be carried out provided there are no significant oxygen reactions in the water during the absorption time experiment.

6.6.3 DETERMINATION OF RATE COEFFICIENTS FOR THE OZONATION KINETICS OF WASTEWATER

The main objective of any kinetic study should be the scale-up of the treated system. This also applies to the ozonation of a real wastewater. Kinetic data obtained in

Kinetics of the Ozonation of Wastewaters

laboratory reactors should be used to design reactors of greater size. However, in addition to problems related to hydrodynamic conditions, the rate coefficients determined from laboratory ozonation experiments are not exactly reaction rate constants but rather parameters involving effects of physical properties (D_B, C_{O3}^*) and mass transfer data ($k_L a$). It is convenient to determine rate coefficients from laboratory experiments carried out in reactors that, at least, maintain geometric similarity as in real size reactors and show the same flow pattern for both the gas and wastewater phases. The type of ozone contactor usually employed for this purpose is the bubble column. These reactors have geometrical and hydrodynamic properties similar to those of the real ozone contactors: the gas phase is fed through porous plates situated at the bottom and wastewater and gas circulate currently or countercurrently. In addition, the gas phase is considered as in plug flow while perfect mixing is assumed for the wastewater phase (see Appendix A1). A general method for scale-up would consist of the following steps:

- Determine the reaction rate and mass transfer coefficients in small laboratory bubble columns.
- Establish the mathematical model of the kinetics, that is, the mass balances of species present for COD and ozone.
- Solve the mathematical model taking into account the type of flow of water and gas phases through the ozone reactor.
- Apply the model to ozone contactors of higher size and compare the experimental and calculated results.

Bubble columns of different size are considered to be of geometrical hydrodynamic similarity when their height/diameter ratio and superficial gas velocity are the same. The study presented here is based on a previous work[91] that used semibatch bubble columns with the continuous gas phase in plug flow and the batch wastewater phase in perfect mixing. Two possible cases are considered: ozonation of wastewater in both fast and slow kinetic regimes. In both of these cases, ozonation is represented by Reaction (6.5). Chapter 11 discusses kinetic modeling by considering both Reaction (6.5) and Reaction (6.6).

6.6.3.1 Fast Kinetic Regime (High COD)*

The absence of dissolved ozone, reaction factors higher than unity, and Hatta numbers higher than 3 (see also Table 5.5) characterize the fast kinetic regime. Conditions regarding the reaction factor and the dissolved ozone concentration can be checked, *a priori*, to conclude that the kinetic regime is fast. However, it is recommended that Condition (4.47) for the fast pseudo first-order kinetic regime be fulfilled since the absorption rate equation is easier to use mathematically. This, however, cannot be checked since the rate coefficient is unknown. As a consequence, the assumption from the outset is that the fast pseudo first-order kinetic regime holds for the

---

* Part of Section 6.6.3.1 is reprinted with permission from Reference 91. Copyright 1995, International Ozone Association.

experimental conditions applied. For this kinetic regime the ozone absorption rate is [see also Equation (5.54)]

$$N_{O3}a = aC_{O3}^* \sqrt{\varepsilon_{O3}D_{O3}COD} \qquad (6.20)$$

Equation (6.20) also represents the term G'_{O3} in the ozone mass balance in the gas phase (see Section 5.3.2). If plug flow conditions hold for the gas phase and the wastewater phase is well mixed in the bubble column, Equation (5.35) is used. If the accumulation term is neglected, once Equation (6.20) is considered, Equation (5.35) becomes

$$-v_g dC_{gb} = a\frac{RT}{He}\beta\sqrt{\varepsilon_{O3}D_{O3}COD}\,C_{gb}Sdz \qquad (6.21)$$

Equation (6.21) can be integrated between the bottom and the top of the column to yield:

$$\ln\frac{C_{geb}}{C_{gob}} = \varsigma h_T \sqrt{COD} \qquad (6.22)$$

where

$$\varsigma = \frac{aRT\beta S\sqrt{\varepsilon_{O3}D_{O3}}}{v_g He} \qquad (6.23)$$

From Equation (6.22) a plot of $\ln(C_{geb}/C_{gob})$ against $(COD)^{0.5}$ should lead to a straight line that allows the rate coefficient to be determined. Then, Condition (4.47) has to be checked to confirm the kinetic regime of ozonation. In practice, however, a more complex procedure has to be followed primarily because of the changing nature with ozonation time of the matrix constituting the wastewater and, as a consequence, the change of kinetic regime with time. Thus, COD gives only a measure of the total concentration of the wastewater at a given time that depends on the nature of the wastewater. This changes with time because ozone reacts in multiple series-parallel reactions to remove and produce compounds of different reactivity. For example, if phenols are the main initial constituents of the wastewater, ozone would react very fast at the beginning of ozonation, producing carboxylic acids that eventually also react, but slowly, with ozone. Therefore, the rate coefficient of ozonation could change with time and the kinetic regime. The problem associated with the change in kinetic regime can be overcome if the general ozone absorption rate Equation (4.25) is used instead of Equation (6.20). This implies, however, the modification of Equation (6.22), which will now become:

$$\ln\frac{C_{geb}}{C_{gob}} = \frac{k_L aRT\beta Sh_T}{v_g He}\frac{Ha_1}{\sinh(Ha_1)}\left[\cosh(Ha_1) - \frac{1}{\cosh(Ha_1) + \frac{\beta k_L}{aD_{O3}}Ha_1\sinh(Ha_1)}\right] \qquad (6.24)$$

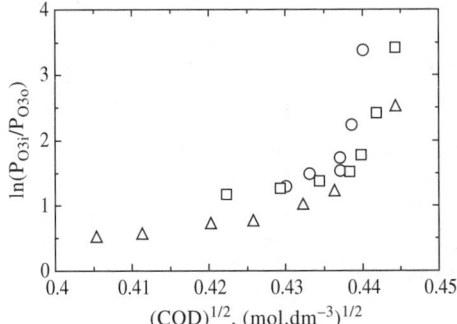

FIGURE 6.12 Verification of Equation (6.22) for different superficial gas velocities during the ozonation kinetics of a distillery wastewater. Superficial gas velocity × 10^3, ms^{-1}: ○ = 2.21, □ = 4.42, △ = 7.74. (From Beltrán, F.J. et al., Modelling industrial wastewater ozonation in bubble contactors. 1. Rate coefficient determination, *Ozone Sci. Eng.*, 17, 355–378, 1995. Copyright 1995, International Ozone Association. With permission.)

A trial-and-error procedure should then be used to obtain the value of the rate coefficient. If the rate coefficient is a function of time, the method should be applied to every reaction time. This procedure was applied in a previous work to the ozonation of distillery wastewater in a small laboratory bubble column.[91] In this study,[91] verification of Equation (6.22) led to Figure 6.12, and for the first minutes of ozonation, no ozone in the gas at the column outlet was detected. Then Equation (6.22) was used only for times when the gas exiting the column contained ozone. In any case, for the appropriate time interval, the experimental points could not be fitted to a straight line as shown in Figure 6.12. The kinetic regime was not always of fast pseudo first-order and the more general Equation (4.25) was applied in a trial-and-error procedure. Finally, the rate coefficients were expressed as a function of the reaction time.[91]

6.6.3.2 Slow Kinetic Regime (Low COD)*

In the ozonation of any wastewater, the slow kinetic regime could be checked for the presence of ozone, reaction factors approximately equal to or lower than unity, and Hatta numbers much lower than 0.3 (see also Table 5.5). In this kinetic regime it is likely that indirect reactions of ozone compete with the direct reactions (see Section 7.1). However, because the method applied has some degree of empiricism, it is sometimes assumed that all reactions (direct or indirect) are represented by the unique Reaction (6.5). Because of the presence of dissolved ozone, the following equations constituted the model to determine the rate coefficient:

---

* Part of Section 6.6.3.2 is reprinted with permission from Reference 91. Copyright 1995, International Ozone Association.

- The ozone mass balance in the gas phase as in the preceding case [Equation (5.35)].
- The ozone mass balance in the bulk of the wastewater phase:

$$\int_0^{k_T} k_L a\left(\frac{RT}{He}C_{gb} - C_{O3b}\right) S dz = -r_{O3}(1-a\delta_L)\beta V + \beta V \frac{dC_{O3b}}{dt} \quad (6.25)$$

where r_{O3} represents the ozone chemical reaction rate.
- The total mass balance:

$$\beta V(C_{O3} + (1/z)\Delta COD) = (m_i - m_o)\Delta t \quad (6.26)$$

where ΔCOD represents the amount of COD consumed in a period of time Δt.

As can be deduced, the left side of Equation (6.25) physically means that the molar rate of ozone that reaches the bulk wastewater (at $x = \delta_L$) is partially consumed by chemical reactions:

$$-r_{O3} = \varepsilon_{O3} C_{O3b} COD \quad (6.27)$$

and partially accumulated in the wastewater, ($\beta V dC_{O3}/dt$). The factor $(1 - a\delta_L)$ represents the bulk water to total volume ratio. The rate coefficient is determined from Equation (5.34) and Equation (6.25). In addition, Equation (5.35) is also used to obtain a relationship between the ozone concentration in the gas at the reactor outlet and the height of the column. For the slow kinetic regime this is Equation (6.14). Equation (6.25) can be solved by numerical integration; the ozone accumulation rate term dC_{O3b}/dt can be obtained from the first derivative with a polynomial function that can be used to fit C_{O3b} with the reaction time. Other data such as the film layer, δ_L, and the wastewater holdup, β, can be estimated from Equation (4.7) and Equation (5.44), respectively. In this kinetic regime, the rate coefficient likely remains independent of reaction time because ozone is reacting with refractory compounds and their nature does not vary as much as it does when fast-reacting compounds are present in the water (as in the preceding case). Finally, values of the stoichiometric ratio, z, can be obtained from the total mass-balance Equation (6.26). The procedure developed here was applied to the ozonation of a tomato wastewater of very low COD (<100 mg/L). The results obtained led to a time-independent rate coefficient of 17 $M^{-1}sec^{-1}$.[91]

When treating the kinetic modeling of wastewater ozonation processes in Chapter 11, the values of the ozone rate coefficient are used for the scale-up of ozone contactors. The practical application of these coefficients is shown.

REFERENCES

1. Lazarova, V. et al., Advanced wastewater disinfection technologies: state of the art and perspectives, *Water Sci. Technol.*, 40, 203–213, 1999.
2. Paraskeva, P., Lambert, S.D., and Graham, N.J.D., Influence of ozonation conditions on the treatability of secondary effluents, *Ozone Sci. Eng.*, 20, 133–150, 1998.
3. Reckhow, D., Singer, P., and Trussell, R., Ozone as a coagulant aid, in AWWA Seminar Proc., *Ozonation Recent Advances and Research Needs*, No. 20005, American Water Works Association, Denver, CO, 17–86, 1986.
4. Jekel, M., Flocculation effects of ozone, *Ozone Sci. Eng.*, 16, 55–66, 1994.
5. Dussert, B.W. and Kovacic, S.L., Impact of drinking water preozonation on activated carbon quality and performance, *Ozone Sci. Eng.*, 1–12, 1997.
6. Croll, B.T., The installation of GAC and ozone surface water treatment plants in Anglian water, *Ozone Sci. Eng.*, 1–18, 1996.
7. Lin, C. et al., Enhanced biodegradation of petrochemical wastewater using ozonation and BAC advanced treatment system, *Water Res.*, 35, 699–704, 2001.
8. Roy-Arcand, L. and Archibald, F.S., Ozonation as a treatment for mechanical and chemical pulp mill effluents, *Ozone Sci. Eng.*, 18, 363–384, 1996.
9. Kamiya, T. and Hirotsuji, J., New combined system of biological process and intermittent ozonation for advanced wastewater treatment, *Water Sci. Technol.*, 38, 145–153, 1998.
10. Huysmans, A. et al., Ozonation of activated sludge in the recycle stream, *J. Chem. Technol. Biotechnol.*, 76, 321–324, 2001.
11. Rice, R.G. and Browning, M.E., *Ozone Treatment of Industrial Wastewater*, Noyes Data Corporation, Park Ridge, NJ, 1981.
12. Masten, S.J. and Davies, H.R., The use of ozonation to degrade organic contaminants in wastewaters, *Environ. Sci. Technol.*, 28, 180A–185A, 1994.
13. Arslan, I., Balcioglu, I.A., and Tuhkanen, T., Advanced oxidation of synthetic dyehouse effluent by O_3, H_2O_2, and H_2O_2/UV processes, *Environ. Technol.*, 20, 921–931, 1999.
14. Paraskeva, P. and Graham, N.J.D., Ozonation of municipal wastewater effluents, *Water Environ. Res.*, 74, 569–581, 2002.
15. Rice, R.G., Ozone in the United States of America — state of the art, *Ozone Sci. Eng.*, 99–118, 1999.
16. Beaman, M.S. et al., Role of ozone and recirculation in the stabilization of landfill and leachates, *Ozone Sci. Eng.*, 121–132, 1998.
17. Arslan, I. and Balcioglu, I.A., Advanced oxidation of raw and biotreated textile industry wastewater with O_3, H_2O_2/UV-C, and their sequential application, *J. Chem. Technol. Biotechnol.*, 76, 53–60, 2001.
18. Alaton, I.A., Balcioglu, I.A., and Bahnemann, D.W., Advanced oxidation of a reactive dyebath effluent: comparison of O_3, H_2O_2/UV-C, and TiO_2/UV-A, *Water Res.*, 36, 1143–1154, 2002.
19. Toffani, G. and Ruchard, Y., Use of ozone for the treatment of a combined urban and industrial effluent: a case history, *Ozone Sci. Eng.*, 17, 345–354, 1995.
20. Mao, H. and Smith, D., Influence of ozone application methods on efficacy of ozone decolorization of pulp mill effluents, *Ozone Sci. Eng.*, 17, 205–236, 1995.
21. Preis, S. et al., Advanced oxidation processes against phenolic compounds in wastewater treatment, *Ozone Sci. Eng.*, 17, 399–418, 1995.

22. Luck, F. et al., Destruction of pollutants in industrial rinse waters by advanced oxidation processes, *Water Sci. Technol.*, 35, 287–292, 1997.
23. Watkins, B.D. et al., Ozonation of swine manure wastes to control odors and reduce the concentrations of pathogens and toxic fermentation metabolites, *Ozone Sci. Eng.*, 19, 425–437, 1997.
24. Höfl, C. et al., Oxidative degradation of AOX and COD by different advanced oxidation processes: a comparative study with two samples of a pharmaceutical wastewater, *Water Sci. Technol.*, 35, 257–264, 1997.
25. Rico, B. et al., Organic matter removal in boiler feed water treatment of a power plant with ozone, *Ozone Sci. Eng.*, 19, 471–480, 1997.
26. Filippi, A., Ozone is the most effective disinfectant for dental treatment units: results after 8 years of comparison, *Ozone Sci. Eng.*, 19, 527–532, 1997.
27. Churchley, J.H., Ozone for dye waste color removal: four years operation at Leeks STW, *Ozone Sci. Eng.*, 20, 111–120, 1998.
28. Orta de Velásquez, M.T. et al., Improvement of wastewater coagulation using ozone, *Ozone Sci. Eng.*, 20, 151–162, 1998.
29. Liberti, L. and Notarnicola, M., Advanced treatment and disinfection for municipal wastewater reuse in agriculture, *Water Sci. Technol.*, 40, 235–245, 1999.
30. López, A. et al., Textile wastewater reuse: ozonation of membrane concentrated secondary effluent, *Water Sci. Technol.*, 40, 99–105, 1999.
31. Egemen, E., Evaluation of ozonation and cryptic growth for biosolids management in wastewater treatment, *Water Sci. Technol.*, 39, 153–158, 1999.
32. Beltrán, F.J. et al., Effects of single and combined ozonation with hydrogen peroxide or UV radiation on the chemical degradation and biodegradability of debittering table olive industrial wastewaters, *Water Res.*, 33, 723–732, 1999.
33. Kos, L. and Perkowski, J., Application of ozone and hydrogen peroxide in textile wastewater treatment, *Fibres and Textiles in Western Europe*, April–June, 61–64, 1999.
34. Beltrán, F.J., García-Araya, J.F., and Álvarez, P., Integration of continuous biological and chemical (ozone) treatment of domestic wastewater. 1. Biodegradation and post-ozonation, *J. Chem. Technol. Biotechnol.*, 74. 877–883, 1999.
35. Beltrán, F.J., García-Araya, J.F., and Álvarez, P., Integration of continuous biological and chemical (ozone) treatment of domestic wastewater. 2. Ozonation followed by biological oxidation, *J. Chem. Technol. Biotechnol.*, 74, 884–890, 1999.
36. Beltrán, F.J., García-Araya, J.F., and Álvarez, P., Wine distillery wastewater degradation. 1. Oxidative treatment using ozone and its effect on the wastewater biodegradability, *J. Agric. Food Chem.*, 47, 3911–3918, 1999.
37. Beltrán, F.J., García-Araya, J.F., and Álvarez, P., Wine distillery wastewater degradation. 1. Improvement of aerobic biodegradation by means of an integrated chemical (ozone)-biological treatment, *J. Agric. Food Chem.*, 47, 3919–3924, 1999.
38. Beltrán, F.J. et al., Improvement of domestic wastewater primary sedimentation through ozonation, *Ozone Sci. Eng.*, 21, 605–614, 1999.
39. Liberti, L., Notarnicola, M., and López, A., Advanced treatment for municipal wastewater reuse in agriculture. III. Ozone disinfection, *Ozone Sci. Eng.*, 22, 151–166, 2000.
40. Zwiener, C. and Frimmel, F.H., Oxidative treatment of pharmaceuticals in water, *Water Res.*, 34, 1881–1885, 2000.
41. Korhonen, M.S., Metsärinne, S.E., and Tuhkanen, T.A., Removal of ethylenediaminetetraacetic acid (EDTA) from pulp mill effluents by ozonation, *Ozone Sci. Eng.*, 22, 279–286, 2000.
42. Weemaes, M. et al., Ozonation of sewage sludge prior to anaerobic digestion, *Water Sci. Technol.*, 42, 175–178, 2000.

43. Perkowski, J., Kos, L., and Ledakowicz, S., Advanced oxidation of textile wastewaters, *Ozone Sci. Technol.*, 22, 535–550, 2000.
44. Beltrán, F.J., García-Araya, J.F., and Álvarez, P., Continuous flow integrated chemical (ozone) activated sludge system treating combined agroindustrial-domestic wastewater, *Env. Prog.*, 19, 28–35, 2000.
45. Rivas, F.J., Beltrán, F.J., and Gimeno, O., Joint treatment of wastewater from table olive processing and urban wastewater. Integrated ozonation-aerobic oxidation, *Chem. Eng. Technol.*, 23, 177–181, 2000.
46. Rivas, F.J. et al., Two step wastewater treatment: sequential ozonation-aerobic biodegradation, *Ozone Sci. Eng.*, 22, 617–636, 2000.
47. Carini, D. et al., Ozonation as pre-treatment step for the biological batch degradation of industrial wastewater containing 3-methyl-pyridine, *Ozone Sci. Eng.*, 23, 189–198, 2001.
48. Beltrán, F.J. et al., Treatment of high strength distillery wastewater (cherry stillage) by integrated aerobic biological oxidation and ozonation, *Biotechnol. Prog.*, 17, 462–467, 2001.
49. Beltrán, F.J., García-Araya, J.F., and Álvarez, P., pH Sequential ozonation of domestic and wine distillery wastewater, *Water Res.*, 35, 929–936, 2001.
50. Baig, S. and Liechti, P.A., Ozone treatment for biorefractory COD removal, *Water Sci. Technol.*, 43, 197–204, 2001.
51. Tosik, R. and Wiktorowski, S., Color removal and improvement of biodegradability of wastewater from dyes production using ozone and hydrogen peroxide, *Ozone Sci. Eng.*, 23, 295–302, 2001.
52. Ledakowicz, S. and Solecka, M., Influence of ozone and advanced oxidation processes on biological treatment of textile wastewater, *Ozone Sci. Eng.*, 23, 327–332, 2001.
53. Lu, S.G. et al., A pilot scale study of tertiary treatment of Jizhuangzi wastewater treatment plant by continuous preozonation-microflocculation-filtration process, *Environ. Technol.*, 22, 331–337, 2001.
54. Verenich, S. et al., Comparison of ozonation and wet oxidation for the destruction of lipophilic wood extractives from paper mill circulation water, *Ozone Sci. Eng.*, 23, 401–409, 2001.
55. Sigge, G.O., Use of ozone and hydrogen peroxide in the post-treatment of USAB treated alkaline fruit cannery effluent, *Water Sci. Technol.*, 44, 69–74, 2001.
56. El-Din, M.G. and Smith, D.W., Maximizing the enhanced ozone oxidation of Kraft pulp mill effluents in an impinging-jet bubble column, *Ozone Sci. Eng.*, 23, 479–493, 2001.
57. Xu, P. et al., Wastewater disinfection by ozone: main parameters for process design, *Water Res.*, 36, 1043–1055, 2002.
58. Pérez, M. et al., Removal of organic contaminants in paper pulp effluents by AOPs: an economic study, *J. Chem. Technol. Biotechnol.*, 77, 525–532, 2002.
59. Radetski, C.M. et al., Ozonation of textile wastewater: physico-chemical and phytotoxic aspects, *Environ. Technol.*, 23, 537–545, 2002.
60. Ahn, K.H. et al., Reduction of sludge by ozone treatment and production of carbon source for denitrification, *Water Sci. Technol.*, 46, 121–125, 2002.
61. Sigge, G.O. et al., Combining UASB technology and advanced oxidation (AOPs) to treat food processing wastewaters, *Water Sci. Technol.*, 45, 329–334, 2002.
62. Haapea, P., Korhonen, S., and Tuhkanen, T., Treatment of industrial landfill leachates by chemical and biological methods: ozonation, ozonation + hydrogen peroxide, hydrogen peroxide, and biological post-treatment for ozonated water, *Ozone Sci. Eng.*, 24, 369–378, 2002.

63. Arslan-Alaton, I. and Balcioglu. A.K., Biodegradability assessment of ozonated raw and biotreated pharmaceutical wastewater, *Arch. Environ. Contam. Toxicol.*, 43, 425–431, 2002.
64. Zenaitis, M.G., Sandhu, H., and Duff, S.J.B., Combined biological and ozone treatment of log yard run-off, *Water Res.*, 36, 2053–2061, 2002.
65. Deleris, S. et al., Minimization of sludge production in biological processes: an alternative solution for the problem of sludge disposal, *Water Sci. Technol.*, 46, 63–70, 2002.
66. Balcioglu, I.A. and Ötker, M., Treatment of pharmaceutical wastewater containing antibiotics by O_3 and O_3/H_2O_2 processes, *Chemosphere*, 50, 85–95, 2003.
67. Alvares, A.B.C., Diaper, C., and Parsons, A., Partial oxidation by ozone to remove recalcitrance from wastewaters — A review, *Environ. Technol.*, 22, 409–427, 2001.
68. Hoigné, J. and Bader, H., Rate constants of reactions of ozone with organic and inorganic compounds in water. II. Dissociating organic compounds, *Water Res.*, 17, 185–194, 1983.
69. Beltrán, F.J., Encinar, J.M., and García-Araya, J.F., Absorption kinetics of ozone in aqueous o-cresol solutions, *Can. J. Chem. Eng.*, 70, 141–147, 1992.
70. Hoigné, J. et al., Rate constants of reactions of ozone with organic and inorganic compounds in water. III. Inorganic compounds and radicals, *Water Res.*, 19, 993–1004, 1985.
71. Yao, C.C.D. and Haag, W.R., Rate constants for direct reactions of ozone with several drinking water contaminants, *Water Res.*, 25, 761–773, 1991.
72. Ridgway, D., Sharma, R.N., and Eanlay, T.R., Determination of mass transfer coefficients in agitated gas liquid reactors by instantaneous reactions, *Chem. Eng. Sci.*, 44, 2935–2942, 1989.
73. Beltrán, F.J., García-Araya, J.F., and Acedo, B., Advanced oxidation of atrazine in water. I. Ozonation, *Water Res.*, 28, 2153–2164, 1994.
74. Beltrán, F.J. et al., Oxidation of polynuclear aromatic hydrocarbons in water. 1. Ozonation, *Ind. Eng. Chem. Res.*, 34, 1596–1606, 1995.
75. Beltrán, F.J., García-Araya, J.F., and Álvarez, P.M., Sodium dodecylbenzenesulfonate removal from water and wastewater. 1. Kinetics of decomposition by ozonation, *Ind. Eng. Chem. Res.*, 39, 2214–2220, 2000.
76. Beltrán, F.J., Encinar, J.M., and Alonso, M.A., Nitroaromatic hydrocarbon ozonation in water. 1. Single ozonation, *Ind. Eng. Chem. Res.* 37, 25–31, 1998.
77. Hoigné, J. and Bader, H., Rate constants of reactions of ozone with organic and inorganic compounds in water. I. Non-dissociating organic compounds, *Water Res.*, 17, 173–183, 1983.
78. Beltrán, F.J. et al., Aqueous degradation of VOCs in the ozone combined with hydrogen peroxide or UV radiation processes. 1. Experimental results, *J. Environ. Sci. Health, Part A Environ. Eng.*, A34, 649–671, 1999.
79. APHA, *Standard Methods for the Examination of Water and Wastewater*, Method 5310 Total Organic Carbon, 19th ed., American Public Health Association, Washington, D.C., 1995.
80. APHA, *Standard Methods for the Examination of Water and Wastewater*, Method 5220 Chemical Oxygen Demand, 19th ed., American Public Health Association, Washington, D.C., 1995.
81. Vogel, F. et al., The mean oxidation number of carbon (MOC) — a useful concept for describing oxidation processes, *Water Res.*, 34, 2689–2707, 2000.
82. Kang, Y.W., Cho, M., and Huang, K., Correction of hydrogen peroxide interference of standard chemical oxygen demand test, *Water Res.*, 33, 1247–1251, 1999.

83. Croué, J.P. et al., Effect of preozonation on the organic halide formation potential of an aquatic fulvic acid, *Ind. Eng. Chem. Res.*, 28, 1082–1089, 1989.
84. Stumm, W. and Morgan, J.J., *Aquatic Chemistry*, 2nd ed., John Wiley & Sons, New York, 1981, 419.
85. Mantzavinos, D. et al., Wet air oxidation of polyethylene glycol: mechanisms, intermediates, and implications for integrated chemical–biological wastewater treatment, *Chem. Eng. Sci.*, 51, 4219–4235, 1996.
86. Scott, J.P. and Ollis, D.F., Integration of chemical and biological oxidation processes for water treatment: review and recommendations, *Environ. Prog.*, 14, 88–103, 1995.
87. Saayman, G.B. et al., Chemical control of filamentous sludge bulking in a full-scale biological nutrient removal of activated sludge plant, *Ozone Sci. Eng.*, 20, 1–16, 1998.
88. Beltrán, F.J., García-Araya, J.F., and Álvarez, P., Impact of chemical oxidation on biological treatment of a primary municipal wastewater. 2. Effects of ozonation on the kinetics of biological oxidation, *Ozone Sci. Eng.*, 19, 513–526, 1997.
89. Beltrán, F.J. et al., Kinetic study of the ozonation of some industrial wastewaters, *Ozone Sci. Eng.*, 14, 303–327, 1992.
90. Beltrán, F.J., García-Araya, J.F., and Álvarez, P.M., Domestic wastewater ozonation: a kinetic model approach, *Ozone Sci. Eng.*, 23, 219–228, 2001.
91. Beltrán, F.J. et al., Modelling industrial wastewater ozonation in bubble contactors. 1. Rate coefficient determination, *Ozone Sci. Eng.*, 17, 355–378, 1995.
92. Beltrán, F.J., García-Araya, J.F., and Encinar, J.M., Henry and mass transfer coefficients in the ozonation of wastewaters, *Ozone Sci. Eng.*, 19, 281–296, 1997.
93. Sotelo, J.L. et al., Henry's law for the ozone water system, *Water Res.*, 1239–1246, 1989.
94. Gurol, M.D. and Nekouinaini, S., Effect of organic substances on mass transfer in bubble aeration, *J. Water Pollut. Control Fed.*, 57, 235–240, 1985.

7 Kinetics of Indirect Reactions of Ozone in Water

At pH lower than 12, indirect ozone reactions develop in the slow kinetic regime of ozone absorption. They are characterized by the presence of dissolved ozone and reaction factors and Hatta numbers lower than or close to unity and 0.3, respectively. These reactions are typical of drinking water ozonation where the concentrations of pollutants are very low (as high as part per million level but usually in the part per billion level). As has been shown previously, some wastewater ozonation can develop in this kinetic regime — specifically, wastewater with low COD level (<200 mg/L). As presented in Section 7.1, in the slow kinetic regime the two kinds of ozone action — direct and indirect reactions (the latter through free radicals) — can compete to remove any compound B present in the water. Indirect reactions result from the ozone decomposition mechanism that can be initiated by the reaction of ozone with the hydroxyl ion, which constituted the first and limiting step of the ozone mechanism leading to hydroxyl radicals (see Section 2.5.1). Indirect reactions or reactions due to hydroxyl radicals can be favored through some other initiation reactions of ozone decomposition (i.e., reactions of ozone with hydrogen peroxide, direct ozone photolysis, or some catalytic-induced reaction) that constitute the so-called ozone-involved advanced oxidation processes (AOPs), as shown in the following chapters. In this section, as a first approximation to the AOPs, ozonation is considered to be the ozone process carried out in the absence of initiators such as hydrogen peroxide or UV radiation or solid catalysts. Also note that at pH < 12 the ozone decomposition reaction is slow, so even if the direct reactions are fast, ozone decomposition will not take place.

In the slow kinetic regime, since ozone can react directly with the compounds present in water or through free radicals, it is convenient to establish some guidelines in order to determine which of these reactions predominates. This is useful for kinetic study and modeling purposes because the equations used (the mass-balance equations) can be simplified in their ozone absorption rate term. We begin with a comparative study about the relative importance of the direct reactions of ozone and its decomposition reaction in water.

7.1 RELATIVE IMPORTANCE OF THE DIRECT OZONE–COMPOUND B REACTION AND THE OZONE DECOMPOSITION REACTION*

In Section 5.2 and Section 5.3, the kinetic regimes of the ozone decomposition reaction and any ozone–compound B direct reaction were discussed together with the potential concentration profiles that ozone and B could have in the water phase. It was seen that the pH value was a crucial parameter for the kinetic regime of the ozone decomposition reaction. Thus, for pH < 12, this reaction is slow and it develops in the bulk water. For the ozone–direct reactions, on the contrary, other parameters such as the reaction rate constant and the concentration of the target compound B can also be fundamental for establishing the kinetic regime. Overall, however, when comparing the decomposition and some direct ozone reaction (when B is a dissociating compound), pH is also fundamental because it affects the constant rate value of the direct reaction. Thus, significant variations in the second-order rate constant of the reaction between ozone and compound B, k_D, lead to drastic changes in the kinetic regime of direct ozonation that can go from instantaneous to even slow. It is evident from these comments that for instantaneous, fast, and even moderate direct reactions, if ozone is consumed in the film layer, the ozone decomposition reaction can be neglected because of the absence of ozone in the bulk water to decompose into free radicals. The absence of dissolved ozone during fast direct reactions is, then, the main proof that confirms the lack of competition. If there is no dissolved ozone in bulk water, there will be no ozone decomposition reaction. Conversely, for pH > 12, the ozone decomposition reaction could be a moderate or even fast reaction and this reaction will compete with the fast direct reactions or it will be the only ozone-consuming reaction if the direct reactions are slow. However, for pH < 12, if dissolved ozone is detected, the ozone decomposition reaction could be the predominant reaction among other possible direct reactions — a situation usually encountered in drinking water ozonation. Competition can be confirmed by calculating the Hatta numbers of the ozone–B direct reaction, by knowing the pH of the water, or by checking the presence of dissolved ozone.

7.1.1 Application of Diffusion and Reaction Time Concepts

The ozone direct reaction and the ozone decomposition reaction can also be compared using the diffusion and reaction time concepts, t_D and t_R, defined in Section 4.2.4. The use of these parameters is based on the surface renewal theories[1] (i.e., Danckwerts theory). Note that for a given ozonation contactor and hydrodynamic conditions, only t_R depends on the chemical reaction rate of the ozone reactions. Thus, when comparing the ozone direct reaction and the ozone decomposition reaction, t_D is constant for both reactions.

Two situations are discussed, based on the relative values of t_D and t_R for each of the reactions considered. These situations correspond to fast and slow kinetic

---

* Part of Section 7.1 is printed with permission from Beltrán, F.J., Theoretical aspects of the kinetics of competitive ozone reactions in water, *Ozone Sci. Eng.*, 17, 163–181, 1995. Copyright 1995, International Ozone Association.

Kinetics of Indirect Reactions of Ozone in Water

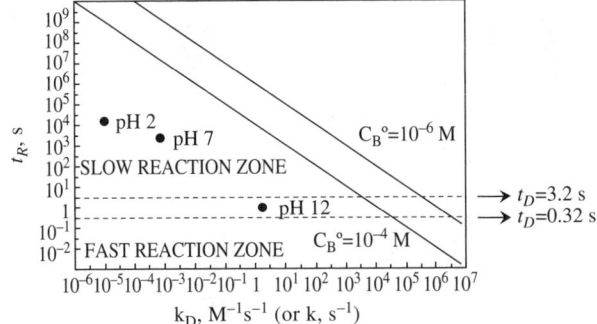

FIGURE 7.1 Variation of reaction time of an ozone gas liquid reaction with direct rate constant. Black circles correspond to the ozone decomposition reaction at different pH levels. (From Beltrán, F.J., Theoretical aspects of the kinetics of competitive ozone reactions in water, *Ozone Sci. Eng.*, 17, 163–181, 1995. Copyright 1995, International Ozone Association. With permission.)

regimes (see Section 5.2 and Section 5.3). As shown in Section 5.2 for the case of the ozone decomposition reaction, a plot of t_R determined from the rate constant of the reactions considered and the concentration of B as a parameter can be prepared. This will allow us to compare the relative importance of the direct and decomposition reactions of ozone.[2] Thus, Figure 7.1, taken from a previous work,[2] shows the conditions at which these reactions develop in the slow or fast kinetic regimes. Figure 7.1 shows two values of t_D that correspond to typical values of the individual mass-transfer coefficient k_L.[3] According to Figure 7.1, the ozone decomposition reaction will compete with any possible ozone–B direct reaction when both reactions simultaneously develop in the slow or fast reaction zones defined according to experimental conditions. For example, for $t_D = 3.2$ s and a concentration of B of 10^{-6} M, both reactions will compete if pH < 12 and k_D is about 5×10^5 $M^{-1}s^{-1}$ or when pH > 11 and $k_D > 5 \times 10^5$ $M^{-1}s^{-1}$.

In another example, taken from Beltrán,[2] a similar plot can be prepared, but plotting, in this case, t_R against the pH. This means of comparison could be useful for the case of the ozonation of dissociating compounds such as phenols where the apparent rate constant, k_D, varies with pH [see Equation (3.22) in Section 3.1]. In Figure 7.2, this plot has been prepared[2] for the ozonation of o-chlorophenol (OCP) and atrazine (ATZ), two compounds of very different reactivity toward ozone. Thus, for $t_D = 3.2$ s, the ozone–ATZ reaction would compete with the ozone decomposition reaction at any pH values except at pH > 11. At these latter conditions, only the decomposition of ozone will take place. In contrast, the reaction between ozone and OCP is the only one to develop at pH between 3 and 11. Thus, the reaction between the hydroxyl radical and OCP does not need to be considered in the corresponding kinetic study. Note that in practical cases, the removal rate of B is the main objective. Thus, the reaction rate terms present in the mass balance of B correspond to the ozone–B direct reaction and the hydroxyl radical–B reaction. However, in order to decide if both reaction rate terms have to be considered, since the hydroxyl radical–B reaction depends on the development of the ozone decomposition reaction, the

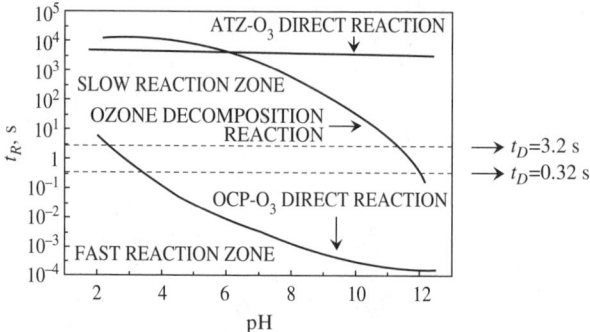

FIGURE 7.2 Reaction time of ozone decomposition and direct reactions of ozone with *o*-chlorophenol (OCP) and atrazine (ATZ) at different pH levels.(From Beltrán, F.J., Theoretical aspects of the kinetics of competitive ozone reactions in water, *Ozone Sci. Eng.*, 17, 163–181, 1995. Copyright 1995, International Ozone Association. With permission.)

comparison between the latter reaction and the ozone–B reaction must be established. Note also that when both the hydroxyl radical–B and ozone–B direct reactions compete, the importance of one of them could be negligible and, then, the corresponding reaction rate term is also eliminated from the kinetic equation. This is the case for the direct reaction of ozone–ATZ when pH > 7. In this case, although the direct reaction also develops (see Figure 7.2), its contribution to the removal of ATZ can be neglected compared to the effect of the hydroxyl radical reaction (see Section 7.2). Therefore, in the kinetic study, the reaction rate term due to the ATZ–ozone reaction can be neglected.

7.2 RELATIVE RATES OF THE OXIDATION OF A GIVEN COMPOUND*

A quantitative method to determine the relative importance of the direct ozonation and free radical oxidation of any given compound B during ozonation can be made by determining the ratio between both oxidation rates. The procedure is applied to the cases where ozone reactions develop in the slow kinetic regime, that is, the Hatta number of all ozone reactions is lower than 0.3 or the reaction time is much higher than the diffusion time. Whatever the ozone kinetic regime, the ratio between the oxidation rates of *B* due to free radical oxidation and direct reaction with ozone is:

$$\frac{r_R}{r_D} = \frac{k_{HO}C_{HO}}{zk_D C_{O3}} \tag{7.1}$$

The concentration of hydroxyl radicals C_{HO} in Equation (7.1) is given by Equation (7.2):

---

* Part of Section 7.2 is printed with permission from Beltrán, F.J., Estimation of the relative importance of free radical oxidation and direct ozonation/UV radiation rates of micropollutants in water, *Ozone Sci. Eng.*, 21, 207–228, 1999. Copyright 1999, International Ozone Association.

Kinetics of Indirect Reactions of Ozone in Water

$$C_{HO} = \frac{2k_{i2}C_{HO_2^-}C_{O3}}{\sum k_s C_s} \qquad (7.2)$$

where $2k_{i2}C_{HO_2^-}C_{O3}$ represents the reaction rate of initiation of free radicals which, in the case of ozonation, is a function of the concentrations of the ionic form of hydrogen peroxide [generated through Reaction (2.18) in Table 2.4] and ozone. By substituting in Equation (7.1), the ratio of oxidation rates attained is:

$$\frac{r_R}{r_D} = \frac{k_{HO} 2k_{i2} C_{HO_2^-}}{zk_D \sum k_s C_s} \qquad (7.3)$$

The problem with Equation (7.3) is that the concentration of hydrogen peroxide is unknown (note that hydrogen peroxide is not added but generated). However, the initiation rate term can be substituted, for practical purposes, by the rate of the reaction between ozone and the hydroxyl ion [Reaction (2.1) or Reaction (2.18)], which constitutes the first reaction in the ozone decomposition mechanism. In this method, the concentration of hydrogen peroxide is not needed. In fact, the ozone–hydroxyl ion reaction has long been considered the initiation rate of the ozone decomposition mechanism for yielding the superoxide ion and the hydroperoxide radicals [also Reaction (2.1)]:

$$O_3 + OH^- \xrightarrow{k_{i1}} HO_2 \bullet + O_2^- \bullet \qquad (7.4)$$

Thus, if Reaction (7.4) is considered the initiation reaction, the ratio between the oxidation rates in Equation (7.1) becomes a function of pH, rate constants, and inhibiting character of the water, $\sum k_s C_s$, which can be calculated as shown later (see also Section 7.3.1.1):

$$\frac{r_R}{r_D} = \frac{k_{HO} 2k_{i1} C_{OH^-}}{zk_D \sum k_s C_s} \qquad (7.5)$$

Equation (7.5) in logarithmic form is:

$$\log \frac{r_R}{r_D} = \log \frac{k_{HO}}{zk_D} + \log \frac{2k_{i1} C_{OH^-}}{\sum k_s C_s} \qquad (7.6)$$

Following Equation (7.6), a plot of the left side against the logarithm of the rate constant ratio k_{HO}/k_D leads to a straight line of slope unity. For any compound B of known kinetics with ozone and hydroxyl radical (that is, known values of z, k_D, and k_{HO}), the relative importance of the direct ozonation and free radical oxidation rates can be estimated at different pH and inhibiting character of the water used. In Figure 7.3, this plot is presented for different pH values and at a given hydroxyl radical

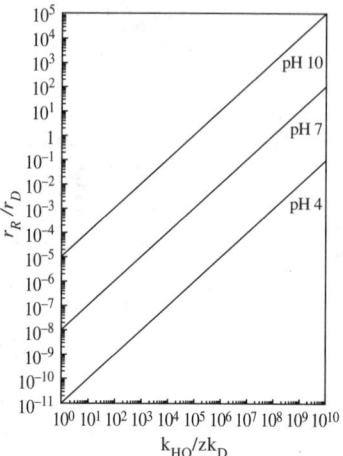

FIGURE 7.3 Comparison between hydroxyl radical and direct ozonation rates of micropollutants in water as a function of reaction rate constant ratio and different pH values in single ozonation. Conditions: 20°C, $\Sigma k_{HOSi} C_{Si} = 10^3$ sec^{-1}. (From Beltrán, F.J., Estimation of the relative importance of free radical oxidation and direct ozonation/UV radiation rates of micropollutants in water, *Ozone Sci. Eng.*, 21, 207–228, 1999. Copyright 1999, International Ozone Association. With permission.)

inhibiting value $\Sigma k_s C_s$. Examples for using Figure 7.3 are straightforward. More details are given on this procedure in an earlier work.[4]

7.3 KINETIC PARAMETERS

In the ozonation process of a given pollutant B, when the ozone reactions are in the slow kinetic regime of absorption, the mass-balance equation of B applied to a small volume of reaction (which is perfectly mixed) in a semibatch system is as follows:

$$-\frac{dC_B}{dt} = zk_D C_B C_{O3} + k_{HOB} C_B C_{HO} \quad (7.7)$$

where the terms $zk_D C_B C_{O3}$ and $k_{HOB} C_{HO} C_B$ represent the contributions of the direct and hydroxyl radical reactions, respectively, to the disappearance of B. In addition, the mass balance of ozone in the water phase at the same conditions is

$$\frac{dC_{O3}}{dt} = k_L a \left(C_{O3}^* - C_{O3} \right) - r_{O3} \quad (7.8)$$

where the ozone decomposition rate r_{O3} has different contribution values because of the ozone reactions with target compound B, the hydroxyl ion, hydroperoxide ion, and superoxide ion and hydroxyl radicals (see mechanism in Table 2.4 or Table 2.5):

Kinetics of Indirect Reactions of Ozone in Water

$$-r_{O3} = k_D C_B C_{O3} + k_{ap} C_{O3} = k_D C_B C_{O3} + C_{O3}\left(k_{i1} C_{OH^-} + k_{i2} C_{HO_2^-} + k_2 C_{O_2^-} + k_6 C_{HO}\right) \quad (7.9)$$

Note that because of the slow kinetic regime, the ozonation gas–liquid reaction is a two-sequential-step series process, where the mass-transfer rate through the film layer is equal to the ozone chemical reaction rate in the bulk water at steady state. Comparing the fast ozonation processes, from Equation (7.7) to Equation (7.9) it is evident that some new unknown parameters appear. These are the rate constant of the reaction between the hydroxyl radical and B, k_{HOB}, the rate constant of the decomposition reaction, k_d, and the concentration of hydroxyl radicals.

Ozone is mainly consumed through reactions with the hydroxyl ion, hydroperoxide ion, hydroxyl radical (ozone acts as promoter of its own decomposition), the superoxide ion radical, and through the direct reaction with B. Rate constants of all these reactions are known from the literature or can be calculated as was shown for the case of the rate constant of the direct reactions (see also Section 3.1 and Section 5.3).[5–7] However, ozone is also consumed through other reactions that can have significant importance such as the initiating reactions, which are different from Reaction (2.1) or Reaction (2.18) (see Reactions in Table 2.4 and Table 2.5). Thus, the rate constants of these reactions must also be known. In addition, because the concentration of hydroxyl radicals is a function of the rate of inhibiting reactions [the reaction between the hydroxyl radical and some scavenger species, denominator of Equation (7.2)], the rate constants of these reactions are also needed. The kinetic study of the ozone reactions in the slow kinetic regime will be addressed to determine all these parameters.

7.3.1 THE OZONE DECOMPOSITION RATE CONSTANT

We can use the general Equation (7.9), used by Staehelin and Hoigné[8] in the mechanism of reactions given in Table 2.4, to determine the apparent pseudo first-order rate constant of the ozone decomposition, k_{app}. Classic methods of homogeneous kinetics can be applied (see Section 3.1). Rate constants of ozone reactions (with OH^-, HO_2^-, $HO\bullet$, and $O_2^-\bullet$) are common to any ozonation process and their values are already known (see Table 2.4 or Table 2.5). However, others such as those for Reaction (7.10) and Reaction (7.11) below are unknown.

$$O_3 + I \xrightarrow{k_{i3}} O_3^- \bullet + I^+ \quad (7.10)$$

$$HO\bullet + S \xrightarrow{k_{HOS}} Products \quad (7.11)$$

Thus, k_{OHS} and k_{i3} are system dependent and have to be determined for each case. In fact, reactions of ozone with initiating compounds [Reaction (7.10)] and those of the hydroxyl radical with inhibiting compounds or scavengers [Reaction (7.11)] will depend on the nature of the water treated. Since in a real case the exact content of the water is not known, a general procedure should be applied to determine these rate constants, as presented below.

Reaction (7.10) and Reaction (7.11) develop in surface waters where numerous substances may play the role of initiators and inhibiting species of the ozone decomposition reaction. However, as experimental results suggest, these reactions are also present during the ozonation of laboratory-prepared waters. For example, in a study on ozone decomposition with phosphate-buffered distilled water,[2] the apparent rate constant of the ozone decomposition was found to be 8.3×10^{-5} and 4.8×10^{-4} sec^{-1} at pH 2 and 7, respectively. At the same conditions, however, the rate constant of the first reaction of the mechanism [Reaction (2.1) or Reaction (7.4)] is 7×10^{-11} and 7×10^{-6} sec^{-1}, respectively. The large differences in the values shown (for each pH) was due not to the other known reactions that initiate and propagate the mechanism but rather to the presence of different substances. In fact, these substances are responsible for the differences observed in the apparent rate constant values of the ozone decomposition reaction when studied in different types of water.[8]

Due to the unknown nature of the initiating and inhibiting species present in water, the true values of k_{i3} and k_{HOS} cannot be known, but the values of their products with the concentrations of these species can be expressed. For the sake of simplicity, the concentrations of these substances are assumed to be constant in the procedure that follows.

From the basic mechanism of ozone decomposition (see Table 2.4 or Table 2.5) by applying the pseudo steady-state conditions, the concentrations of hydroxyl and superoxide ion radicals can be expressed as follows:

$$C_{HO} = \frac{(2k_{i1}10^{pH-14} + k_{i3})C_{O3}}{k_t} \quad (7.12)$$

and

$$C_{O_2^{\cdot-}} = \frac{2k_{i1}10^{pH-14} + k_6 C_{HO}}{k_3} \quad (7.13)$$

where

$$k_t = \sum k_{HOS} C_s \quad (7.14)$$

which when substituted in the ozone chemical rate Equation (7.9) lead to:

$$r_{O3} = (3k_{i1}10^{pH-14} + k_{i3})C_{O3} + 2k_6 \frac{2k_{i1}10^{pH-14} + k_{i3}}{k_t} C_{O3}^2 = k_A C_{O3} + k_B C_{O3}^2 \quad (7.15)$$

In a homogeneous perfectly mixed batch reactor, the mass balance of ozone in water is given by Equation (7.8) with the absorption rate term removed and the ozone decomposition rate term given by Equation (7.15). The experimental concentrations of ozone at any time can then be fitted to Equation (7.15) to obtain the values of the rate constants k_A and k_B and, hence, the values of k_{i3} and k_t. With these values, the initiating and inhibiting character of the water for the ozone decomposition can

Kinetics of Indirect Reactions of Ozone in Water

be established. Note that k_t involves all possible contributions of inhibiting substances.

7.3.1.1 Influence of Alkalinity

As observed previously, the concentration of hydroxyl radicals will depend greatly on the inhibiting character of the water treated (k_t). In many cases, carbonates are used as scavenger substances of hydroxyl radicals in ozonation studies[9,10] to check the importance of the free radical oxidation (indirect way of ozone action). In fact, these substances are used because in the case of natural (surface or ground) water, they are the main natural scavengers.[8] The contributing term of these substances to the inhibiting character of the ozonated water is due to the following reactions[11] (see also Table 2.5):

$$HCO_3^- + HO\bullet \xrightarrow{k_{c1}=8.5 \times 10^6 M^{-1}s^{-1}} CO_3^-\bullet + H_2O \qquad (7.16)$$

$$CO_3^= + HO\bullet \xrightarrow{k_{c2}=3.9 \times 10^8 M^{-1}s^{-1}} CO_3^-\bullet + OH^- \qquad (7.17)$$

The rate constants of these reactions are not very high when compared to other hydroxyl radical reactions with organic pollutants.[12] However, since the rate of reaction is proportional to both the rate constant and concentration of reactants, the carbonate–bicarbonate inhibiting effect is usually high because there is a concentration of these ions in natural waters. Thus, the k_t term for carbonate–bicarbonate ions is a function of pH and can be determined as follows:

$$k_t = k_{c1} C_{HCO3} + k_{c2} C_{CO3} = k_c C_{HCO3t} \qquad (7.18)$$

where C_{HCO3t} represents the total concentration of bicarbonates in water, with

$$k_c C_{HCO3t} = \left(k_{c1} + k_{c2} 10^{pH-pK_2}\right) \frac{10^{pH-pK_1}}{1 + 10^{pH-pK_1} + 10^{2pH-pK_1-pK_2}} C_{HCO3t} \qquad (7.19)$$

and pK_1 and pK_2 the pK of equilibrium of carbonates in water. Thus, at neutral pH and 20°C, k_t is 1233 sec^{-1}, which corresponds to an alkalinity of 10^{-4} M in total carbonates. This value is of the same order of magnitude as that from a given inhibiting pollutant at a concentration of 10^{-6} M whose reaction with the hydroxyl radical has a rate constant value of 10^9 M^{-1}sec^{-1}. Strictly speaking, however, the inhibiting term due to the alkalinity of water is not exactly that given by Equation (7.19). In fact, the carbonate ion radical, $C_{O3}\bullet^-$, generated in Reaction (7.16) and Reaction (7.17), reacts with hydrogen peroxide to regenerate the hydroperoxide radical or the superoxide ion radical:[13]

$$C_{O3}^-\bullet + H_2O_2 \xrightarrow{k_{CH1}=4.3 \times 10^5 M^{-1}s^{-1}} HO_2\bullet + HCO_3^- \qquad (7.20)$$

and

$$C_{O3}^{-}\bullet + HO_2^{-} \xrightarrow{k_{CH2}=5.6\times 10^7 M^{-1}s^{-1}} HO_2\bullet + CO_3^{=} \quad (7.21)$$

that in the presence of ozone eventually yields the hydroxyl radical (see Table 2.4). According to this, the carbonate–bicarbonate ions would not be an absolute inhibiting species of the ozone decomposition in ozonation processes where hydrogen peroxide is formed. In addition, the carbonate ion radical also reacts with the organic matter present in water through selective reactions (similar to the case of the direct ozone reactions) and, in this way, terminates the radical chain.[14–16] A compilation of rate constant values of the reactions between the carbonate ion radical and different substances can be found elsewhere.[17] From the above observation, it can be assumed that there is a fraction of carbonate-bicarbonate ions that, while reacting with the hydroxyl radical [Reaction (7.16) and Reaction (7.17)], eventually regenerates it through Reaction (7.20) and Reaction (7.21). Thus, the fraction of carbonate ion radicals that reacts with hydrogen peroxide as compared to other reactions is:

$$w = \frac{k_{CH}C_{H2O2t}}{k_{CH}C_{H2O2t} + \sum k_{CM}C_M} \quad (7.22)$$

where

$$k_{CH}C_{H2O2t} = \left(k_{CH1} + k_{CH2}10^{pH-pK}\right)\frac{C_{H2O2t}}{1+10^{pH-pK}} \quad (7.23)$$

where C_{H2O2t} and pK are the total concentration of hydrogen peroxide and pK value of its equilibrium in water, and k_{CM} is the rate constant value of any reaction between a given compound M present in water and the carbonate ion radical that terminates the radical chain.

7.3.2 Determination of the Rate Constant of the OH–Compound B Reaction

The contribution of free radical reactions to the oxidation rate of pollutants (B) in water during ozonation can be established if both the rate constant k_{OHB} and the concentration of the hydroxyl radical are known. For the latter, in the absence of B, the appropriate expression is given in Equation (7.12). In the presence of B, depending on the nature of the role of this substance in the ozone reaction mechanism, the concentration of the hydroxyl radical will also depend on k_{HOB} and C_B (in the case of B as inhibitor of ozone decomposition). The term $k_{HOB}C_B$ will be part of the inhibiting character of the water given by $\Sigma k_{HOS}C_S$. Thus, the rate constant k_{HOB} is a crucial parameter to know. Reactions of hydroxyl radicals are usually defined as nonselective, which could mean that the rate constant k_{HOB} is always similar regardless of the nature of B, although this is not correct because k_{HOB} can vary up to three orders of magnitude. For example, for an organochlorine compound such as trichloroethane, k_{HOB} is $2 \times 10^7 \, M^{-1} \text{sec}^{-1}$ [18] while for phenol it is about $10^{10} \, M^{-1} \text{sec}^{-1}$.[12] Thus, k_{HOB} must also be determined.

The best way to determine k_{HOB} is from data of the disappearance rate of B. In an ozonation system, the chemical disappearance rate of B is theoretically due to the reaction with ozone (direct reaction) and with the hydroxyl radical. In a semibatch or batch well-agitated reactor, the accumulation rate of B is

$$-\frac{dC_B}{dt} = zk_D C_{O3} C_B + k_{HOB} C_{HO} C_B \tag{7.24}$$

The system is simplified if the direct reaction can be neglected, a situation that is likely to be present when the ozonation develops in the slow kinetic regime. Then the disappearance rate of B will be a function of the concentration of hydroxyl radicals, C_{HO}, which depends on the initiating and inhibiting character of the water system [see Equation (7.12)]. It is evident from this information that the main difficulty in determining k_{HOB} is the unknown concentration of hydroxyl radicals. Two methods can be followed: the absolute and the competitive.

7.3.2.1 The Absolute Method

The absolute method leads to the direct determination of k_{HOB}. In fact, in a semibatch well-mixed ozonation system, by assuming the concentration of hydroxyl radical constant (as would correspond to a short live species), the integration of Equation (7.24), once the direct rate term has been neglected and variables have been separated, yields:

$$\ln \frac{C_B}{C_{B0}} = k_{HOB} C_{HO} t \tag{7.25}$$

A plot of the left side of this equation against time should give a straight line of slope $k_{HOB} C_{HO}$. Then a value of C_{HO} is needed to find k_{HOB}. According to Equation (7.12), the exact nature of the water used and the role of the substances (as initiators and/or inhibitors) present must be known. This is rather difficult because ozone decomposition is very sensitive to the action of substances present even at very low concentrations. However, using a procedure similar to that shown previously, values of k_{i3} and k_t that would correspond to the ozone–water system treated could be determined in the absence of B, and, consequently, the concentration of hydroxyl radicals (see also Section 7.4). Two possible situations arise depending on the inhibiting or promoting nature of B. If B promotes the ozone decomposition reaction, that is, B reacts with the hydroxyl radical to give the superoxide ion radical that eventually regenerates the hydroxyl radical (see mechanism in Table 2.4), the rate constant k_{HOB} would be

$$k_{HOB} = \frac{mk_t}{r_i} \tag{7.26}$$

where m is the slope of the straight line mentioned previously ($k_{HOB}C_{HO}$) and r_i is the initiation rate of free radicals, which is given by the numerator of Equation (7.12). On the other hand, if B inhibits the ozone decomposition, that is, if the reaction of B with the hydroxyl radical terminates the radical chain, the product $k_{HOB}C_B$ would be part of the denominator of Equation (7.27):

$$C_{HO} = \frac{\left(2k_{i1}10^{pH-14} + k_{i3}\right)C_{O3}}{k_{HOB}C_B + k_t} \quad (7.27)$$

Then, the rate constant k_{HOB} is

$$k_{HOB} = \frac{mk_t}{r_i - mC_B} \quad (7.28)$$

In addition, when $k_{HOB}C_B \gg k_t$, a zero-order kinetics would develop so that the rate of B accumulation would be constant and nondependent on C_B. In this case, the rate constant could not be obtained from the absolute method. Because of these limitations — unknown values of k_{i3}, k_t, etc. — the absolute method is more suitable when other AOPs are used, as will be shown in the following chapters.

7.3.2.2 The Competitive Method

This method is similar to that shown in Section 5.3.4 for the determination of the rate constant of the direct reactions. In this case, the limitation is that the contributions of the direct reactions ozone–B and ozone–R (for reference compound, see Section 5.3.4) have to be negligible. Fortunately, in slow kinetic regimes this is likely to be the case. Then, from ozonation results in a semibatch or batch well-agitated reactor, the ratio of accumulation rates of B and R is the ratio of their chemical reaction rates with hydroxyl radicals:

$$\frac{dC_B}{dC_R} = \frac{k_{HOB}C_B}{k_{HOR}C_R} \quad (7.29)$$

The resulting equation does not depend on C_{HO}, and after integration, the ratio of rate constants k_{HOB}/k_{HOR} is obtained from the slope of a plot of $\ln(C_B/C_{B0})$ vs. $\ln(C_R/C_{R0})$. Since k_{HOR} is known, k_{HOB} is finally determined. Also note that in this case the ozonation kinetics of both B and R must not be of zero order. If so, a situation similar to that mentioned above for the absolute method would be present and it would not be possible to determine k_{HOB}. Another possible limitation of this method is that the reference compound needs to have a reactivity toward the hydroxyl radical similar to that of the target compound B. Haag and Yao[19] used this competitive method with batch ozone solutions at pH > 8 to determine k_{HOB} of numerous reactions between the hydroxyl radical and compounds.[19]

7.4 CHARACTERIZATION OF NATURAL WATERS REGARDING OZONE REACTIVITY

Natural water from lakes, rivers, reservoirs, etc. is the source for the preparation of drinking water. Although it has much lower pollution than wastewater, natural water also contains numerous and different compounds, most of them of the organic type, defined as natural organic matter, NOM. This matter may contain a large variety of substances from vegetable plant degradation and animal wastes, and pollutants from agricultural, industrial, and urban activities, which are often grouped into macromolecules that constitute humic substances. Humic substances can also contain metal linked to them as complexes or as simple molecules such as pesticides, etc.[20] Because NOM constitutes the major fraction of the matter present in surface waters, the parameter usually employed to characterize the water is dissolved organic carbon, DOC (see Section 6.3.3).

7.4.1 Dissolved Organic Carbon, pH, and Alkalinity

DOC has also been used to establish the reactivity of natural water with ozone[21] although this relationship has not yet been well established. Thus, different ozone reactivities have been observed in natural water with the same DOC.[21] Another parameter frequently used to characterize ozone reactivity of natural water is the UV absorbance or specific UV absorbance (usually at 254 nm).[22] In this sense, Westerhoff et al.[23] correlated the ozone reactivity, measured as the rate constants of the ozone decomposition and hydroxyl radical–DOC reactions, with the specific ultraviolet absorbance at 254 nm with good results. In this work,[23] other different parameters or, rather, properties of DOC of 17 different natural water samples were correlated with the corresponding rate constants of the ozone direct and indirect decomposition reactions. These were, among others, the aromatic, aliphatic, and carbonyl contents, molecular weight, etc. of hydrophobic organic acids, a fraction of humic substances that usually reaches 50% of the DOC content of the water. From the results obtained, it was found that A_{254} was a good parameter of characterization of ozone reactivity.[23]

Total organic carbon, pH, and alkalinity are also current parameters used to characterize natural water because they are easy to measure and are related to the variables affecting the overall ozone decomposition rate constant, as can be deduced from Equation (2.70).[24,25] Thus, if it is assumed that ozone direct reactions are negligible, which is the general situation in natural water due to the absence of high concentrations of simple organic chemicals, the ozone decomposition rate constant given by Equation (2.70) reduces to the following:

$$k_d = 3k_i C_{OH^-} \frac{\sum k_{Pi} C_{Pi}}{\sum k_{Si} C_{Si}} \qquad (7.30)$$

As can be deduced from Equation (7.30), the ozone reactivity will depend on the pH (C_{OH^-}) presence of promoters C_{Pi}, that is, TOC, and scavengers C_{Si} that can be represented by the alkalinity of the carbonate–bicarbonate content of the water, Alk. These parameters have been correlated to yield relationships as follows:

$$k_d = a\left(C_{OH^-}\right)^b (TOC)^c (Alk)^d \qquad (7.31)$$

or

$$\log k_d = \log a - b(pH) + c\log(TOC) + d\log(Alk) \qquad (7.32)$$

where TOC and alkalinity are measured in mg of C per liter and mg of $CaCO_3$ per liter, respectively. As examples, Equation (7.33) and Equation (7.34) below were used in previous research works[24,25] to relate the ozone decomposition rate constant to the parameters mentioned:

$$k_d = 2.908 \times 10^{-8}\, 10^{0.66\,pH}\, TOC^{0.61} (alk/10)^{-0.42},\ sec^{-1} \qquad (7.33)$$

at the conditions $7 < pH < 9$, $0.3 < TOC(mgL^{-1}) < 4.3$, $25 < alk(mgCaCO_3 L^{-1}) < 150$ and

$$k_d = 3.26 \times 10^{-8}\, 10^{0.24\,pH}\, TOC^{1.08}\, alk^{-0.19}\, A^{0.7537},\ sec^{-1} \qquad (7.34)$$

From a strictly kinetic point of view, however, characterization of natural waters can also be accomplished by defining the rate constant or parameters derived from the application of mass balances of ozone and hydroxyl radicals in water. The procedure shown in Section 7.3.1 is along this line. Thus, rate constants k_A and k_B experimentally determined from ozone decomposition in natural waters can also be taken as characterization parameters of the water. In fact, rate constants k_A and k_B give a measure of the effects of pH and presence of initiating compounds, on the one hand, and the effect of promoting and inhibiting compounds, on the other hand.

7.4.2 THE OXIDATION–COMPETITION VALUE

As pointed out above, Hoigné and Bader[26] defined the oxidation–competition values of natural water as the amount of ozone needed to reach a 63% conversion of a given pollutant, probe, or reference compound in the natural or raw water. These parameters are then dependent on the nature of the reference compound of known indirect kinetics with ozone, which means that the rate constant of its reaction with the hydroxyl radical must be known. In addition, this reference compound must have a negligible direct reaction with ozone to avoid interference in the kinetic procedure. The procedure to determine the oxidation–competition value of the water treated, Ω_B, is in fact a competitive method where the probe compound and the matter present in water compete for the available hydroxyl radicals that come from the ozone decomposition. The procedure is based on the assumption that the ozone reacting system in the natural water is exclusively due to indirect reactions so that the following reaction steps are considered:[21,26]

- Decomposition of ozone in hydroxyl radicals:

$$O_3 \longrightarrow z_r HO\bullet + \ldots \tag{7.35}$$

with z_r as the stoichiometric coefficient

- Reaction of hydroxyl radicals with the reference or probe compound B:

$$HO\bullet + B \xrightarrow{k_{HOB}} P \tag{7.36}$$

- Reactions of any other natural substances with hydroxyl radicals:

$$HO\bullet + S \xrightarrow{k_{HOS}} P \tag{7.37}$$

The procedure is also based on the following conditions:

- The direct reaction of ozone with B is negligible.
- The contribution of B in reactions with hydroxyl radicals is also negligible compared to the total consumption of hydroxyl radicals in Reaction (7.37), that is:

$$k_{HOB}C_B \ll \sum_i k_{HOS_i} C_{S_i} \tag{7.38}$$

With these conditions, the reactivity of the natural water with ozone due to indirect reactions (which is usually the most common situation in natural water) is simulated. Determination of the oxidation–competition value comes from the application of mass balances of ozone, hydroxyl radical, and B and stoichiometric rules. Thus, for a perfectly mixed batch reactor, these equations are (see also Appendix A1):

- For ozone:

$$r_{O3} = \frac{dC_{O3}}{dt} \tag{7.39}$$

where r_{O3} is intrinsically negative when expressed as a function of the rate constant and concentrations due to the negative value of the ozone stoichiometric coefficient in Reaction (7.35), which is -1.[27]

- For B:

$$r_B = \frac{dC_B}{dt} = -k_{HOB} C_{HO} C_B \quad (7.40)$$

where the minus sign is also due to the negative stoichiometric coefficient of B in Reaction (7.36), which is also −1.

- For the hydroxyl radicals:

$$r_{HO\bullet} = \frac{dC_{HO\bullet}}{dt} = r_f - C_{HO} \sum_i k_{HOS_i} C_{S_i} \quad (7.41)$$

with the two right terms of Equation (7.41) representing the formation and decomposition rates of hydroxyl radicals, respectively, the latter including the contribution of Reaction (7.36). At steady-state conditions, the accumulation rate term of hydroxyl radicals is zero, that is, $dC_{HO}/dt = 0$. Thus, the rate of hydroxyl radical formation is, in fact, the rate of initiation of the radical chain [Reaction (7.35)] which can be expressed as the rate of ozone decomposition due to the absence of direct reactions and once the stoichiometric coefficients are accounted for:

$$r_f = r_i = -z_r r_{O3} \quad (7.42)$$

From Equation (7.42) and Equation (7.39), the concentration of hydroxyl radicals in Equation (7.41) is:

$$C_{HO} = \frac{-z_r \dfrac{dC_{O3}}{dt}}{\sum_i k_{HOS_i} C_{S_i}} \quad (7.43)$$

which, when substituted in Equation (7.40) and after variable separation and simplification, leads to:

$$\frac{dC_B}{C_B} = \frac{z_r k_{HOB}}{\sum_i k_{HOS_i} C_{S_i}} dC_{O3} \quad (7.44)$$

where the oxidation–competition value is:[21,26]

$$\Omega_B = \frac{\sum_i k_{HOS_i} C_{S_i}}{z_r k_{HOB}} \quad (7.45)$$

Kinetics of Indirect Reactions of Ozone in Water

The Hoigné and Bader oxidation–competition value for natural water is in fact the fraction of hydroxyl radical consumed by the probe compound times the ratio between the stoichiometric coefficient of ozone–hydroxyl radical initiation reaction and the concentration of the probe compound:

$$\Omega_B = f \frac{C_B}{z_r} \quad (7.44)$$

where

$$f = \frac{k_{HOB} C_B}{\sum_i k_{HOS_i} C_{S_i}} \quad (7.45)$$

Once Ω_B, given by Equation (7.45), is substituted in Equation (7.44), the resulting differential equation can be integrated with the following initial condition:

$$C_{O3} = C_{O3_0} \quad C_B = C_{B_0} \quad (7.48)$$

Equation (7.49), deduced from this procedure, gives the relationship between the concentration profile of B and the amount of ozone consumed:[26]

$$\ln \frac{C_B}{C_{B_0}} = -\frac{C_{O3_0} - C_{O3}}{\Omega_B} \quad (7.49)$$

According to Equation (7.49), a plot of the left side against the ozone consumption at different times should yield a straight line of slope $-1/\Omega_B$. This kind of plot, presented in Figure 7.4 for an imaginary natural water and probe compound, has been reported by Hoigné and Bader for true natural waters.[21,26,27] However, preparation of this plot is not necessary, as observed by Hoigné and Bader,[21,26] because from Equation (7.49) it is deduced that at 63% conversion of B its left term is -1. This means that the amount of ozone consumed to reach a 63% conversion of B is the oxidation–competition value of the natural water. Values of this parameter for different natural waters have been reported in different works[21,26,28] and have shown good linearity with the UV absorbance of the natural waters.[26]

The oxidation–competition value can also be determined from ozone decomposition experiments in natural water in steady-state continuous plug flow and stirred tank reactors (PFR and CSTR).[28] In the former case, balance equations are the same as in the perfectly mixed batch reactor with the difference that the actual reaction time is replaced by the hydraulic or spatial time, τ:

$$\tau = \frac{V}{v} \quad (7.50)$$

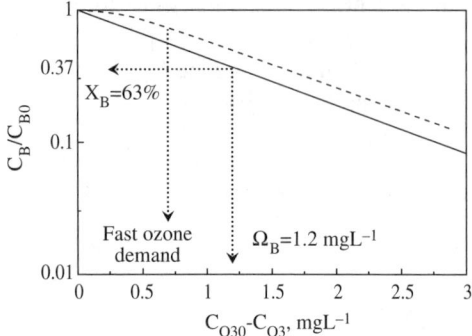

FIGURE 7.4 Checking Equation (7.49) for the determination of the Hoigné and Bader oxidation–competition value of an arbitrary water in a batch or plug flow reactor. The dotted line shows the typical case of a natural water with an initial period due to the fast ozone demand (x and y axes show arbitrary values).

The final equation, however, is also Equation (7.49) since it is not directly time-dependent. For a CSTR, the procedure differs, but the final equation is something similar, as shown by Equation (7.59) below.

Theoretical consideration of a CSTR as the ozone contactor is also a recommended option to compare the efficiency of ozonation processes in different reactor types (i.e., to compare the PFR and CSTR). This comparison can also be made as far as the Hoigné and Bader oxidation–competition value is concerned. For a CSTR, given the ozone-reacting system of Step (7.35) to Step (7.37), the mass-balance equations for ozone, B, and hydroxyl radicals at steady-state conditions are as follows:

- For ozone:

$$v\left(C_{O3_0} - C_{O3}\right) + Vr_{O3} = 0 \qquad (7.51)$$

- For the probe compound, B:

$$v\left(C_{B_0} - C_B\right) + Vr_B = 0 \qquad (7.52)$$

- For the hydroxyl radicals:

$$-vC_{HO} + Vr_{HO} = 0 \qquad (7.53)$$

where V is the reactor volume, v is the total volumetric flow rate, and C_{O3_0} and C_{B_0} are the concentrations of ozone and B, just at the reactor inlet. In Equation (7.53), r_{HO} represents the net formation rate of hydroxyl radicals, which is given by:

Kinetics of Indirect Reactions of Ozone in Water

$$r_{HO} = r_i - C_{HO} \sum_i k_{HOS_i} C_{S_i} \tag{7.54}$$

Now, if the stoichiometric ratio between ozone and hydroxyl radical in Equation (7.35) [see also Equation (7.42)] and the spatial time [Equation (7.50)] are considered, the concentration of hydroxyl radicals can be expressed in an explicit way from Equation (7.52) as follows:

$$C_{HO} = -\frac{\tau z_r r_{O_3}}{1 + \tau \sum_i k_{HOS_i} C_{S_i}} \tag{7.55}$$

where the ozone reaction rate, r_{O_3}, can be expressed as a function of the ozone consumption using the ozone mass-balance equation to yield:

$$C_{HO} = \frac{z_r}{1 + \tau \sum_i k_{HOS_i} C_{S_i}} \left(C_{O_{3_0}} - C_{O_3} \right) \tag{7.56}$$

Since the reaction rate of the oxidation of B is:

$$r_B = -k_{HOB} C_{HO} C_B \tag{7.57}$$

substitution of r_B in the B mass-balance Equation (7.52) yields

$$C_{HO} = \frac{C_{B_0} - C_B}{C_B} \frac{1}{\tau k_{HOB}} \tag{7.58}$$

If it is assumed that the term $\tau \sum k_{HOS_i} C_{S_i}$ is much greater than 1, elimination of the concentration of hydroxyl radicals from Equation (7.56) and Equation (7.58) once Equation (7.45) has been accounted for leads to

$$\frac{C_{B_0}}{C_B} - 1 = \frac{C_{O_{3_0}} - C_{O_3}}{\Omega_B} \tag{7.59}$$

According to Equation (7.59), a plot of its left side against the ozone consumption term $(C_{O_{3_0}} - C_{O_3})$ for different spatial times must lead to a straight line of slope equal to $1/\Omega_B$ (see Figure 7.5 for an arbitrary case). However, this type of plot is not necessary as deduced from Equation (7.59) because Ω_B is the amount of ozone consumed to reach 50% conversion of B. For purposes of comparison, note that for the plug flow or batch reactor, Equation (7.49) can be expressed in a form similar to Equation (7.59) since Equation (7.49) can be written as follows:

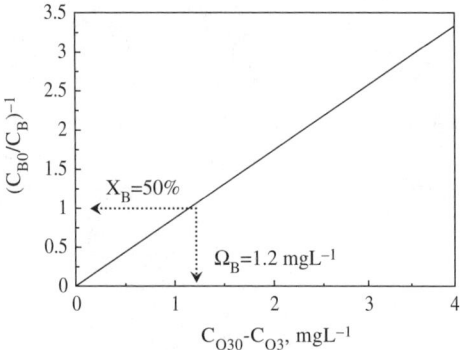

FIGURE 7.5 Checking Equation (7.59) for the determination of the Hoigné and Bader oxidation–competition value of an arbitrary water in a continuous-stirred tank reactor (x and y axes show arbitrary values).

$$\frac{C_{B_0}}{C_B} = \exp\left[\frac{C_{O3_0} - C_{O3}}{\Omega_B}\right] \qquad (7.60)$$

As deduced from the above observations, the consumption of ozone in the plug flow reactor (or batch reactor) corresponding to the oxidation–competition value is lower than that of the CSRT.

7.4.3 THE R_{CT} CONCEPT

Finally, another possible parameter to characterize ozone reactivity in any natural water with respect to is the R_{CT} value proposed by Elovitz and von Gunten.[29] This parameter was defined as the ratio between the time-integrated concentrations of hydroxyl radicals and ozone during an ozone decomposition experiment in a natural water in the presence of a probe compound B:[29]

$$R_{CT} = \frac{\int C_{HO} dt}{\int C_{O3} dt} \qquad (7.61)$$

For a batch or plug flow reactor, use of Equation (7.61) in the integrated Equation (7.40) allows the B concentration profile to be expressed as a function of the ozone-integrated concentration, which is known from experimental results as:[29]

$$\ln \frac{C_B}{C_{B_0}} = -k_{HOB} R_{CT} \int C_{O3} dt \qquad (7.62)$$

Figure 7.6 shows for an arbitrary case the straight line that Equation (7.62) represents. The slope of this line is $k_{HOB}R_{CT}$. Note that in this case Condition (7.38) and the

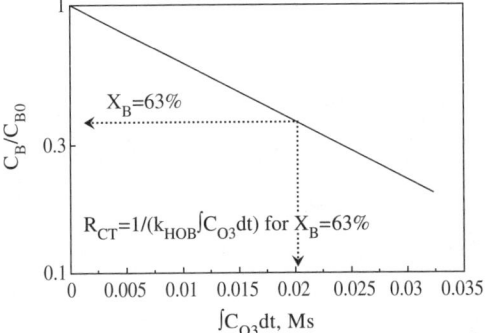

FIGURE 7.6 Checking Equation (7.62) for the determination of the Elovitz and von Gunten R_{CT} value for an arbitrary water in plug flow or batch reactors (x and y axes show arbitrary values).

negligible ozone–B direct reaction must hold. Elovitz and von Gunten[29] showed that the R_{CT} parameter was constant for batch ozone decomposition experiments regardless of the reaction time. They proposed R_{CT} as a parameter that directly relates the concentrations of the hydroxyl radical and ozone:

$$R_{CT} = \frac{C_{HO}}{C_{O3}} \qquad (7.63)$$

For a CSTR, from Equation (7.58) and Equation (7.63) the following is obtained:

$$\frac{C_{B_0}}{C_B} - 1 = k_{HOB} R_{CT} C_{O3} \tau \qquad (7.64)$$

and this represents the equation of a straight line in a plot of C_{B0}/C_B against $C_{O3}\tau$ (see also Figure 7.7 for an arbitrary case). Again, the plug flow reactor (or batch reactor) is better than the CSTR for determining the R_{CT} with the lowest amount of ozone, as can be deduced from Equation (7.62) and Equation (7.64). Values of R_{CT} have been reported by Elovitz and von Gunten for several natural and prepared water types.[29–32] Since the R_{CT} allows the concentration of hydroxyl radical (always an unknown concentration) to be removed from mass-balance equations, it is a useful parameter for kinetic modeling purposes, as has already been reported[31,32] (see also Chapter 11).

A final question should be considered when Ω_B and R_{CT} parameters are used to characterize natural waters: the direct ozone demand of the water due to the presence of fast-reacting compounds [e.g., nitrites[33]]. Thus, it is common practice to obtain plots similar to those presented in Figure 7.4 as examples (dotted lines) where two reaction periods are observed: the first one of about 60 to 120 sec corresponding to the ozone direct demand and the second one corresponding the indirect ozone reactions.

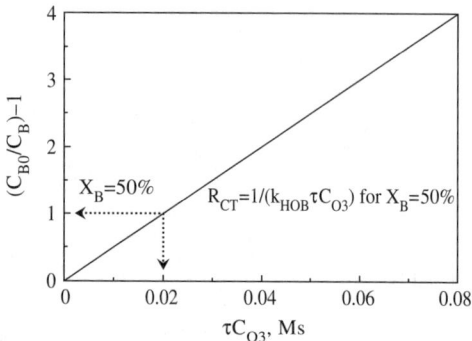

FIGURE 7.7 Checking Equation (7.64) for the determination of the Elovitz and von Gunten R_{CT} value of an arbitrary water in continuous-stirred tank reactors (x and y axes show arbitrary values).

References

1. Astarita, G., *Mass Transfer with Chemical Reaction*, Elsevier, Amsterdam, 1967, 8–10.
2. Beltrán, F.J., Theoretical aspects of the kinetics of competitive ozone reactions in water, *Ozone Sci. Eng.*, 17, 163–181, 1995.
3. Charpentier, J.C., Mass transfer rates in gas liquid absorbers and reactors, in *Advances in Chemical Engineering*, Vol. 11, Academic Press, New York, 1981, 3–133.
4. Beltrán, F.J., Estimation of the relative importance of free radical oxidation and direct ozonation/UV radiation rates of micropollutants in water, *Ozone Sci. Eng.*, 21, 207–228, 1999.
5. Hoigné, J. and Bader, H., Rate constants of the reactions of ozone with organic and inorganic compounds. I. Non dissociating organic compounds, *Water Res.*, 17, 173–183, 1983.
6. Hoigné, J. and Bader, H., Rate constants of the reactions of ozone with organic and inorganic compounds. II. Dissociating organic compounds, *Water Res.*, 17, 185–194, 1983.
7. Yao, C.C.D. and Haag, W.R., Rate constants of direct reactions of ozone with several drinking water contaminants, *Water Res.*, 25, 761–773, 1991.
8. Staehelin, S. and Hoigné, J., Decomposition of ozone in water in the presence of organic solutes acting as promoters and inhibitors of radical chain reactions, *Environ. Sci. Technol.*, 19, 1206–1212, 1985.
9. Acero, J.L. and von Gunten, U., Influence of carbonate on the ozone/hydrogen peroxide based advanced oxidation process for drinking water, *Ozone Sci. Eng.*, 22, 305–308, 2000.
10. Westerhoff, P. et al., Applications of ozone decomposition models, *Ozone Sci. Eng.*, 19, 55–73, 1997.
11. Weeks, J.L. and Rabani, J., The pulse radiolysis of deaerated carbonate solutions. 1. Transient optical spectrum and mechanism. 2. pK for OH radicals, *J. Phys. Chem.*, 82, 138–141, 1966.
12. Buxton, G.V. et al., Critical review of data constants for reactions of hydrated electrons, hydrogen atoms and hydroxyl radicals (•OH/•O⁻) in aqueous solution, *J. Phys. Chem. Ref. Data*, 17, 513–886, 1988.
13. Behar, D., Czapski, G., and Duchovny, I., Carbonate radical in flash photolysis and pulse radiolysis of aqueous carbonate solutions, *J. Phys. Chem.*, 74, 2206–2211, 1970.

14. Chen, S. and Hoffman, M.Z., Rate constants for the reactions of the carbonate radical with compounds of biochemical interest in neutral aqueous solution, *Radiat. Res.*, 56, 40–47, 1973.
15. Chen, S. and Hoffman, M.Z., Reactivity of the carbonate radical in aqueous solution. Tryptophan and its derivatives, *J. Phys. Chem.*, 78, 2099–2102, 1974.
16. Chen, S., Hoffman, M.Z., and Parsons, G.H., Jr., Reactivity of the carbonate radical toward aromatic compounds in aqueous solution, *J. Phys. Chem.*, 79, 1911–1912, 1975.
17. Neta, P., Huie, R.E., and Ross, A.B., Rate constants for reactions of inorganic radicals in aqueous solution, *J. Phys. Chem. Ref. Data*, 17, 1027–1284, 1988.
18. Beltrán, F.J. et al., Contribution of free radical oxidation to eliminate volatile organochlorine compounds in water by ultraviolet radiation and hydrogen peroxide, *Chemosphere*, 32, 1949–1961, 1996.
19. Haag, W.R. and Yao, C.C.D., Rate constants for reactions of hydroxyl radicals with several drinking water contaminants, *Environ. Sci. Technol.*, 26, 1005–1013, 1992.
20. Schnitzer, M. and Khan, S.U., *Humic Substances in the Environment*, Marcel Dekker, New York, 1972.
21. Haag, W. and Yao, C.C.D., Ozonation of US drinking water sources: HO concentration and oxidation-competition values, in *Ozone in Water and Wastewater*, Proceedings of 11th Ozone World Congress, Vol. 2, S-17-119-125, San Francisco, CA, 1993.
22. Legube, B. et al., Effect of ozonation on the organic-halide formation potential of fulvic acids, *Sciences de l'eau*, 6, 435–448, 1987.
23. Westerhoff, P. et al., Relationships between the structure of natural organic matter and its reactivity towards molecular ozone and hydroxyl radicals, *Water Res.*, 33, 2265–2276, 1999.
24. Yurteri, C. and Gurol, M. Ozone consumption in natural waters: effect of background organic matter, pH and carbonate species, *Ozone Sci. Eng.*, 10, 277–290, 1988.
25. Laplanche, A. et al., Modelization of micropollutant removal in drinking water treatment by ozonation or advanced oxidation processes (O_3/H_2O_2), *Ozone Sci. Eng.*, 17, 97–117, 1995.
26. Hoigné, J. and Bader, H., Ozonation of water: oxidation–competition values of different types of waters used in Switzerland, *Ozone Sci. Eng.*, 1, 357–372, 1979.
27. Fogler, H.S., *Elements of Chemical Reaction Engineering*, 3rd ed., Prentice-Hall, Upper Saddle River, NJ, 1999, 77–81.
28. Hoigné, J., Chemistry of aqueous ozone and transformation of pollutants by ozonation and advanced oxidation processes, in J. Hrubec, Ed., *The Handbook of Environmental Chemistry*. 2nd ed., Vol. 5, Part C: Quality and Treatment of Drinking Water, Springer-Verlag, Berlin, 1998, 83–141.
29. Elovitz, M.S. and von Gunten, U., Hydroxyl radical/ozone ratios during ozonation processes. I. The R_{CT} concept, *Ozone Sci. Eng.*, 21, 239–260, 1999.
30. Elovitz, M.S. and von Gunten, U., Hydroxyl radical/ozone ratios during ozonation processes. II. The effect of temperature, pH, alkalinity, and DOM properties, *Ozone Sci. Eng.*, 22, 123–150, 2000.
31. Acero, J.L., Stemmler, K., and von Gunten, U., Degradation kinetics of atrazine and its degradation products with ozone and OH radicals: a predictive tool for drinking water treatment, *Environ. Sci. Technol.*, 34, 591–597, 2000.
32. Acero, J.L. et al., MTBE oxidation by conventional ozonation and the combination ozone/hydrogen peroxide: efficiency of the processes and bromate formation, *Environ. Sci. Technol.*, 35, 4252–4259, 2001.
33. Hoigné, J. and Bader, H., Ozonation of water: kinetics of oxidation of ammonia by ozone and hydroxyl radicals, *Environ. Sci. Technol.*, 12, 79–84, 1978.

8 Kinetics of the Ozone/Hydrogen Peroxide System

An advanced oxidation process (AOP) (see also Chapter 2) is characterized by the production of hydroxyl radicals, which are the main species responsible for oxidation. AOPs can be classified according to the experimental conditions applied. Thus, there are AOPs carried out at ambient conditions of temperature and pressure and at severe experimental conditions. Belonging to the first group are the ozone-involving AOPs such as ozone at high pH, ozone/hydrogen peroxide (O_3/H_2O_2), ozone/UV radiation (O_3/UV), ozone/catalyst (O_3/Cat), etc.,[1,2] and others such as the Fenton, photofenton, and photocatalytic processes,[3-5] to name some of the best-known processes. To the second group belong wet air and supercritical oxidation.[6,7] What might be called a third group are those initiated at ambient conditions but that develop with the generation of high temperature and pressures, such as the cavitation processes.[8] In this chapter, we discuss ozone–hydrogen peroxide advanced oxidation kinetics.

Staehelin and Hoigné[9] extensively studied the reaction between ozone and hydrogen peroxide at different conditions of pH and concentration of H_2O_2 in both pure and natural water. In homogeneous systems, they found a second-order kinetics for the ozone decomposition (first order with respect to ozone and hydrogen peroxide) with the particular characteristic that only the ionic form of hydrogen peroxide reacted with ozone. Only for experiments at pH > 4 did they note an appreciable rate of ozone decomposition that increased one order of magnitude with an increase of one unit in pH. The stoichiometric equation they proposed was:

$$O_3 + HO_2^- \xrightarrow{k_{i2}=2.8 \times 10^6 M^{-1} s^{-1}} O_3^- \bullet + HO_2 \bullet \qquad (8.1)$$

This reaction constitutes the initiation step of a radical chain mechanism that eventually leads to the formation of hydroxyl radicals.[10] In fact, this is also considered the initial step of ozone decomposition in water due to the hydroxide ion as recently reported by Hoigné[11] and previously proposed by Tomiyashu et al.[10] Therefore, the mechanism of radical reactions presented in Table 2.4 is also the mechanism of the O_3/H_2O_2 system. At pH 7, low concentrations of hydrogen peroxide (10^{-5} to 10^{-4} M) yield significant ozone decomposition rates and, hence, a high concentration of hydroxyl radicals. This is due to the high value of the rate constant of Reaction (8.1). For this reason, the O_3/H_2O_2 system is one of the most studied and applied AOPs in the laboratory, pilot plant works, and even in practical drinking water treatment. Here, unless otherwise indicated, the kinetics of this process is considered as a heterogeneous gas–liquid reaction.

8.1 THE KINETIC REGIME OF THE O_3/H_2O_2 PROCESS

The kinetic regime of ozone absorption in water containing H_2O_2 is defined by its corresponding Hatta number [Equation (4.40)], which in this case is

$$Ha_2 = \sqrt{\frac{k_{i2} 10^{pH-pK} C_{H2O2t} D_{O3}}{k_L^2 (1 + 10^{pH-pK})}} \quad (8.2)$$

As observed from Equation (8.2), Ha_2 depends on both pH and concentration of hydrogen peroxide. For example, at pH 7, if $k_L = 10^{-4}$ msec^{-1}, an appropriate value for gas–liquid contactors[12] for total concentrations of hydrogen peroxide (C_{H2O2t}) of 10^{-3}, 10^{-2}, and 0.1 M Ha_2 is 0.006, 0.06, and 0.6, respectively. This means that as the kinetic regime goes from slow to moderate, the corresponding ozone absorption rate law will be different (see Table 5.5). In a previous work,[13] the reaction time t_R (see Section 4.2.4) of Reaction (8.1) was plotted against the total concentration of hydrogen peroxide at different pH (see Figure 8.1) in order to establish the conditions needed for slow and fast kinetic regimes. From Figure 8.1 it can be seen that at the experimental conditions usually found in drinking water treatment (pH < 10 and $C_{H2O2t} < 10^{-4}$ M), the kinetic regime would be slow. However, in some cases the kinetic regime is different. These situations are presented below.

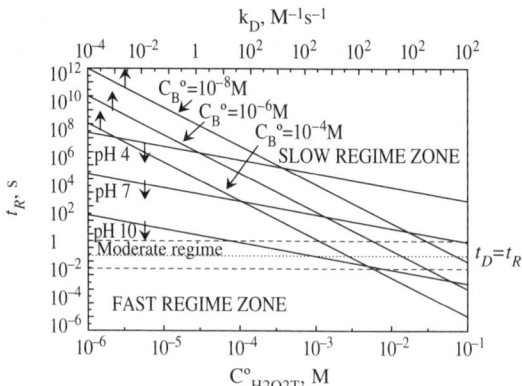

FIGURE 8.1 Variation of reaction time with direct rate constant and concentration of micropollutant, for ozone–B direct reaction, and with hydrogen peroxide concentration and pH for Reaction (8.1). C_B^o is the concentration of B in bulk water. (Reprinted with permission from Beltrán, F.J., Theoretical aspects of the kinetics of competitive first order reactions of ozone in the O_3/H_2O_2 and O_3/UV oxidation processes, *Ozone Sci. Eng.*, 19, 13–38, 1997. Copyright 1997, International Ozone Association.)

8.1.1 SLOW KINETIC REGIME

If it is assumed that a given pollutant B during the O_3/H_2O_2 oxidation is only removed from water by free radical oxidation, which is the usual case in drinking water ozonation. The rate of disappearance of B is:

$$-r_B = k_{HOB} C_{HO} C_B \tag{8.3}$$

where a simplified expression for the concentration of hydroxyl radicals, obtained from the mechanism of reactions (see Table 2.4) after the application of a steady-state situation, is given by:

$$C_{HO} = \frac{2k_{i2} C_{H2O2t} 10^{pH-11.8}}{(1+10^{pH-11.8}) \sum k_{OHS} C_S} \tag{8.4}$$

Note that the hydrogen peroxide concentration does not appear in the denominator of Equation (8.4) because of its role in promoting ozone decomposition through Reactions (2.27) and (2.28). Also, in the denominator of Equation (8.4), when carbonate is present in the water, the term w [see Equation (7.22)] has to be included so that the concentration of hydroxyl radicals is:

$$C_{HO} = \frac{2k_{i2} C_{H2O2t} 10^{pH-11.8}}{(1+10^{pH-11.8})\left[(1-w)k_C C_{HCO3t} + \sum k_{OHS} C_S\right]} \tag{8.5}$$

The concentration of B, however, is determined from the corresponding mass balance that varies with the type of reactor or, to be more exact, that varies with the type of flow of the gas and water phases through the reactor. In a semibatch well-mixed reactor, the equation to be solved is

$$-\frac{dC_B}{dt} = r_B \tag{8.6}$$

and, in a continuous well-mixed system at steady state,

$$v_0 (C_{B0} - C_B) + r_B V = 0 \tag{8.7}$$

8.1.2 FAST-MODERATE KINETIC REGIME

As shown in Figure 8.1, for some special experimental conditions the kinetic regime of the ozone–hydrogen peroxide reaction could be moderate or even fast. In these cases, the concentration of dissolved ozone is zero or really negligible so that neither Equation (8.6) nor Equation (8.7) is appropriate to determine the remaining concentration of B. Rigorously, this should be obtained from the ozone absorption rate law corresponding to a moderate or fast reaction (see equations in Table 5.5), which is a rather complex equation where the reaction factor, E, needs to be known:

$$-z\frac{dC_B}{dt} = N_{O3} = k_L a C_{O3}^* E \qquad (8.8)$$

In theory, this case is similar to that studied for the ozone direct reactions with a fast kinetic regime. However, the ozone absorption rate equation is even more complex than that for direct ozonation kinetics [see Equation (4.44)]. As was shown in Chapter 4, the ozone absorption rate equation, N_{O3}, comes from the application of Fick's law at the interface ($x = 0$), and its determination requires that the concentration profile of ozone through the film layer is known. The profile concentration is obtained from the solution of the system of differential equations constituted by the microscopic mass balances of ozone and B. The problem now is that the chemical reaction rate term in the ozone equation involves all the reactions that ozone undergoes in water, including reactions with free radicals (HO•, O_2•⁻, etc.). Thus, the solution of the differential equations also needs the expressions for the concentrations of free radicals. It is evident that the mathematical model that results is very complicated. As a consequence, a different, simpler method is recommended. This method was first reported by Glaze and Kang,[14] and, among other aspects, allows the optimal ozone–hydrogen peroxide ratio to be determined.

8.1.3 CRITICAL HYDROGEN PEROXIDE CONCENTRATION

While the kinetic regime of the ozone–hydrogen peroxide reaction is slow, the increase in the concentration of the latter leads to increases in the ozonation rate of any pollutant in water. If the reaction is fast, however, the opposite situation is observed. In this latter case, the concentration of ozone in bulk water is zero or very small. In other words, the rate of ozone accumulation in water is zero ($dC_{O3}/dt = 0$). In the ozone/hydrogen peroxide system, from the macroscopic ozone balance with $dC_{O3}/dt = 0$, the rate term due to the reaction between ozone and hydrogen peroxide is as follows:[13]

$$2k_{i2}10^{pH-11.8}C_{H2O2t}C_{O3} = k_L a C_{O3}^* - k_H C_{H2O2t}C_{HO} - wk_C C_{HCO3t}C_{HO} \qquad (8.9)$$

where

$$k_H C_{H2O2T} = \left(k_{H1} + k_{H2}10^{pH-pK}\right)\frac{C_{H2O2T}}{1+10^{pH-pK}} \qquad (8.10)$$

with k_{H1} and k_{H2} as rate constants of Reactions (2.27) and (2.28), respectively.

The left side of Equation (8.9) is the numerator of the concentration of hydroxyl radicals as shown in Equation (8.5). Thus, if Equation (8.9) is substituted in Equation (8.5), a modified equation for the concentration of hydroxyl radicals is obtained:

Kinetics of the Ozone/Hydrogen Peroxide System

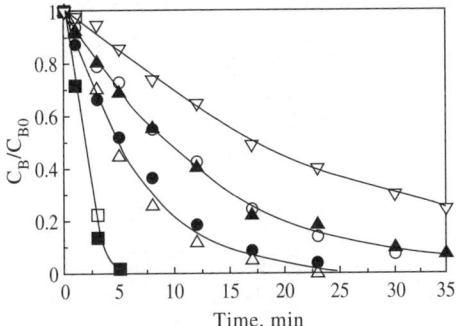

FIGURE 8.2 Variation of normalized remaining concentration of atrazine with time during its O_3/H_2O_2 oxidation. Influence of initial hydrogen peroxide concentration. Conditions: $T = 20°C$; inlet ozone partial pressure = 800 Pa; gas flow rate = 50 dm^3h^{-1}, C_{H2O2Ti}; M: ○ = 0; ● = 10^{-5}; □ = 10^{-4}; ■ = 10^{-3}; ▲ = 10^{-2}; ▽ = 2×10^{-1}. (Reprinted with permission from Beltran, F.J. et al., Aqueous degradation of atrazine and some of its main by-products with ozone and ozone/hydrogen peroxide, *J. Chem. Technol. Biotechnol.*, 71, 345–355, 1998. Copyright 1998, John Wiley & Sons, Inc.)

$$C_{HO} = \frac{k_L a C_{O3}^*}{k_H C_{H2O2t} + k_C C_{HCO3t} + \sum k_{OHS} C_S} \quad (8.11)$$

From this equation, we can see that the concentration of hydroxyl radicals, when the ozone–hydrogen peroxide reaction is moderate or fast, is proportional to the ozone mass transfer efficiency (specifically, it depends on the maximum ozone absorption rate $k_L a C_{O3}^*$). Hydrogen peroxide also changes its role to become a true inhibiting species of ozone decomposition through free radical reactions. In other words, the increasing presence of hydrogen peroxide concentration stops or slows the disappearance rate of B. This has been observed often and an example is shown in Figure 8.2. In fact, in different studies, a critical concentration of hydrogen peroxide is reported above, which the oxidation rate of B diminishes.[14–16] Paillard et al.[16] studied the ozone/hydrogen peroxide oxidation of oxalic acid and other refractory organic compounds in a batch reactor to determine the optimum ratio between the concentrations of ozone and hydrogen peroxide. They found the ratio of 2 mol of ozone per mol of hydrogen peroxide as the optimum one. In the heterogeneous ozone/hydrogen peroxide oxidation, in semibatch well-agitated reactors, the rate of disappearance of B due to hydroxyl radical oxidation, when the kinetic regime is moderate fast, will now be:

$$-\frac{dC_B}{dt} = k_{HOB} C_B \frac{k_L a C_{O3}^*}{k_H C_{H2O2t} + k_C C_{HCO3t} + \sum k_{OHS} C_S} \quad (8.12)$$

8.2 DETERMINATION OF KINETIC PARAMETERS

From the kinetic point of view, the rate constant k_{HOB} is the key parameter when the ozone/hydrogen peroxide system is applied. This rate constant can also be determined from absolute and competitive methods, provided the direct reactions are negligible and compounds are not volatile.

8.2.1 THE ABSOLUTE METHOD

The best time to apply the absolute method is during the moderate–fast reaction-state in water where Equation (8.12) holds. This equation represents a first-order kinetics with respect to B so that its integration after variable separation yields a linear relationship between the logarithm of B and reaction time. Thus, according to the integrated Equation (8.12), a plot of the $\ln C_B$ against time leads to a straight line of slope $k_{HOB} C_{HO}$. In a kinetic study, the experiment is usually carried out in the absence of carbonates in ultrapure water. Then the term $\Sigma k_{OHS} C_S$ in the denominator of the expression for the concentration of hydroxyl radicals reduces to $k_{HOB} C_B$. Thus, the denominator is only a function of the inhibiting contributions of B and hydrogen peroxide. In most practical situations, for the kinetic regime to be moderate or fast, a high concentration of hydrogen peroxide is used. As a consequence, the term $k_{HOB} C_B$ can also be neglected so the denominator of Equation (8.11) that will now depend only on hydrogen peroxide concentration. In addition, at a high concentration of hydrogen peroxide, the denominator probably remains constant and known during the ozonation time since the concentration of hydrogen peroxide diminishes only slightly. Then, from the slope of the straight line plotted according to the integrated Equation (8.12), the rate constant, k_{HOB}, can be obtained:

$$k_{HOB} = slope \frac{k_H C_{H2O2t}}{k_L a C^*_{O3}} \qquad (8.13)$$

With some modification, this procedure was applied by Glaze and Kang[14] to the experimental results of the semibatch ozone/hydrogen peroxide oxidation of trichloroethylene and tetrachloroethylene. At the conditions they investigated, it was found that Equation (8.13) was applicable for a 1.2×10^{-4} M concentration of hydrogen peroxide. It is evident that in their system, the individual liquid-phase mass-transfer coefficient should have been very low so that Ha_2 is at least higher than 0.3 (moderate kinetic regime). Another possible way to determine k_{HOB}, as Glaze and Kang[14] reported, is based on experiments carried out in the presence of high concentrations of carbonates. If the inhibiting terms due to hydrogen peroxide and B can be neglected in the denominator of Equation (8.11), the rate Equation for B can be reduced to:

$$-\frac{dC_B}{dt} = k_{HO} C_B \frac{k_L a C^*_{O3}}{k_C C_{HCO3t}} \qquad (8.14)$$

According to this equation, the reciprocal of the pseudo first-order rate constant, $k_{HOB}C_{HO}$, varies linearly with the increasing concentration of bicarbonate, as follows:

$$\frac{1}{k_{HOB}C_{HO}} = \frac{k_c C_{HCO3}}{k_L a C_{O_3}^* k_{HOB}} \tag{8.15}$$

Then, once the apparent pseudo first-order constant is obtained at different high carbonate concentrations, a plot of the inverse of this constant against the concentration of carbonates should yield a straight line, as given by Equation (8.15). From the slope of the line, the rate constant k_{HOB} can be calculated, provided the mass transfer coefficient and Henry's law constant are known. With this procedure, k_{HOB} for the hydroxyl radical–trichloroethylene reaction was calculated to be $2.1 \times 10^9\ M^{-1}\text{sec}^{-1}$.[14]

8.2.2 THE COMPETITIVE METHOD

The competitive kinetic method can be applied to any kinetic regime of ozonation in the presence of hydrogen peroxide carried out in a semibatch well-mixed reactor. This is so because in this type of reactor, the disappearance rate of B is always given by Equation (8.6) regardless of the kinetic regime. Then, if the ozonation of B is carried out in the presence of a reference compound of known hydroxyl radical kinetics (provided it does not react directly with ozone), the competitive procedure shown in Section 5.3.4 also holds. According to this method, the rate constant k_{HOB} is calculated from the value of k_{HOR}, the rate constant of the hydroxyl radical–R reaction, and the slope of the plot $\ln(C_B/C_{B0})$ with $\ln(C_R/C_{R0})$. This method can also be applied to both homogeneous and heterogeneous ozonation results. The literature reports some examples where this procedure was also applied.[17–19]

8.2.3 THE EFFECT OF NATURAL SUBSTANCES ON THE INHIBITION OF FREE RADICAL OZONE DECOMPOSITION

Another interesting aspect that can be studied when the kinetic regime of ozone absorption is moderate or fast is related to the influence of natural substances to scavenge hydroxyl radicals and limit the free radical ozone decomposition — hence the removal of a given pollutant from natural water. It is evident that, in this case, ozonation is carried out in the presence of carbonates that abound in this type of water. If, in addition, it admits the presence of other inhibiting substances S, the concentration of hydroxyl radicals for moderate–fast kinetic regimes will be given by Equation (8.11). In the denominator, the contribution of hydrogen peroxide could also be negligible when compared to the other inhibiting terms (carbonates, other natural substances, etc.), depending on the concentration of these substances and the specific conditions for the reaction to be moderate or fast. For example, this is the case of a system of low mass-transfer efficiency (low $k_L a$) where the hydrogen peroxide concentration for the moderate or fast kinetic regime is low, as is its inhibiting contribution ($k_H C_{H2O2t}$). Thus, if a probe compound, B, is used to follow the ozonation kinetics, since the corresponding k_{HOB} is known, the application of the absolute method will allow the contribution $\sum k_{HOS}C_S$ be calculated from experimental ozonation data. Glaze and Kang[14] estimated values of $\sum k_{HOS}C_S$ in natural water, which represented in some cases as much as 98% of total hydroxyl radical scavenged. Values of $\sum k_{HOS}C_S$ are very important for the design of ozone contactors, especially in the treatment of drinking water since it helps to know the time needed for a given pollutant to be removed in the ozonation process.

8.3 THE OZONE/HYDROGEN PEROXIDE OXIDATION OF VOLATILE COMPOUNDS

In some cases, the compounds present in water have a high vapor pressure, that is, they are volatile compounds. In these cases, removal of these compounds can be achieved by aeration of water, and the use of ozone could be a waste of time and money. Thus, the rate of removal of B will have another contribution — volatility. Note that the effect of volatile compounds can also be applied to single ozonation systems. The rate of the removal of a volatile B compound in a well-mixed semibatch reactor can be expressed as

$$-\frac{dC_B}{dt} = k_D C_{O3} C_B + k_{HOB} C_{HO} C_B + (k_L a)_B (C_B - C_B^*) \qquad (8.16)$$

where $(k_L a)_B$ and C_B^* represent the volumetric mass transfer coefficient for B in the water and its concentration at the water–gas interface. In Equation (8.16), the terms on the right side represent the contributions of the direct reaction ozone–B, the hydroxyl radical–B reaction, and volatility to the removal rate. The concentration of B at the interface is calculated from the corresponding Henry's law:

$$P_B = He_B C_B^* \qquad (8.17)$$

If the direct reaction contribution is negligible, the determination of k_{HOB} will present some difficulty because of the unknown partial pressure or concentration of B at the gas–water interface (P_{Bi}) in the rate equation. The problem, however, could be solved with the use of the mass-balance equation of B in the gas phase. If the gas phase is also well mixed and its resistance to mass transfer negligible, the mass balance of B becomes

$$(k_L a)_B \left(C_B - \frac{P_B}{He_B} \right) V\beta - v_g \frac{P_B}{RT} = (1-\beta) V \frac{v_g}{RT} \frac{dP_B}{dt} \qquad (8.18)$$

where β is the liquid holdup in the gas–liquid reactor, v_g is the gas flow rate, and R is the perfect gas constant. The simultaneous solution of Equation (8.16) and Equation (8.18) and those of ozone in the gas and water phases will give the concentration profiles of B and ozone in both phases at given values of the rate constants and mass-transfer coefficients.

Kinetics of the Ozone/Hydrogen Peroxide System

Another simpler way to solve this case is to use volatility coefficients to measure the contribution of volatility to the removal rate of B. These coefficients can be experimentally determined from the aeration of aqueous solutions of B in the absence of ozone. As shown in a previous work,[20] volatilization of compounds from water is usually a first-order kinetics so that the logarithm of the concentration of B with time follows a straight line with the volatility as the slope of coefficient k_v. Once k_v is known at different gas flow rate conditions, experiments of ozonation are carried out. In these cases, the resulting pseudo first-order rate constant is the sum of the volatility coefficient and the product $k_{HOB}C_{HO}$. Thus, once k_v has been accounted for, k_{HOB} can be calculated from the procedures presented in the previous sections.

8.4 THE COMPETITION OF THE DIRECT REACTION

Chapter 7 discussed the factors that influence the competition between the direct reaction of ozone and a given compound B and the decomposition reaction of ozone (which leads to hydroxyl radicals). This kind of study can also be carried out when the first ozone reaction leading to hydroxyl radicals is due to the action of hydrogen peroxide [Reaction (8.1)]. Note that Reaction (8.1) is also the reaction reported as leading to hydroxyl radicals in simple ozonation processes (see Table 2.4). However, in simple ozonation, the first reaction of the mechanism (not the initiation reaction) is the ozone–hydroxyl ion reaction [Reaction (2.18) in Table 2.4]. This study can be oriented in two ways:[13,21,22]

1. The competition between Reaction (8.1) and the direct reaction of ozone with B from the standpoint of the kinetic regime of ozone absorption.
2. The competition between the direct reaction of ozone with B and the reaction between the hydroxyl radical (coming from the ozone–hydrogen peroxide reaction) and B.

8.4.1 COMPARISON BETWEEN THE KINETIC REGIMES OF THE OZONE– COMPOUND B AND OZONE–HYDROGEN PEROXIDE REACTIONS*

The ozone reactions to be compared are:

- The direct reaction ozone–compound B:

$$O_3 + zB \xrightarrow{k_D} \text{Products} \tag{8.19}$$

- The direct reaction ozone–hydrogen peroxide:

$$O_3 + HO_2^- \xrightarrow{k_{12}} HO_2 \bullet + O_3^- \bullet \tag{8.1}$$

---

* Most of the text in Section 8.4.1 is reprinted with permission from Beltrán.[13] Copyright 1997, International Ozone Association.

Diffusion and reaction time concepts are used to compare the kinetic regimes of both reactions (see Section 4.2.4). Thus, as reported in a previous work,[13] in Figure 8.1 the reaction time was plotted against the total concentration of hydrogen peroxide [for Reaction (8.1)] and against k_D [for Reaction (8.19)]. Values of t_R for both ozone reactions are also plotted at different pH and concentrations of B. The dotted line that divides the plot in the fast and slow kinetic regimes of ozone absorption was drawn for a value of t_D calculated by considering a value of 2×10^{-4} sec^{-1} for the individual mass-transfer coefficient k_L. This represents a common value for ozonation systems where the gas is feeding through porous plates.[12] Note that the position of the dotted line can vary according to the value of t_D — hence k_L. For poor ozone mass-transfer systems, t_D will increase, as will the number of experimental conditions for the reactions to be fast. Thus, if k_L is 10^{-5} sec^{-1}, t_D becomes 13 sec and Reaction (8.1) would be moderate at pH ≥ 7 with $C_{H2O2t} \geq 5 \times 10^{-4}$ M. Then, from Figure 8.1, experimental conditions for competition of Reaction (8.1) and Reaction (8.19) can be established. For example, at pH 7, if the concentration of B varies between 10^{-4} and 10^{-8} M, the direct ozone–B Reaction (8.19) will compete with Reaction (8.1) for the available ozone, provided the rate constant k_D is lower than 10^3 M^{-1}sec^{-1} because at these conditions both reactions develop in the slow kinetic regime. From Figure 8.1 it can also be deduced that the addition of hydrogen peroxide at concentrations lower than 10^{-2} M to a water containing B at concentrations higher than 10^{-6} M will be a waste of reagent and hence money because Reaction (8.1) will not develop if the rate constant of the direct reaction $k_D \geq 10^6$ M^{-1}sec^{-1}. This is because the available ozone will be exclusively consumed in the proximity of the gas–water interface through the direct reaction ozone–B that develops in the moderate–fast kinetic regime. Then, ozone would not be able to reach the bulk water where Reaction (8.1) would take place. An example of this situation is found when phenols are ozonated at pH ≥ 7. Conversely, always considering pH 7, we find the use of hydrogen peroxide is advisable when the reaction between ozone and the compound B to be removed from water has a rate constant $k_D < 10^4$ M^{-1}sec^{-1} and the concentration of B is <10^{-6} M, a situation that could be found in some surface waters. In these situations, the addition of hydrogen peroxide at concentrations higher than 10^{-3} M (or even at lower concentration if pH is higher than 7) is recommended to remove B through free radical reactions. Next, Reaction (8.1) develops in the moderate–fast kinetic regime and will consume the ozone transferred from the gas near the interface. It is evident that the direct reaction will not take place.

The way t_R and t_D are plotted in Figure 8.1 against k_D and the hydrogen peroxide concentration can be considered a standard plot to compare the kinetic regimes of Reaction (8.1) and Reaction (8.19). However, for the case where B is a dissociating compound with pH (e.g., phenols), a more convenient plot to represent the comparison between the kinetic regimes of both reactions has also been reported,[13] and it is shown in Figure 8.3. In this case, t_R is plotted against pH with concentrations of B and hydrogen peroxide taken as parameters. This new plot is necessary because of the strong effect of pH on the rate constant values of both reactions that can be expressed as

Kinetics of the Ozone/Hydrogen Peroxide System

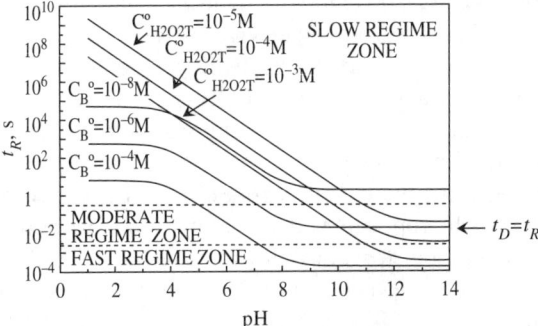

FIGURE 8.3 Variation of reaction time with pH, for the ozone–B direct reaction (case of o-chlorophenol ozonation, a dissociating organic micropollutant), and, at different hydrogen peroxide concentration, for Reaction (8.1). C_B^o and C_{H2O2T} represent the concentration of B and hydrogen peroxide in bulk water, respectively. (Reprinted with permission from Beltrán, F.J., Theoretical aspects of the kinetics of competitive first order reactions of ozone in the O_3/H_2O_2 and O_3/UV oxidation processes, *Ozone Sci. Eng.*, 19, 13–38, 1997. Copyright 1997, International Ozone Association.)

For Reaction (8.19),

$$t_R = \frac{1}{\left[k_{Dn}(1-\alpha) + k_{Dd}\alpha\right]C_B} \tag{8.20}$$

For Reaction (8.1),

$$t_R = \frac{1+10^{pH-11.8}}{2.8 \times 10^{pH-5.8}C_{H2O2t}} \tag{8.21}$$

where in Equation (8.20), k_{Dn} and k_{Dd} are the rate constants of the direct reactions of ozone with the nondissociating and dissociating species of B, respectively, and α is the degree of dissociation, a function of pH and pK of B:

$$\alpha = \frac{1}{1+10^{pK-pH}} \tag{8.22}$$

This plot is particularly useful for phenol compounds. For example, let us take the case of the ozonation of o-chlorophenol. The rate constant of the reaction of dissociating and nondissociating species of this compound with ozone at 20°C are 1300 and $1.4 \times 10^6\ M^{-1}sec^{-1}$, respectively, and its pK is 8.3.[23] Then, for the ozone–hydrogen peroxide–o-chlorophenol system, the addition of hydrogen peroxide would only be recommended for concentrations of o-chlorophenol lower than $10^{-7}\ M$ and/or when the water is weakly acidic. Only in these cases will the oxidation rate increase with the addition of hydrogen peroxide.

Determination of the kinetic regimes of both Reaction (8.1) and Reaction (8.19) from Figure 8.1 or Figure 8.3 is not enough cause to disregard one of the two reactions in the kinetic study of the ozone–hydrogen peroxide oxidation of a given compound B. Note that what is known from these plots are the kinetic regimes of the absorption of ozone corresponding to both reactions. Although this allows the neglect of one of the reactions in extreme cases, as indicated above, it does not give decisive information when the kinetic regimes of both reactions are the same, for example, when B is o-chlorophenol with a concentration of 10^{-8} M, pH 7 and C_{H2O2t} is 10^{-5} M. In this case, both reactions are slow although the direct reaction has a reaction time 100 times lower than that of Reaction (8.1). It is evident that most of the ozone will be consumed in the direct reaction, but in the kinetic study both reactions should be considered unless other information supports the fact that the contribution of Reaction (8.1) to the removal of B can be neglected. This information comes from the comparison of the oxidation rates of B due to its direct reactions with ozone and hydroxyl radicals.

8.4.2 COMPARISON BETWEEN THE RATES OF THE OZONE–COMPOUND B AND HYDROXYL RADICAL–B REACTIONS*

A definitive conclusion about the relative importance of the direct and free radical types of oxidation when both Reaction (8.1) and Reaction (8.19) develop in the same kinetic regime can be achieved from the estimation of oxidation rates of both reactions. Here, both reactions are assumed to develop in the slow kinetic regime because this case covers most of the situations that can be found in practical cases in water treatment. In the slow kinetic regime, the ozone absorption rate can be expressed as follows:

$$N_{O3} = k_L a \left(C_{O3}^* - C_{O3} \right) = \beta \left[k_D C_B + k_{i2} C_{HO_2^-} \right] C_{O3} \tag{8.23}$$

where $k_L a$ and β are the volumetric mass-transfer coefficient and liquid holdup, respectively. As deduced from Equation (8.23), the importance of Reaction (8.1) and Reaction (8.19) can be established, provided the dissolved ozone concentration is known. This concentration can be estimated from Equation (8.24) following the film theory concept[21] from:

$$C_{O3} = \frac{C_{O3}^*}{\cosh Ha + \dfrac{\beta}{a\delta_L} Ha \tanh Ha} \tag{8.24}$$

where Ha is the Hatta number defined, in this case, as

---

* Most of the text in Section 8.4.2 is reprinted with permission from Beltrán.[13] Copyright 1997, International Ozone Association.

TABLE 8.1
Calculated Ozone Dissolved Concentration in Bulk Water and Importance of Reaction (8.1) in the O_3/H_2O_2 Oxidation of Organic Micropollutants[a]

Case #	C_{H2O2T} M	$C_{O3} \times 10^5$ M[b]	$C_{O3} \times 10^5$ M[c]	$C_{O3} \times 10^5$ M[d]	$N_{O3} \times 10^7$ $Msec^{-1}$[e]	% Due to Reaction (8.1)
1	0	4.9			0.15	
2	10^{-5}	4.9	4.9	4.8	0.36	59.6
3	10^{-4}	4.9	4.1	4.1	1.9	93.7
4	10^{-3}	4.9	1.6	1.6	7.2	99.3
5	10^{-2}	4.9	0.23	0.23	10.0	99.3
6	10^{-1}	4.9	0.02	0.02	11.1	100
7[f]	10^{-4}	49.3	41.3	40.8	19	93.7
8[g]	10^{-5}	4.9	0.23	0.23	10	99.9
9	0	3.4			3.4	
10	10^{-5}	3.4	4.9	3.3	3.5	4.2
11	10^{-4}	3.4	4.1	3.0	4.3	30.5
12	10^{-3}	3.4	1.6	1.4	7.6	81.4
13	10^{-2}	3.4	0.23	0.22	10	97.8
14	10^{-1}	3.4	0.02	0.02	11	100
15[f]	10^{-4}	33.8	41.3	29.6	43	30.5
16[g]	10^{-5}	3.4	0.23	0.23	10	97.7

[a] Cases 1 to 8 correspond to ATZ oxidation. Cases 9 to 16 correspond to MCPP oxidation. Other data used: $\beta = 0.95$, $a = 100$ m^{-1}, $k_L = 2 \times 10^{-4}$ msec^{-1}, $D_{O3} = 1.3 \times 10^{-9}$ m^{2}sec^{-1}, k_D(ATZ) = 6.3 M^{-1}sec^{-1}, $z = 1$, k_D(MCPP) = 101 M^{-1}sec^{-1}, $z = 1$, $C_{O3}^* = 5 \times 10^{-5}$ M, pH = 7 $m = C_B$(ATZ) = 5×10^{-5} M, C_B(MCPP) = 10^{-4} M unless otherwise indicated.
[b] From Equation (8.24) with $k_T = k_D C_B$
[c] From Equations (8.24) and (8.26) with $k_T = k_{i2} C_{HO_2^-}$.
[d] From Equations (8.24) and (8.25).
[e] From Equation (8.23).
[f] $C_{O3}^* = 5 \times 10^{-4}$ M.
[g] pH = 10.

Source: Reprinted with permission from Beltrán, F.J., Theoretical aspects of the kinetics of competitive first order reactions of ozone in the O_3/H_2O_2 and O_3/UV oxidation processes, *Ozone Sci. Eng.*, 19, 13–38, 1997. Copyright 1997, International Ozone Association.

$$Ha = \frac{\sqrt{k_T D_{O3}}}{k_L} \qquad (8.25)$$

and k_T is the sum of the contribution of both reactions

$$k_T = k_D C_B + k_{i2} C_{HO_2^-} \qquad (8.26)$$

Thus, for different experimental conditions (C_B, C_{O3}^* or P_{O3}, pH, C_{O3}, etc.), the percentage contribution of both terms in the ozone absorption rate [right side of Equation (8.23)] is obtained. Table 8.1 referred to previously[13] shows data corresponding to the ozonation of two herbicides, atrazine and mecoprop. We can see that for hydrogen peroxide concentrations >0.1 M, the concentration of dissolved ozone is similar to that found if only the ozone–hydrogen peroxide reaction takes place. Accordingly, the contribution of Reaction (8.1) will increase with the addition of hydrogen peroxide, increase of pH, and/or decrease of C_B. The exception is for $C_{H2O2T} = 0.1$ M; the kinetic regime is then diffusional, the ozone absorption rate equals the maximum physical ozone absorption rate ($k_L a C_{O3}^*$), and the whole process rate is reduced, or does not increase any more. This negative effect of hydrogen peroxide is due to the change in kinetic regime corresponding to its reaction with ozone as discussed above and observed in previous works.[14–16] Some other conclusions from Table 8.1 are related to the importance of k_D. Thus, in the case of atrazine, Reaction (8.1) contributes to the oxidation rate more than 90% in most cases, while the contribution is significantly reduced in the case of mecoprop because k_D is much higher. In this latter case, more rigorous conditions are then required so that the contribution of Reaction (8.1) is important.

Data in Table 8.1 pose another question. The fact that Reaction (8.1) is more important for consuming ozone than the direct reaction ozone–B does not necessarily mean that the oxidation of B through free radicals is the main method of removal from water. To determine this, it is necessary to compare the oxidation rate contributions of reactions between ozone and the hydroxyl radical with B. Although the whole reaction mechanism should be considered, preliminary estimations of the initial reaction rates can be made as follows.

In the slow kinetic regime, the initial oxidation rate of B is

$$-r_{Bo} = zk_D C_B C_{O3} + k_{HOB} C_{HO} C_B \qquad (8.27)$$

where the terms on the right side of Equation (8.27) represent the contribution of the direct and free radical means of oxidation. Depending on the step that controls the process rate, the concentration of hydroxyl radicals is given by Equation (8.5) for chemical reaction control, $t_R > 10 t_D$, and Equation (8.11) for diffusion control ($t_R = 10 t_D$). Note that while $\Sigma k_{HOS} C_S$ is always an inhibiting term for the formation of hydroxyl radicals, the term $k_H C_{H2O2t}$ plays a similar role only while the diffusion controls the process. Table 8.2 taken from Reference 13 gives data on initial rates of B oxidation (where B is ATZ or MCPP) for the experimental conditions presented in Table 8.1. For the calculations, carbonates have been considered the inhibiting substances (with $C_S = 0$) with $k_C C_{HCO3T}$ given by Equation (7.19). For chemical reaction rate control, it can be seen from Table 8.2 that the increase in hydrogen peroxide concentration and/or pH leads to increases in the oxidation rate of B. In contrast, for ozone diffusion control conditions, a significant decrease of $-r_{Bo}$ is observed regardless of the nature of B. It is logical that percentages of the free radical means of oxidation are lower in the case of MCPP as a result of the higher value of k_D compared to values of atrazine oxidation. At first glance, one can conclude

TABLE 8.2
Calculated Initial Oxidation Rate of Organic Micropollutants and Importance of Free Radical Pathway in Their O_3/H_2O_2 Oxidation[a]

Case No.	$-r_{Bo} \times 10^7$, Msec^{-1} [b]	% Free Radical Pathway	$-r_{Bo} \times 10^7$, Msec^{-1} [c]
1			0.15
2	0.58	74.4	0.15
3	3.7	96.7	0.15
4	14	99.6	0.15
5	20	99.9	0.15
6	2.5	100	0.15
7	3.7	96.7	1.5
8	17	99.9	0.15
9			3.4
10	3.6	7.5	3.4
11	5.4	44.8	3.4
12	13	89.0	3.4
13	19	98.8	3.4
14	0.7	100	3.4
15	54	44.8	34
16	11	97.8	3.4

Note: Comparison with direct ozonation alone.

[a] Conditions and case numbers as in Table 8.1.
[b] From Equations (8.27) and (8.4) or (8.11) with C_B given in column 5 of Table 8.1.
[c] Direct ozonation alone: $-r_{Bo} = zk_D C_B C_{O3}$, with C_{O3} given in column 3 of Table 8.1.

Source: Reprinted with permission from Beltrán, F.J., Theoretical aspects of the kinetics of competitive first order reactions of ozone in the O_3/H_2O_2 and O_3/UV oxidation processes, *Ozone Sci. Eng.*, 19, 13–38, 1997. Copyright 1997, International Ozone Association.

that the addition of hydrogen peroxide is appropriate to increase the oxidation rate of both compounds. However, these values should also be compared to those obtained from direct ozonation. Thus, as observed from Table 8.2, the oxidation rate of atrazine with the ozone/hydrogen peroxide system would be much higher than the corresponding one due only to direct ozonation regardless of conditions (even for the case of ozone diffusion control, run 6). A very different situation is observed in the case of MCPP. Here, direct ozonation rates are of a magnitude similar to that found from the O_3/H_2O_2 system when chemical reaction controls the process rate. However, in the ozone diffusion control regime, direct ozonation would be the best means of oxidation. Then, at these conditions, addition of hydrogen peroxide is not recommended. Note that this discussion can be applied to the oxidation of any other compound B present in water, but the nature of scavengers or inhibiting substances should be known. As follows from Equation (8.5) and Equation (8.11), $\sum k_{HOS} C_S$ is an important term regardless of the kinetic regime that could decide whether or not the free radical oxidation would take place.

8.4.3 RELATIVE RATES OF THE OXIDATION OF A GIVEN COMPOUND*

As has been shown in Section 7.2, for the case of ozonation alone, a more straightforward way to compare the relative importance of both rates of oxidation of a given compound is from the ratio of the rates of direct and free radical oxidation. Thus, Equation (8.27) is also applied in this case, with the difference that the concentration of hydroxyl radical is given by Equation (8.5) or Equation (8.11), depending on the rate controlling step. In the case of chemical reaction control, at any time, the concentration of any species is constant through the film layer and equal to the concentration in the bulk water. Equation (8.5) is then applied for the concentration of hydroxyl radicals. Then, the ratio between oxidation rates of B due to the direct and free radical oxidation becomes:

$$\frac{r_R}{r_D} = \frac{k_{HOB} 2 k_{i2} 10^{pH-pK} C_{H2O2t}}{z k_D (1+10^{pH-pK}) \sum k_{OHS} C_S} \tag{8.28}$$

where $\sum k_{HOS} C_S$ also includes the scavenging term due to carbonates, if present in the water. Equation (8.28) in logarithmic form is:

$$\log \frac{r_R}{r_D} = \log \frac{k_{HOB}}{z k_D} + \log \frac{2 k_{i2} 10^{pH-pK} C_{H2O2t}}{(1+10^{pH-pK}) \sum k_{HOS} C_S} \tag{8.29}$$

which is similar to Equation (7.6) for the case of ozonation alone. According to Equation (8.29), a plot of the logarithm of the reaction rate ratio against the logarithm of the reaction rate constant ratio leads to a straight line of slope unity. The ordinate of this line depends on the pH of the water, any inhibiting factor, and the hydrogen peroxide concentration applied. From plots such as those presented in Figure 8.4,[22] we can estimate the experimental conditions of pH, hydrogen peroxide concentration, etc. needed to increase the oxidation rate of any pollutant B in a significant way. Also, with these kinds of plots, the combined ozonation process (O_3/H_2O_2) can be considered a recommended option or not for the oxidation removal of any pollutant B of known kinetics with ozone and hydroxyl radicals (known z, k_D, and k_{HOB}). Details of this procedure with practical examples are given elsewhere.[22]

---

* Most of the text in Section 8.4.3 is reprinted with permission from Beltrán.[22] Copyright 1999, International Ozone Association.

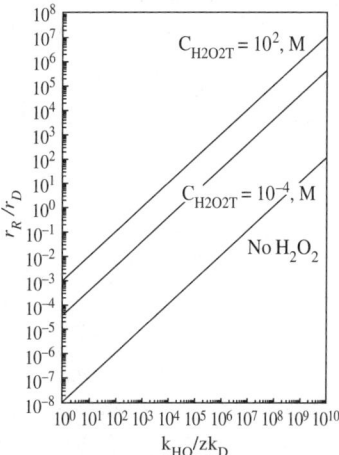

FIGURE 8.4 Comparison between hydroxyl free radical and direct ozonation rates of micropollutants in water as a function of reaction rate constant ratio and different hydrogen peroxide concentration in combined ozonation with hydrogen peroxide. Conditions: 20°C, $\Sigma k_{HOS} C_S = 10^3$ sec^{-1}. (Reprinted with permission from Beltrán, F.J., Estimation of the relative importance of free radical oxidation and direct ozonation/UV radiation rates of micripollutants in water, *Ozone Sci. Eng.*, 21, 207–228, 1999. Copyright 1999, International Ozone Association.)

References

1. Glaze, W.H., Kang, J.W., and Chapin, D.H., The chemistry of water treatment processes involving ozone, hydrogen peroxide, and ultraviolet radiation, *Ozone Sci. Eng.*, 9, 335–342, 1987.
2. Legube, B. and Karpel Vel Leitner, N., Catalytic ozonation: a promising advanced oxidation technology for water treatment, *Catal. Today*, 53, 61–72, 1999.
3. Goldstein, S., Meyerstein, D., and Czapski, G., The Fenton reagents, *Free Rad. Biol. Med.*, 15, 435–445, 1993.
4. Bhatkhande, D.S., Pangarkar, V.G., and Beenackers, A.C.M., Photocatalytic degradation for environmental applications — a review, *J. Chem. Technol. Biotechnol.*, 77, 102–116, 2001.
5. Legrini, O., Oliveros, E., and Braun, A.M., Photochemical processes for water treatment, *Chem Rev.*, 93, 671–698, 1993.
6. Kolaczkowski, S.T. et al., Wet air oxidation: a review of process technologies and aspects in reactor design, *Chem. Eng. J.*, 73, 143–160, 1999.
7. Shaw, R.W. et al., Supercritical water: a medium for chemistry, *Chem. Eng. News*, December 23, 26–39, 1991.
8. Weavers, L.K. and Hoffmann, M.R., Sonolytic decomposition of ozone in aqueous solution: mass transfer effects, *Env. Sci. Technol.*, 32, 3941–3947, 1998.
9. Staehelin, S. and Hoigné, J., Decomposition of ozone in water: rate of initiation by hydroxide ions and hydrogen peroxide, *Environ. Sci. Technol.*, 16, 666–681, 1982.

10. Tomiyasu, H., Fukutomi, H., and Gordon, G., Kinetics and mechanism of ozone decomposition in basic aqueous solution, *Inorg. Chem.*, 24, 2962–2966, 1985.
11. Hoigné, J., Chemistry of aqueous ozone and transformation of pollutants by ozonation and advanced oxidation processes, in *The Handbook of Environmental Chemistry. Vol. 5, Part C, Quality and Treatment of Drinking Water,* 2nd ed., J. Hrubec, Springer-Verlag, Berlin, 1998, 83–141.
12. Charpentier, J.C. Mass transfer rates in gas liquid absorbers and reactors, in *Advances in Chemical Engineering*, Vol. 11. Academic Press, New York, 1981, 3–133.
13. Beltrán, F.J., Theoretical aspects of the kinetics of competitive first order reactions of ozone in the O_3/H_2O_2 and O_3/UV oxidation processes, *Ozone Sci. Eng.*, 19, 13–38, 1997.
14. Glaze, W.H. and Kang, J.W., Advanced oxidation processes. Test of a kinetic model for the oxidation of organic compounds with ozone and hydrogen peroxide in a semibatch reactor, *Ind. Eng. Chem. Res.*, 28, 1580–1587, 1989.
15. Beltran, F.J. et al., Aqueous degradation of atrazine and some of its main by-products with ozone and ozone/hydrogen peroxide, *J. Chem. Technol. Biotechnol.*, 71, 345–355, 1998.
16. Paillard, H., Brunet, R., and Dore, M., Optimal conditions for applying an ozone–hydrogen peroxide oxidizing system, *Water Res.*, 22, 91–103, 1988.
17. Chramosta, N. et al., Etude de la degradation de triazines par O_3/H_2O_2 et O_3 cinètique et sous-produits d'ozonation, *Water Supply*, 11, 177–185, 1993.
18. Acero, J.L., Stemmler, K., and von Gunten, U., Degradation kinetics of atrazine and its degradation products with ozone and OH radicals: a predictive tool for drinking water treatment, *Env. Sci. Technol.*, 34, 591–597, 2000.
19. Acero, J.L. et al., MTBE oxidation by conventional ozonation and the combination ozone/hydrogen peroxide: efficiency of the processes and bromate formation, *Env. Sci. Technol.*, 35, 4252–4259, 2001.
20. Beltrán, F.J. et al., Aqueous degradation of VOCs in the ozone combined with hydrogen peroxide or UV radiation processes. 2. Kinetic modelling, *J. Environ. Sci. Health Part A Environ. Eng.*, A34, 673–693, 1999.
21. Beltrán, F.J. Theoretical aspects of the kinetics of competitive ozone reactions in water, *Ozone Sci. Eng.*, 17, 163–181, 1995.
22. Beltrán, F.J., Estimation of the relative importance of free radical oxidation and direct ozonation/UV radiation rates of micropollutants in water, *Ozone Sci. Eng.*, 21, 207–228, 1999.
23. Beltrán, F.J. et al., Degradation of *o*-clorophenol with ozone in water, *Trans. Inst. Chem. Eng., Part B Process Saf. Environ. Prot.*, 71, 57–65, 1993.

9 Kinetics of the Ozone–UV Radiation System

The photolysis of aqueous ozone has been the subject of many works aimed at establishing the mechanism of reactions and the kinetics of the photolytic process. Peyton and Glaze[1] reported that the photolysis of dissolved ozone directly yields hydrogen peroxide, then the photolysis of the hydrogen peroxide formed and/or its reaction with ozone initiates the mechanism of free radical reactions leading to the hydroxyl radical:

$$O_3 + H_2O \xrightarrow{h\nu} H_2O_2 \qquad (9.1)$$

$$H_2O_2 \xrightarrow{h\nu} 2HO\bullet \qquad (9.2)$$

and Reaction (8.1), etc. (see Table 2.3 or Table 2.4).

According to these observations, the O_3/UV system is the most complete ozone system involving an advanced oxidation process since there could be up to three possible initiation reactions for the generation of hydroxyl radicals. These reactions are possibly due to the initiating species present in the water [see Reaction (7.4) and Reaction (7.10)], the free radicals formed from the photolysis of hydrogen peroxide, Reaction (9.2), and the free radicals formed from the ozone–hydrogen peroxide Reaction (8.1).

In addition, in the UV/O_3 system there are also three other possible types of direct oxidation/photolysis: direct ozonation, direct oxidation with hydrogen peroxide (although kinetically unfavorable in most cases), and direct photolysis. Since direct photolysis can constitute a significant kind of oxidation, this type of treatment is first examined from the kinetics point of view. The radiation applied can be visible or ultraviolet, although for kinetic studies involving ozone processes, the commonly used radiation is the 254 nm wavelength because at this wavelength ozone is at its maximum absorption efficiency.

9.1 KINETICS OF THE UV RADIATION FOR THE REMOVAL OF CONTAMINANTS FROM WATER

Many substances that absorb radiation can decompose through what is known as photolytic reactions or simply direct photolysis. The photolytic process can also be due to direct and indirect mechanisms. In the direct mechanism, the target substance absorbs part of the incident radiation and decomposes. The second mechanism is called photosensitization, and the decomposition is due to the reaction of the target compound with another substance, called a photosensitizer, which has absorbed the radiation and is in an excited state.[2] In this work on kinetics, only UV radiation due to the direct mechanism will be considered.

The rates of photolytic degradation can vary widely depending on the energy of radiation, the molar absorptivity of the target compound, and the quantum yield. In ozone/UV processes, the wavelength is usually fixed as 254 nm (energy is 112.8 kcalmol^{-1}), high enough to break down numerous chemical bonds. The conditions for the photolytic reaction to develop are that the target substance absorbs, at least, a fraction of the incident UV radiation and undergoes a decomposition reaction. The molecular structure of a given compound provides the best information whether this compound will absorb UV radiation. The quantum theory indicates that each molecule has a minimum energy at a given temperature. It is then said that the molecule is in the ground state. When radiation affects the molecule, a given increment of energy is absorbed and the molecule goes to an excited state. Then the molecule can undergo different mechanisms to return to the ground state. These mechanisms are called fluorescence, vibrational relaxation, internal conversion, phosphorescence, intersystem crossing, and photolytic decomposition.[3] The latter mechanism is the one responsible for the removal of substances through direct photolysis.

9.1.1 THE MOLAR ABSORPTIVITY

The magnitude of radiation that any given substance absorbs is measured by the Beer–Lambert law:

$$A = \log \frac{I_o}{I} = \varepsilon C L \tag{9.3}$$

where A is the absorbance or the ratio between the logarithm of intensities of incident, I_o, and transmitted radiation, I; L is the path of radiation; and ε is the molar absorptivity o and molar extinction coefficient, a parameter that depends on the chemical structure and represents a measure of the amount of absorbed radiation. The molar absorptivity of any compound corresponding to 254 nm radiation can be directly measured with one spectrophotometer from the absorbance of aqueous solutions at different concentrations of the target compound. This will allow a calibration curve to be prepared according to Equation (9.3). Thus, a plot of A with the concentration should lead to a straight line with slope εL.

9.1.2 THE QUANTUM YIELD

The quantum yield, Φ, is the parameter that expresses the fraction of the absorbed radiation employed for the photolytic decomposition reaction. It is also defined as the number of moles of the irradiated substance decomposed per mole of photon absorbed (one mole of photon is called one Einstein). For any given radiation, the energy associated with one Einstein is calculated from Planck's constant, $h = 6.63 \times 10^{-34}$ Js, the frequency of radiation, v, and Avogadro's number, $N_{AV} = 6 \times 10^{23}$ photons/Einstein:

$$E = hv\, N_{AV} \qquad (9.4)$$

For the particular case of 254 nm UV radiation ($v = 1.181 \times 10^{15}$ s^{-1}photon^{-1}), the energy associated with one Einstein is 4.69×10^5 J.

Knowledge of the quantum yield is necessary for further kinetic study of the photolytic process. The quantum yield of the target substance can be determined in photolysis experiments at specific conditions, as shown later. For the treatment of data, however, the characteristics of the incident radiation (intensity, flux of radiation, etc.) and the geometry of the photoreactor will be needed. The specific definition of these parameters will depend on the photolytic kinetic model applied.

9.1.3 Kinetic Equations for the Direct Photolysis Process

The photolysis rate of a given compound, B, depends not only on the photolytic properties of the target substance and UV wavelength of radiation (ε, Φ, etc.) but also on the nature of radiation source, lamp, and reactor type. The overall kinetic equation of a photolytic reaction can be expressed as the product of the quantum yield and the local rate of absorbed radiation per unit of time and volume, I_a:

$$-r_{BUV} = \Phi_B I_a \qquad (9.5)$$

For a monochromatic radiation, I_a can be defined as follows:

$$I_a = \mu q \qquad (9.6)$$

where μ is the attenuation coefficient and q is the density flux of radiation. The attenuation coefficient can be related to the molar absorptivity:[4]

$$\mu = 2.303 \sum \varepsilon_i C_i \qquad (9.7)$$

where the subindex i refers to any substance that absorbs radiation.

On the other hand, the following simplified mechanism can represent the UV photolysis of any substance in water:[5,6]

$$B \xrightarrow{hv, k_{1UV}} B* \qquad (9.8)$$

$$B* \xrightarrow{k_{2UV}} B \qquad (9.9)$$

$$B^* \xrightarrow{k_{3UV}} \text{Products} \qquad (9.10)$$

According to this mechanism, in an elementary volume of reaction, dV, the disappearance rate of B due to the direct photolysis or photolytic decomposition is:

$$-r_{BUV}\big|_{dV} = k_{1UV} F_B I_a - k_2 C_{B^*} \qquad (9.11)$$

where F_B is the fraction of absorbed radiation that B absorbs:

$$F_B = \frac{\varepsilon_B C_B}{\sum \varepsilon_i C_i} \qquad (9.12)$$

k_1 and k_2 are the rate constants of steps (9.8) and (9.9), respectively, the latter involving all pathways except the photochemical reaction; C_B^* is the concentration of B in the excited form; and ε_i and C_i, respectively, are the molar absorptivity and concentration of any i substance that absorbs radiation at the given wavelength. If the stationary state situation is applied to C_B^*, Equation (9.11) becomes:

$$-r_{BUV}\big|_{dV} = \Phi_B F_B I_a \qquad (9.13)$$

where the quantum yield is defined as:

$$\Phi_B = k_1 \frac{k_3}{k_2 + k_3} \qquad (9.14)$$

with k_3 the rate constant of the photochemical Reaction (9.10). If the rate equation is applied to the whole photoreactor volume, V, the disappearance rate of B becomes:

$$r_B = \frac{1}{V} \int r_{BUV}\big|_{dV} dV \qquad (9.15)$$

If Equation (9.6), Equation (9.13), and Equation (9.14) are considered, Equation (9.15) finally becomes:

$$r_B = \frac{1}{V} \int \Phi_B F_B \mu q \big|_{dV} dV \qquad (9.16)$$

where the flux density of radiation, q, depends on the position and characteristics of the radiation source (intensity of emitted radiation, geometry, etc.) and can be

Kinetics of the Ozone–UV Radiation System

determined from an energy radiation balance whose mathematical complexity depends on the kinetic model applied.

The different photochemical kinetic models are divided into two main groups: incidence and emission models. Incidence models give rise to a mathematical algorithm assuming the existence of a given radiant energy distribution in the vicinity of the reaction. Emission models are based on the source emission. Because of their simplicity, in this book only source emission models are considered. Further information concerning any of these models can be found elsewhere.[7] Two of the source emission models have been used extensively in ozone/UV or H_2O_2/UV processes:

- The linear source with emission in parallel planes to the lamp axis (LSPP model)
- The point source with spherical emission (PSSE model)

In addition, a third, more empirical model, should be mentioned:

- The Lambert's law model (LL model)

Table 9.1 gives the main characteristics of these models and equations for determining the flux density of radiation and photolytic removal of a given compound B.

Of the three models presented, as far as kinetics is concerned, the LL model is highly recommended because of its mathematical simplicity although it is based on an empirical situation and two parameters are needed for its application. Also note that the LSPP and PSSE models, although based on more realistic assumptions, consider a homogeneous radiation system and do not take into account the distortions that the radiation field presents in heterogeneous radiation systems as the UV/O_3 process. Thus, according to these models all radiation emitted by the source is entirely absorbed by the solution without end effects, reflection and/or refraction that is far away from the actual situation in a heterogeneous system. Especially in dilute aqueous systems, important fractions of the incident radiation can pass through the aqueous solution without being absorbed and be reflected at the wall. Some researchers, however, have considered the use of homogeneous photoreactor models by introducing some correction factors. Jacob and Dranoff[12] noted some of these limitations and introduced such a correction factor, the function of position, to account for deviations. Otake et al.[13,14] proposed a modified attenuation coefficient that contains the absorption effects of the liquid phase and the reflection, refraction, and transmission effects that reult from the gas phase. Yokota et al.[15] also proposed a modified attenuation coefficient function for the liquid phase, bubble diameter, and gas holdup. Thus far, no model has been proposed to account for the source of the presence of reflecting surfaces and scattering. According to the preceding discussion, given the heterogeneous character of the ozone/UV radiation system, the kinetics of photolytic processes is usually analyzed using the LL model. The reader can find further information on heterogeneous photoreactor models in an excellent review by Alfano et al.[16]

TABLE 9.1
Source Emission Kinetic Models for Photoreactors[a]

Model	q, Einstein m^{-2}s^{-1}		Photolytic Decomposition Rate of B, r_B [b]		Reference
LSSP[c]	$q = \dfrac{q_0 R_0}{r} \exp\left[\mu(r - R_0)\right]$	(9.17)	$\Phi_B F_B \dfrac{2\pi R_0 q_0 L}{V}\left[1 - \exp\left[-\mu(R_1 - R_0)\right]\right]$	(9.18)	8
PSSE[d]	$q = \dfrac{E_0}{4\pi} \displaystyle\int_{z'}^{z'+L} \dfrac{\exp\left[-\mu(r-R_0)\dfrac{\rho}{r}\right]}{\rho^2} dz'$	(9.19)	$\Phi_B F_B \dfrac{E_0 \mu}{4} \displaystyle\int_0^L \int_{R_0}^{R_1} \int_{z'}^{z'+L} \dfrac{\exp\left[-\mu\left[1 - \dfrac{R_0}{r}\right]\sqrt{r^2 + (z - z')^2}\right]}{r^2 + (z - z')^2} drdzdz'$	(9.20)	9
LL[e]	$q = \dfrac{I_0}{\mu}\left[1 - \exp(-\mu L)\right]$	(9.21)	$\Phi_B F_B I_0 \left[1 - \exp(-\mu L)\right]$	(9.22)	4, 10, 11

[a] Equations correspond to homogeneous systems.
[b] Calculated from Equation (9.16) after considering Equations (9.17), (9.19), or (9.21).
[c] Radiation source assumed as a consecutive line of points each emitting radiation in all directions in a plane perpendicular to the lamp axis, with q_0 as the density flux of radiation at the internal wall of the photoreactor ($r = R_0$) (see Figure 9.1).
[d] Radiation source assumed as emitting radiation in all space directions with E_0 as the radiant energy of the lamp per unit of length (see Figure 9.2).
[e] I_0 and L are the intensity of incident radiation and effective path of radiation through the photoreactor, respectively; q_0, E_0, or I_0 and L are calculated by actinometry experiments (see Appendix A4).

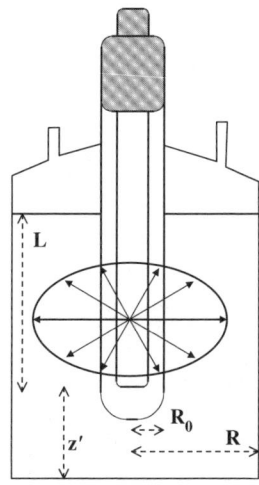
FIGURE 9.1 Scheme of photo-reactor for the LSPP model.

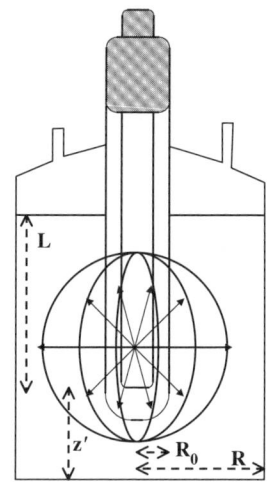
FIGURE 9.2 Scheme of photo-reactor for the PSSE model.

9.1.4 Determination of Photolytic Kinetic Parameters: The Quantum Yield

Quantum yield values of substances present in the water to be photolytically decomposed are necessary to quantify the magnitude of their photolytic removal rate. The quantum yield is determined from experiments in the photoreactor where I_o and L are already known. Procedures to determine Φ_B depend on the competitive character of intermediates and other substances present in the aqueous medium that absorb the incident radiation (see also Appendix A4).

9.1.4.1 The Absolute Method

If the target compound whose Φ_B has to be determined is treated with radiation of a given wavelength, without the presence of any other competing substance and without the interference of intermediate compounds resulting from the photolysis mechanism, Equation (9.22) can directly be applied. If a batch (or semibatch, in ozone systems) photoreactor is considered, the photolytic disappearance and accumulation rates of B in the water phase coincide, so that

$$-\frac{dC_B}{dt} = r_B = \Phi_B F_B I_0 [1 - \exp(-\mu L)] \qquad (9.23)$$

In the absence of any competing substance for the UV radiation, $F_B = 1$, so that integration of Equation (9.23) leads to

$$C_{B0} - C_B + \frac{1}{2.303 L \varepsilon_B} \ln \left[\frac{1 - \exp(-2.303 L \varepsilon_B C_{B0})}{1 - \exp(2.303 L \varepsilon_B C_b)} \right] = I_\theta \Phi_B t \quad (9.24)$$

According to Equation (9.24), a plot of the left side against time should yield a straight line of slope $I_0 \Phi_B$. Knowing the intensity of incident radiation allows us to determine the quantum yield. Equation (9.23) can be simplified when the concentration of the target compound is so low that the exponential term results are lower than 0.2. The resulting equation is:

$$-\frac{dC_B}{dt} = \mu L \Phi_B F_B I_0 = 2.303 \varepsilon_B L \Phi_B I_0 C_B \quad (9.25)$$

Equation (9.25) is an apparent first-order kinetics that, integrated after variable separation, yields:

$$\ln \frac{C_B}{C_{B_0}} = -2.303 \varepsilon_B L \Phi_B I_0 t \quad (9.26)$$

According to Equation (9.26), experimental points [$\ln(C_B/C_{B0})$ with time] of a direct photolysis experiment should yield a straight line of slope $-2.303\varepsilon_B L \Phi_B I_0$. If the molar absorptivity, intensity of incident radiation, and the effective pathway of the photoreactor (see Appendix A4) are known, the quantum yield can be determined from the slope of the indicated straight line. For example, this procedure can be applied to the photolysis of trichloroethylene[17] or similar organochlorine compounds, photoreactions that do not lead to competing intermediates for the incident radiation. In this particular case, however, the overall rate of decomposition of B could be due to two mechanisms: volatilization and photolysis. The procedure, then, would require previous knowledge of a first-order volatility coefficient that can easily be found from volatilization experiments.[17]

Another possible way to simplify Equation (9.23) results when the exponential term is higher than 2. Then, Equation (9.23) becomes:

$$-\frac{dC_B}{dt} = \Phi_B \frac{\varepsilon_B C_B}{\sum \varepsilon_i C_i} I_0 \quad (9.27)$$

Three possible cases arise from this situation: (1) the target compound absorbs all the incident radiation; (2) other compounds (e.g., intermediates) absorb most of the incident radiation; and (3) both the target compound and others absorb, with similar percentages, the incident radiation. In the first case, $\varepsilon_B C_B$ is basically $\sum \varepsilon_i C_i$ so that Equation (9.27) simplifies to:

$$-\frac{dC_B}{dt} = I_0 \Phi_B \quad (9.28)$$

Kinetics of the Ozone–UV Radiation System

Equation (9.28) is a zero-order kinetics so that a plot of the concentration of B with time yields a straight line of slope $-I_0\Phi_B$. Then the quantum yield can be determined, if the intensity of incident radiation (see Appendix A4) is known. When intermediates or other compounds present in water absorb most of the radiation, B is not directly photolyzed and its quantum yield cannot be determined. The third situation is the most complicated because it requires knowledge of the composition of the water and molar absorptivities of the components. In this case the absolute method can rarely be applied.

9.1.4.2 The Competitive Method

The usual way of finding the quantum yield value is through competitive UV radiation experiments where the target compound and another compound taken as the reference compound of known quantum yield are irradiated simultaneously. No conditions regarding the exponential term in Equation (9.23) are needed. With this procedure, radiation is applied to a solution containing the target compound and a reference compound of known quantum yield at the wavelength of the radiation. Then, applying Equation (9.23) to both the target compound and reference compound, division of the resulting equations, and integration between the limits

$$C_B = C_{B0} \quad C_R = C_{R0}$$
$$C_B = C_B \quad C_R = C_R \tag{9.29}$$

leads to:

$$\ln \frac{C_B}{C_{B0}} = \frac{\varepsilon_B \Phi_B}{\varepsilon_R \Phi_R} \ln \frac{C_R}{C_{R0}} \tag{9.30}$$

so that a plot of the logarithm of the dimensionless concentration of the target compound against that of the reference compound should yield a straight line, the slope of which is the ratio of the products of quantum yield and molar absorptivity. Since the quantum yield of the reference compound and molar absorptivities are known, the quantum yield of the target compound can be determined. Table 9.2 shows the values of quantum yield found following the procedures given above.

9.1.5 QUANTUM YIELD FOR OZONE PHOTOLYSIS

Ozone presents a very high molar absorptivity at 254 nm UV radiation both in the gas phase (2950 $M^{-1}cm^{-1}$ [42]) and in water (3300 $M^{-1}cm^{-1}$ [19]). In fact, this property makes the absorbance method one of the analytical procedures for measuring the concentration of ozone. There are different ozone analyzers commercially available that are based on the absorption of radiation at 254 nm wavelength of a gas stream containing ozone. The apparatus works like a spectrophotometer that measures the absorbance of ozone at this wavelength. Then the absorbance is correlated with the ozone concentration using some standard method[42–45] that acts as a calibration

TABLE 9.2
Quantum Yield and/or Rate Constants of the Reaction between the Hydroxyl Radical and Compounds Determined from UV and UV/H_2O_2 Kinetic Studies in Water

Compound	Reacting System Conditions	Quantum Yield and/or Rate Constant × 10^{-9}	Reference No. (Year)
Hydrogen peroxide	254-nm UV low pressure Hg lamp, 10W 25°C, acetic acid (0.01–0.1 M), AK	Φ = 0.5 AK	18 (1957)
Ozone	254-nm Hg resonance lamp, 0–0.6 M acetic acid, 0.05 M $HClO_4$, 14°C, AK	Φ = 0.62 AK	19 (1957)
MCPA	Philips, HPK-125 W, filter solution of 25g/L $CuSO_4$, 290 nm, AK	Φ = 0.53 AK	20 (1986)
Nitroaromatic compounds	Merry-go-round photoreactor, 450-Hg lamp, Corning 0.52 and 7.37 filters for 366 nm and $10^{-3}M$ K_2CrO_4 in 3% K_2CO_3 for 313 nm, AK	Φ = 2.9 × 10^{-5} (nitrobenzene, Φ = 0.0022 (2-nitrotoluene, 366), etc. AK	21 (1986)
Chloroaromatics	Hanovia 450W Hg lamp, Corning 7.54 filter, main line at 313 nm, CK, phenol as reference compound	Φ = 0.37 (Chlorobenzene) Φ = 0.043 (1,3,5-TCB) Φ = 0.29 (2-CBP)	22 (1986)
Ozone	Low pressure Hg lamp, 20 W, 2–40 Wm^{-2}, pH 2–9, T = 6–30°C	Empirical correlation deduced	23 (1988)
Parathion	Philips, HPK-125 W, > 290 nm, AK, 25°C, pH 4.69–9.59	Φ = 0.007 to 0.0016 depending on pH	24 (1992)
Atrazine	254-nm low pressure Hg lamp, 20°C, 1.6 × 10^{-6} Einstein $L^{-1}s^{-1}$ and UV/H_2O_2, up to 0.075 M, CK, phenol as reference compound	Φ = 0.05 k_{HOB} = 18	25 (1993)
Chloroethanes	Hanau TNN 15/32, 20°C, 253.7 nm, 4.44 × 10^{-6} Einstein s^{-1}, AK, 20°C, H_2O_2 continuously fed, 5.7 × 10^{-4} – 9.9 × 10^{-3}	k_{HOB} = 0.022 (1,1,1-TCE) k_{HOB} = 0.107 (1,1,2-TCE) k_{HOB} = 0.077 (1,1,2,2-TCE)	10 (1994)
1,2-dibromo-3-chloropropane, DBCP	1 to 4 low pressure Hg lamp, 254 nm, 8.4 W/lamp	Φ = 0.49, AK k_{HOB} = 0.14 (CK, PCE reference compound)	11 (1995)
Polynuclear aromatic hydrocarbons and Ozone	254-nm TNN 15/32 Hanau low pressure vapor Hg lamp, pH 7, 20°C, AK, 1.8 W L^{-1}	Φ = 0.0075 (Fluorene) Φ = 0.0069 (Phenanthrene) Φ = 0.052 (Acenaphthene) Φ = 0.64 (ozone)	6 (1995)
Trichloroethylene	254-nm TNN 15/32 Hanau low pressure vapor Hg lamp, pH 7, 20°C, AK, 1.8 W L^{-1}	Φ = 0.868	17 (1995)
Polynuclear aromatic hydrocarbons	254-nm TNN 15/32 Hanau low pressure vapor Hg lamp, pH 7, 20°C, 1.8 W L^{-1}, C_{H2O2T} = 0.4 M, AK	k_{HOB} = 9.9 (Fluorene) k_{HOB} = 13.4 (Phenanthrene) k_{HOB} = 8.8 (Acenaphthene)	26 (1996)

TABLE 9.2 (continued)
Quantum Yield and/or Rate Constants of the Reaction between the Hydroxyl Radical and Compounds Determined from UV and UV/H_2O_2 Kinetic Studies in Water

Compound	Reacting System Conditions	Quantum Yield and/or Rate Constant × 10^{-9}	Reference No. (Year)
Atrazine by-products	254-nm TNN 15/32 Hanau low pressure vapor Hg lamp, pH 7, 20°C, 6.2×10^{-3} Wcm^{-1}, $C_{H2O2T} = 10^{-2}$ M, CK, phenol, reference compound	For CEAT: $\Phi = 0.038$ and $k_{HOB} = 2.9$ For CIAT: $\Phi = 0.042$ and $k_{HOB} = 1.8$	27 (1996)
Trichloroethylene and tricloroethane	254-nm TNN 15/32 Hanau low pressure vapor Hg lamp, pH 7, 20°C, 6.2×10^{-3} Wcm^{-1}, $C_{H2O2T} > 10^{-2}$M, AK	$k_{HOB} = 0.02$ (TCA) $k_{HOB} = 1.8$ (TCE)	28 (1996)
Tomato wastewater	254-nm TNN 15/32 Hanau low pressure vapor Hg lamp, pH 7, 20°C, 6.2×10^{-3} Wcm^{-1}, COD = 930 mg O_2 L^{-1}, AK	$\Phi = 0.7$	29 (1997)
Atrazine	254-nm TNN 15/32 Hanau low pressure vapor Hg lamp, pH 7, 20°C, 5.98×10^{-6} Einstein L^{-1} s^{-1}, $C_{H2O2T} > 0.04$M, AK	$\Phi = 0.04$ $k_{HOB} = 2.08$	30 (1997)
Nitroaromatics	254-nm TNN 15/32 Hanau low pressure vapor Hg lamp, pH 7, 20°C, 3.3×10^{-8} Einstein cm^{-2} s^{-1}, CK, atrazine, reference compound, $C_{H2O2T} = 10^{-2}$ M, AK	Nitrobenzene: $\Phi = 0.007$ and $k_{HOB} = 2.9$ 2,6-Dinitrotoluene: $\Phi = 0.022$ and $k_{HOB} = 0.75$	31 (1998)
1,3,5,trinitrotriaza-cyclohexane (RDX)	Osram 150W Xe short-arc lamp, 0.15 W L^{-1}, 254 nm, AK	$\Phi > 0.13$	32 (1998)
Resorcinol (Re), nitroresorcinol (N) Muconic acid (M)	254-nm TNN 15/32 Hanau low pressure vapor Hg lamp, pH 7, 20°C, 3.3×10^{-8} Einstein cm^{-2} s^{-1}, CK, phenol, reference compound, $C_{H2O2T} = 10^{-2}$M, AK	Re: $\Phi = 0.0105$, $k_{HOB} = 2.7$ N: $\Phi = 0.0079$, $k_{HOB} = 3.4$ M: $\Phi = 0.0087$, $k_{HOB} = 6.9$	33 (1999)
Alachlor	254-nm TNN 15/32 Hanau low pressure vapor Hg lamp, pH 7, 20°C, 1.6×10^{-6} Einstein L^{-1} s^{-1}, CK, atrazine, reference compound, $C_{H2O2T} = 10^{-2}$ M, AK	$\Phi = 0.177$ $k_{HOB} = 32$	34 (2000)
Sodium dodecyl-benzenesulfonate	254-nm TNN 15/32 Hanau low pressure vapor Hg lamp, pH 7, 20°C, 1.6×10^{-6} Einstein L^{-1} s^{-1}, CK, phenol, reference compound, $C_{H2O2T} = 0.1$ M	$k_{HOB} = 11.6$	35 (2000)
MTBE	Low pressure Hg lamp, 254 nm, 5.3×10^{-6} Einstein L^{-1}s^{-1} H_2O_2/MTBE: 4.1–15.1, AK	$k_{HOB} = 3.9$	36 (2000)

TABLE 9.2 (continued)
Quantum Yield and/or Rate Constants of the Reaction between the Hydroxyl Radical and Compounds Determined from UV and UV/H_2O_2 Kinetic Studies in Water

Compound	Reacting System Conditions	Quantum Yield and/or Rate Constant × 10^{-9}	Reference No. (Year)
Acenaphthylene	254-nm TNN 15/32 Hanau low pressure vapor Hg lamp, pH 7, 20°C, 1.6 × 10^{-6} Einstein L^{-1} s^{-1}, AK, $C_{H2O2T} = 0.4\ M$	$\Phi = 0.004$ $k_{HOB} = 8.0$	37 (2000)
Simazine	254-nm TNN 15/32 Hanau low pressure vapor Hg lamp, pH 7, 20°C, 1.9 × 10^{-6} Einstein L^{-1} s^{-1}, CK, atrazine as reference compound, $C_{H2O2T} = 0.1\ M$, AK	$\Phi = 0.062$ $k_{HOB} = 2.1$	38 (2000)
Carbendazim	Philips HPM-12 Hg lamp, 400 W, 250–750 nm, pH:5–11, AK	Φ values varied depending on pH and oxygen concentration	39 (2000)
Polynuclear aromatic hydrocarbons	TNN 15/32 low pressure Hg lamp: 4.13 × 10^{-6} Einstein $L^{-1}s^{-1}$ and TQ150 medium pressure Hg lamp: 9.93 × 10^{-5} Einstein $L^{-1}s^{-1}$, pH 2.5–11.7	Φ values varied depending on pH and hydroxyl radical scavengers	40 (2001)
Debittering table olive wastewater	TNN 15/32 low pressure Hg lamp: 3.7 × 10^{-6} Einstein $L^{-1}s^{-1}$ and TQ150 medium pressure Hg lamp: 3.4 × 10^{-5} Einstein $L^{-1}s^{-1}$, pH 2.5–11.7, COD: 20000	For 254 nm: Φ = 4595 mgO_2 Einstein^{-1} For 200–600 nm: Φ = 529 mgO_2 Einstein^{-1}	41 (2001)

^a Φ = Quantum yield as mol Einstein^{-1}; k_{HOB} = rate constants in $M^{-1}sec^{-1}$; AK = absolute kinetic method; CK = competitive kinetic method.

method. In the gas phase, the standard method is the iodometric one.[43] In water, the most useful method is the karman indigo analysis,[44] but in the absence of compounds that also absorb at 254 nm radiation, the direct absorbance procedure can also be used to measure the ozone concentration, as reported by Schechter.[45]

9.1.5.1 The Ozone Quantum Yield in the Gas Phase

Since ozone gas also absorbs radiation and decomposes, the quantum yield of this photolytic process is a fundamental parameter to establish the actual ozone transferred mass to water in an O_3/UV system when both ozone and UV radiation are simultaneously fed to the photoreactor.

Morooka et al.[23] studied the photolytic decomposition of gaseous ozone and proposed a mechanism of reactions. According to their mechanism, the presence of

Kinetics of the Ozone–UV Radiation System

other gases (oxygen, nitrogen, water vapor) can affect the photolytic rate. From this mechanism, a very complex rate equation was found. Values of the photolytic decomposition rate of ozone gas in the presence of different gases were also determined experimentally. The authors observed that experimental values were higher than the calculated ones although the rate equation they proposed qualitatively explained the experimental results. Finally, the quantum yield was obtained from Equation (9.31), which is basically the LL model equation with $F_{O3g} = 1$:

$$\Phi_{O3g} = \left(-\frac{dC_{O3g}}{dt}\right)_{calc} \frac{h\nu N_{Av}}{2.303\varepsilon_{O3g}\bar{I}C_{O3g}} \tag{9.31}$$

where $\bar{I}$ is the mean UV irradiance in the reactor.

The overall quantum yield values obtained at different temperatures and in the presence of gases were correlated as a function of the composition of the gas phase and temperature to yield:[23]

For oxygen concentrations higher than ambient air:

$$\Phi_{O3g} = 7.9 \times 10^{14} \exp(-23/RT) C_{O3g}^{0.5} \left(C_{O2} + C_{N2}^{0.7}\right)^{-0.7} \tag{9.32}$$

For oxygen concentration of ambient air:

$$\Phi_{O3g} = 7.9 \times 10^{14} \exp(-23/RT) C_{O3g}^{0.5} \left(C_{O2} + C_{N2}^{0.7}\right)^{-0.7} \left(1 + 0.9 C_{H_2O}\right) \tag{9.33}$$

where the numerator of the exponential term is expressed in kJmol⁻¹ and concentrations are given in molm⁻³.

9.1.5.2 The Ozone Quantum Yield in Water

Taube[19] was one of the first who studied the aqueous ozone photolysis at 254 nm radiation with the aim of analyzing the atomic oxygen reactions in solution. The photolytic experiments were carried out in the presence of acetic acid or HCl to stop the radical mechanism of ozone decomposition. Thus, primary quantum yield values of 0,62 and 0.23 mol Einstein⁻¹ at 254 and 313 nm radiation were found.

Also, in this case, the determination of the quantum yield can be accomplished in a way similar to that shown above for the LL model. In most of the cases, the rate Equation (9.23), where $F_B = F_{O3}$ is assumed to be unity, can be simplified because the exponential term is higher than 2. Then the ozone photolysis should follow a zero-order kinetics. From the slope of the straight line that results when experimental concentrations of ozone are plotted against time, the value of $I_o\Phi_{O3}$ can be obtained after least squares analysis and the corresponding quantum yield that resulted is 0.64 mol Einstein⁻¹.[6,19]

9.2 KINETICS OF THE UV/H_2O_2 SYSTEM

Baxendale and Wilson[18] first studied the hydrogen peroxide photolytic decomposition in water in the absence and presence of acids ($HClO_4$, acetic acid), alkalis (NaOH), metal cations (Cu^{2+}), alcohols, etc. They obtained different experimental values of the quantum yield depending on the composition of water and presence or absence of oxygen. From the results in strong acid or alkali conditions, it was deduced that hydroxyl radicals from the photolytic reaction then react with hydrogen peroxide so that the experimental quantum yield was found to be 1 mol Einstein^{-1}, regardless of the pH value. It is evident that this represents the overall quantum yield of the photolytic decomposition reaction. In the presence of acetic acid, a scavenger of hydroxyl radicals, these radicals do not react with hydrogen peroxide, so that the measured rate of its disappearance is exclusively due to the photolytic Reaction (9.2). The quantum yield determined at these conditions was found to be 0.5 mol Einstein^{-1}, that is, half the value of the corresponding amount to the overall process. This represents the primary quantum yield of hydrogen peroxide photolysis. As shown later, this is the value to consider in kinetic models of hydrogen peroxide–UV radiation systems when the initial hydrogen peroxide decomposition is considered. However, the hydroxyl radical concentration is proportional to two times the initial photolytic decomposition rate of hydrogen peroxide since two hydroxyl radicals are formed in the primary step. The following set of reactions explains the main steps of the mechanism of the hydrogen peroxide photolytic decomposition in water:

$$H_2O_2 \xrightarrow{h\nu} 2HO\bullet \quad (9.2)$$

$$HO\bullet + H_2O_2 \longrightarrow HO_2\bullet + H_2O \quad (9.34)$$

$$HO\bullet + HO_2^- \longrightarrow HO_2\bullet + OH^- \quad (9.35)$$

$$2HO\bullet \longrightarrow H_2O + \frac{1}{2}O_2 \quad (9.36)$$

Note that Reaction (9.34) to Reaction (9.36) are also steps of the ozone decomposition mechanism in water (see Chapter 2).

9.2.1 DETERMINATION OF KINETIC PARAMETERS

When the UV/H_2O_2 oxidation system is applied to remove compounds in water, as a consequence of the direct use of this oxidizing system or because ozone is photolyzed, the rate of disappearance of any compound in water can be due to direct photolysis and hydroxyl radical oxidation (the direct action of hydrogen peroxide is usually negligible):

$$r_B = r_{UVB} + r_{Rad} \quad (9.37)$$

Kinetics of the Ozone–UV Radiation System

In addition to the quantum yield of B photolysis, Φ_B, the rate constant of the reaction between the hydroxyl radical and B has to be determined. In preceding chapters, different ways of obtaining these parameters are presented but the UV/H$_2$O$_2$ oxidation system is specifically suitable for determining the rate constant of hydroxyl radical reactions. Again, this determination can be made using absolute and competitive kinetic methods.

9.2.1.1 The Absolute Method

In this case, a possible simplification applies to Equation (9.37) when the direct photolysis of the target compound is negligible or simply does not occur. If this is the case, the disappearance rate of B in a batch photoreactor is exclusively due to its reaction with the hydroxyl radical:

$$-\frac{dC_B}{dt} = k_{HOB} C_{HO} C_B \qquad (9.38)$$

Since the hydroxyl radical concentration can be considered constant, Equation (9.38) represents a first-order kinetic process. Then, integration of Equation (9.38) gives a linear relationship between the $\ln(C_B/C_{Bo})$ and reaction time. The slope of the straight line that would result from this kind of plot is $k_{HOB} C_{HO}$. Thus, the problem reduces to finding the expression for the concentration of hydroxyl radicals as in the O$_3$ or O$_3$/H$_2$O$_2$ processes. For this case, this concentration is expressed as follows:

$$C_{HO\cdot} = \frac{2\Phi_{H2O2} I_0 \left[1 - \exp\left(-2.303 \sum \varepsilon_i C_i\right)\right] F_{H2O2}}{k_H C_{H2O2t} + \sum k_{HOS} C_S} \qquad (9.39)$$

where, in the denominator, the contribution of hydrogen peroxide to scavenging hydroxyl radicals, $k_H C_{H2O2t}$, is present as in the O$_3$/H$_2$O$_2$ oxidation system where mass transfer controls the process rate. Equation (9.39) can further be simplified if the concentration of hydrogen peroxide is high enough so that the rest of scavenging contributions, $\Sigma k_{HOS} C_S$, are negligible compared to $k_H C_{H2O2t}$, and when the exponential term becomes zero because $2.303 \Sigma \varepsilon_i C_{H2O2t} > 2$, and $F_{H2O2} = 1$ because hydrogen peroxide absorbs all the radiation. With these assumptions, the combination of Equation (9.38) and Equation (9.39) leads to:

$$-\frac{dC_B}{dt} = k_{HOB} C_B \frac{2\Phi_{H2O2} I_0}{k_H C_{H2O2t}} = k_T C_B \qquad (9.40)$$

According to this equation a plot of the apparent pseudo first-order rate constant, k_T, obtained from experiments at different high concentrations of hydrogen peroxide against the total concentration of hydrogen peroxide applied should lead to a straight line of slope $2\Phi_{H2O2} I_0 k_{HOB}/k_H$. From this slope the rate constant k_{HOB} can be calculated.

This procedure has been applied in different works.[28,30,38] Table 9.2 gives data on k_{HOB} obtained from this absolute method. Note that Equation (9.40) also holds in cases where B decomposes by direct photolysis since the contribution of direct photolysis to the disappearance rate may likely be negligible due to the high concentrations of hydrogen peroxide applied. In other words, hydrogen peroxide would absorb most (if not all) of the radiation.

If the direct photolysis of B cannot be neglected, the absolute method could still be applied. In these cases, however, a differential method has to be used since the term F_{H2O2T} is different from 1. If not, the hydrogen peroxide is likely to absorb nearly all the radiation and the integral absolute method also holds. The differential method, however, is not recommended because of the inaccuracies related not only to determining the accumulation rate of B ($-dC_B/dt$) at any time but also to quantifying the direct photolysis contribution, especially the term $\Sigma \varepsilon_i C_i$, at any time.

Note that, in natural waters, the scavenging term $\Sigma k_{HOS} C_S$ can be determined in a way similar to that shown for the O_3/H_2O_2 oxidation system. In this case, a model compound of known kinetics with the hydroxyl radical (known k_{HOB}) and a medium concentration of hydrogen peroxide must be added to the water so that the terms $\Sigma k_{HOS} C_S$ and $k_H C_{H2O2t} + k_{HOB} C_B$ will be of similar order of magnitude. Then, from the slope of the straight line of the plot $\ln(C_B/C_{B0})$ against time, the scavenging effect of the substances present in the natural water can be determined:

$$\Sigma k_{HOS} C_S = \frac{k_{HOB}}{k_T} 2 I_0 \Phi_{H2O2} - k_H C_{H2O2T} - k_{HOB} C_{B0} \qquad (9.41)$$

Note that in Equation (9.41) the scavenging term from the model compound B has also been included. For this term, the initial concentration of B, C_{B0}, has been considered despite its variation during the oxidation process. When B is removed other intermediate compounds of unknown nature are formed. These compounds also scavenge hydroxyl radicals; the term $k_{HOB} C_{B0}$ then accounts at all times for the role of these intermediates and the remaining model compound.

9.2.1.2 The Competitive Method

When the appropriate experimental conditions for the absolute method cannot be achieved (i.e., with compounds of very high molar absorptivity and/or quantum yields such as some polynuclear aromatic hydrocarbons) the fastest method to determine k_{HOB} is the competitive method. This is similar to the other competitive methods presented previously, corresponding to different oxidation or radiation systems. For this method, again a reference compound, R, of known kinetics with the hydroxyl radical must be used. This compound, in addition, should present a reactivity toward the hydroxyl radical similar to that of the target compound B. Also, direct photolysis of both the target compound and the reference compound should be negligible. The method is not explained in detail since, as stated above, it is similar to other competitive methods described previously. From a plot of the $\ln(C_B/C_{B0})$ against $\ln(C_R/C_{R0})$, a straight line of slope k_{HOB}/k_{HOR} is obtained. The rate constant k_{HOB} will depend on the accuracy of k_{HOR}, which represents the main drawback of this method. Table 9.2 presents the values of k_{HOB} determined in this way together with the reference compound used.

9.2.2 Contribution of Direct Photolysis and Free Radical Oxidation in the UV/H$_2$O$_2$ Oxidation System

The relative importance of both direct photolysis and free radical oxidation for the removal of B in the UV/H$_2$O$_2$ oxidation system can be estimated from the ratio of their respective rates:[46]

$$\frac{r_R}{r_{UV}} = \frac{k_{HOB} C_{HO} C_B}{I_0 \Phi_B F_B [1 - \exp(-\mu L)]} \quad (9.42)$$

with the concentration of hydroxyl radicals given by Equation (9.39), where the rate of the initiation step is two times the direct photolysis rate of hydrogen peroxide [see Reaction (9.2)] and the inhibiting factor involves the scavenging reactions of hydroxyl radicals with hydrogen peroxide. Equation (9.42), on the other hand, can be simplified in two cases that represent most of the practical situations:

- Low absorbing solution, $\mu L < 0.4$, then

$$r_{UV} = 2.303 I_0 \Phi_B L \varepsilon_B C_B \quad (9.43)$$

- High absorbing solution, $\mu L > 2$, then

$$r_{UV} = I_0 \Phi_B F_B \quad (9.44)$$

According to previous discussion, Equation (9.39) for the concentration of hydroxyl radicals can also be simplified, depending on the absorbing character of the water. However, if one or another simplifications is considered, the ratio between the rates of B degradation becomes the same, as Equation (9.45) indicates:

$$\frac{r_R}{r_{UV}} = \frac{k_{HOB} 2\varepsilon_{H2O2} \Phi_{H2O2} C_{H2O2T}}{\Phi_B \varepsilon_B \left(\sum k_{OHS} C_S + k_H C_{H2O2T} \right)} \quad (9.45)$$

Equation (9.45) can further be simplified in the case when the UV/H$_2$O$_2$ oxidation is carried out in laboratory-prepared water with $k_H C_{H2O2t} \gg k_{OHS} C_S$. If this condition is considered, Equation (9.45) reduces to Equation (9.46):

$$\frac{r_R}{r_{UV}} = \frac{k_{HOB} 2\varepsilon_{H2O2} \Phi_{H2O2}}{\Phi_B \varepsilon_B k_H} \quad (9.46)$$

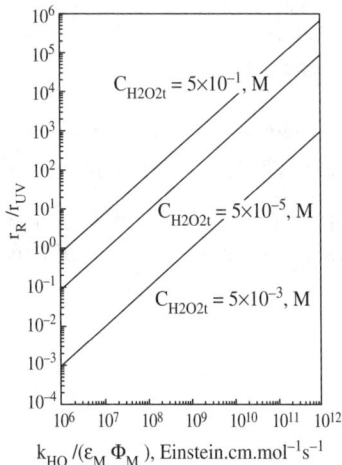

FIGURE 9.3 Comparison between the rates of hydroxyl radical oxidation and direct photolysis of micropollutants in water as a function of the ratio $k_{HOB}/(\Phi_B\varepsilon_B)$ at different hydrogen peroxide concentrations in the combined UV radiation–hydrogen peroxide oxidation. Conditions: 20°C, pH = 7, $\Sigma k_{HOS}C_S = 10^6$. (From Beltrán, F.J., Estimation of the relative importance of free radical oxidation and direct ozonation/UV radiation rates of micropollutants in water, *Ozone Sci. Eng.*, 21, 207–228, 1999. Copyright 1999, International Ozone Association. With permission.)

In this latter case, the rate ratio becomes independent of the concentration of hydrogen peroxide. From Equation (9.46), note that the rate ratio is neither dependent on the intensity of incident radiation nor the effective path of radiation through the photoreactor. If Equation (9.45) is put in logarithmic form, we obtain:

$$\log \frac{r_R}{r_{UV}} = \log \frac{k_{HOB}}{\Phi_B\varepsilon_B} + \log \frac{2\Phi_{H2O2}\varepsilon_{H2O2}C_{H2O2T}}{\sum k_{OHS}C_S + k_H C_{H2O2T}} \quad (9.47)$$

A plot of the logarithm of the reaction rate ratio against the logarithm of the ratio $k_{HOB}/\Phi_B\varepsilon_B$ leads to a straight line of slope unity (see Figure 9.3). The ordinate depends on the values of the concentration and molar absorptivity of hydrogen peroxide (which are also functions of the pH) and the inhibiting effect of the water, including that due to hydrogen peroxide, $k_H C_{H2O2t}$. With the hydroxyl radical kinetics (k_{HOB}) and the photolytic parameters of B (ε_B and Φ_B) we can estimate the relative importance of direct photolysis and free radical oxidation and, hence, the conditions where the addition of hydrogen peroxide is recommended to improve the overall oxidation rate of a given compound B. More details of this study are given in Beltrán.[46]

Kinetics of the Ozone–UV Radiation System

9.3 COMPARISON BETWEEN THE KINETIC REGIMES OF THE OZONE–COMPOUND B AND OZONE–UV PHOTOLYSIS REACTIONS

As shown in Chapter 8, when the ozone/hydrogen peroxide system was studied, the application of the reaction and diffusion time concepts is another possible way to compare the competition of the direct reaction between ozone and B and the ozone direct photolysis reaction. In fact, this comparison can be made using different approaches:

- By comparing the contributions of different reactions to the ozone absorption rate
- By comparing the contributions to the disappearance rate of the target compound
- By estimating the ratio between the oxidation rates of the target compound due to the direct reaction with ozone and to the action of hydroxyl radicals

These aspects will be discussed below.

9.3.1 Comparison between Ozone Direct Photolysis and the Ozone Direct Reaction with a Compound B through Reaction and Diffusion Times*

The concepts of reaction and diffusion times are now applied to ozone photolysis and direct ozone reactions for comparative reasons. The stoichiometry of these reactions was shown in Equation (9.1) and Equation (8.19), respectively. For the case of the direct reaction of ozone with a given compound, the corresponding reaction time has already been defined in Section 8.4.1 [see Equation (8.20)]. The case of Reaction (9.1) is somewhat more complicated due to the expression of the rate of ozone photolysis. For the sake of simplicity, the LL model is applied so that the ozone photolysis rate is given by Equation (9.48):

$$-r_{O3/UV} = I_0 \Phi_{O3} F_{O3} [1 - \exp(-\mu L)] \qquad (9.48)$$

where F_{O3} is the fraction of absorbed radiation that ozone absorbs:

$$F_{O3} = \frac{\varepsilon_{O3} C_{O3}}{\sum \varepsilon_i C_i} \qquad (9.49)$$

---

* Most of Section 9.3.1 is reprinted with permission from Beltrán, F.J., Theoretical aspects of the kinetics of competitive first-order reactions of ozone in the O_3/H_2O_2 and O_3/UV oxidation processes, *Ozone Sci. Eng.*, 19, 13–38, 1997. Copyright 1997, International Ozone Association.

Equation (9.48) can be simplified as already shown for any compound B [Equation (9.43) and Equation (9.44)]. Thus, depending on the absorbance of the solution, two cases can be considered:[47]

- Strong absorbance, $\mu L > 2$:

$$-r_{O3/UV} = I_0 \Phi_{O3} F_{O3} = 2.303 I_0 \Phi_{O3} \frac{\varepsilon_{O3} C_{O3}}{\mu} = k_{01} C_{O3} \qquad (9.50)$$

- Weak absorbance solution, or $\mu L < 0.4$:

$$-r_{O3/UV} = 2.303 L I_0 \Phi_{O3} \varepsilon_{O3} C_{O3} = k_{02} C_{O3} \qquad (9.51)$$

For these two cases the reaction time is defined as follows:

- Strong absorbance, $\mu L > 2$:

$$t_R = \frac{1}{k_{01}} = \frac{\mu}{2.303 I_0 \Phi_{O3} \varepsilon_{O3}} \qquad (9.52)$$

- Weak absorbance solution, or $\mu L < 0.4$:

$$t_R = \frac{1}{k_{02}} = \frac{1}{2.303 L I_0 \Phi_{O3} \varepsilon_{O3}} \qquad (9.53)$$

with

$$\mu = 2.303 (\varepsilon_{O3} C_{O3} + \varepsilon_B C_B) \qquad (9.54)$$

Note that only the photolysis rates of ozone and B are considered since conditions at the start of the process are applied.

There is also a third possible simplification of the photolysis rate equation of ozone. This is the case of strong absorbing solution exclusively due to the ozone molecules, that is, where B is a nonabsorbing compound. For this particular case, the rate of ozone photolysis is a zero-order kinetics defined as follows:

$$-r_{O3/UV} = I_0 \Phi_{O3} = k_0. \qquad (9.55)$$

with a reaction time given by Equation (9.56):

$$t_R = \frac{C_{O3}^*}{2k_0}. \qquad (9.56)$$

Kinetics of the Ozone–UV Radiation System

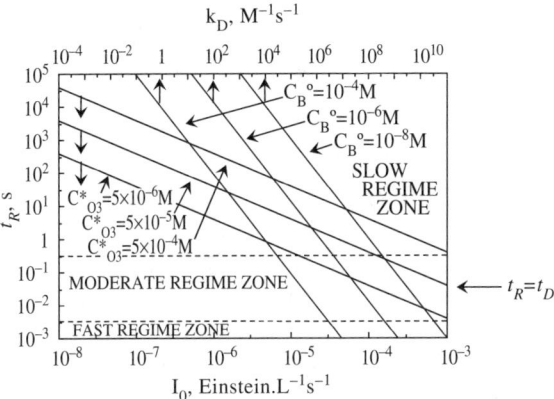

FIGURE 9.4 Variation of reaction time with direct rate constant and concentration of micropollutant, for Reaction (8.19), and with the intensity of UV radiation and ozone solubility, for Reaction (9.1) (case of non-UV absorbing compounds and $\mu L > 2$). C_B^o represents the concentration of B in bulk water. (From Beltrán, F.J., Theoretical aspects of the kinetics of competitive first-order reactions of ozone in the O_3/H_2O_2 and O_3/UV oxidation processes, *Ozone Sci. Eng.*, 19, 13–38, 1997. Copyright 1997, International Ozone Association. With permission.)

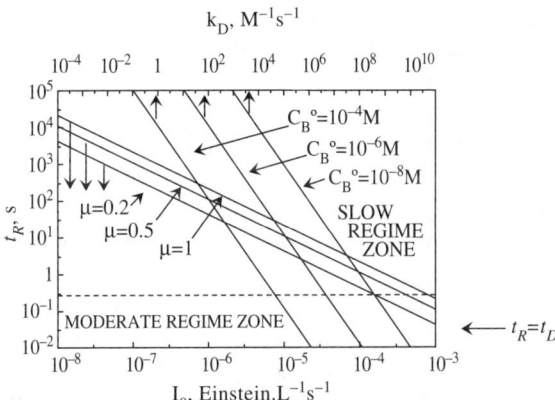

FIGURE 9.5 Variation of reaction time with direct rate constant and concentration of micropollutant, for Reaction (8.19), and with the intensity of UV radiation and absorbance solution, for Reaction (9.10) (case of medium or high UV absorbing compounds and $\mu L > 2$). C_B^o represents the concentration of B in bulk water. (From Beltrán, F.J., Theoretical aspects of the kinetics of competitive first-order reactions of ozone in the O_3/H_2O_2 and O_3/UV oxidation processes, *Ozone Sci. Eng.*, 19, 13–38, 1997. Copyright 1997, International Ozone Association. With permission.)

Now conditions for the competition of the direct and ozone photolysis reactions can be established by plotting the corresponding reaction times against the direct rate constant and the intensity of incident radiation, I_0, respectively, and by taking, depending on the case, C_{O3}^*, μ, or L as parameters. These plots are presented in Figures 9.4 to 9.6 where a value of 5×10^{-2} s has been taken for the diffusion time

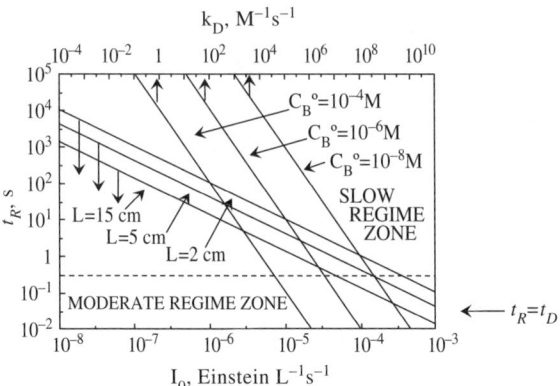

FIGURE 9.6 Variation of reaction time with direct rate constant and concentration of micropollutant, for Reaction (8.19), and with the intensity of incident UV radiation and effective path of radiation through the photoreactor, for Reaction (9.1) (case of non-UV absorbing compounds and μL < 0.4). C_B° represents the concentration of B in bulk water. (From Beltrán, F.J., Theoretical aspects of the kinetics of competitive first-order reactions of ozone in the O_3/H_2O_2 and O_3/UV oxidation processes, *Ozone Sci. Eng.*, 19, 13–38, 1997. Copyright 1997, International Ozone Association. With permission.)

as in the preceding oxidation systems (see Figure 8.1). According to these plots, both Reaction (9.1) and Reaction (8.19) will compete when their kinetic regimes coincide for a given set of experimental conditions. For example, in Figure 9.4 for a nonabsorbing compound B in a strong absorbing solution, if the concentration of B is very low (10^{-8} M) which is typical of surface water, and 10^6 $M^{-1}s^{-1}$ is the rate constant of the direct reaction between B and ozone, for ozone solubilities lower than 5×10^{-4} M, the direct and ozone photolytic reactions develop in the slow kinetic regime. Hence, they will compete for the available ozone. Similar conclusions can be reached for other experimental conditions and for other cases, where the ozone photolysis rate is simplified. More detailed discussion of these results can be found elsewhere.[47]

However, once conditions for the competition have been established, it is also possible that the contribution of one of the reactions could be neglected. This requires the determination of the initial reaction rates of ozone through both reactions that can be made as shown for the ozone/hydrogen peroxide system.

9.3.2 Contributions of Direct Photolysis and Direct Ozone Reaction to the Ozone Absorption Rate*

The case of slow kinetic regime will be considered since it represents the one closest to practical situations (that is, low concentration of micropollutant). The absorption rate is given by Equation (9.57):

---

* Most of Section 9.3.2 is reprinted with permission from Beltrán, F.J., Theoretical aspects of the kinetics of competitive first-order reactions of ozone in the O_3/H_2O_2 and O_3/UV oxidation processes, *Ozone Sci. Eng.*, 19, 13–38, 1997. Copyright 1997, International Ozone Association.

Kinetics of the Ozone–UV Radiation System

$$N_{O3} = \beta \left[k_D C_{O3} C_B + \left(-r_{O3/UV} \right) \right] \quad (9.57)$$

where the ozone photolysis rate term takes the form given by Equation (9.50), Equation (9.51), or Equation (9.55), depending on the nature of the solution and concentration of dissolved ozone. Thus, the following cases can be considered:

1. Strong UV absorption exclusively due to dissolved ozone
2. Strong UV absorption due to the dissolved ozone and a given compound B
3. Weak UV absorption

9.3.2.1 Strong UV Absorption Exclusively due to Dissolved Ozone

The concentration of dissolved ozone for a case where two parallel ozone (gas) liquid slow reactions (the direct reaction and the photolytic reaction) develop can be obtained from Equation (9.58) below, after application of film theory,[48] obtained in a similar manner as Equation (8.24):

$$C_{O3} = \frac{k_L \dfrac{Ha}{\sinh Ha} \left[C_{O3}^* + \dfrac{k_0}{k_D C_B}(1 - \cos Ha) \right] - \dfrac{\beta}{a} k_0}{\dfrac{\beta}{a} k_D C_B + k_L Ha \cosh Ha} \quad (9.58)$$

Note that one of the reactions (the photolytic reaction) is of zero-order kinetics. The ozone absorption rate, on the other hand, is given by Equation (9.59) below, obtained after substitution of $(-r_{O3/UV})$, given by Equation (9.55), in Equation (9.57):

$$N_{O3} = \beta \left[k_D C_{O3} C_B + k_0 \right] \quad (9.59)$$

By taking the values of k_0 [Equation (9.55)] and C_{O3}, Equation (9.58), substituted in Equation (9.59), the contributions of both ozone reactions to the ozone absorption rate at the start of the process can be estimated. Furthermore, the ozone concentration can also be calculated for the cases where only the direct reaction or the zero-order ozone photolysis rate develops. In this way, comparisons can be made between the individual ozone absorption rates and that corresponding to the combined process. In a previous work, comparison details are given for the case of the ozone/UV oxidation of trichloroethane, a compound that does not absorb UV light.[47] It should be noted that the application of Equation (9.58) leads, for some intensities of radiation (high values of I_0), to negative values of the dissolved ozone concentration. In these cases, Equation (9.58) cannot be applied since it has no physical meaning. These cases represent situations where the kinetic regime of the ozone photolysis is moderate or fast because of the high intensity of incident radiation. Recall that the cases presented refer to slow ozone gas liquid reactions, that is, their Hatta number is lower than 0.3.

9.3.2.2 Strong UV Absorption due to Dissolved Ozone and Compound B

The equation for the ozone photolysis rate is given by Equation (9.50), which, substituted in Equation (9.57), leads to the expression for the ozone absorption rate:

$$N_{O3} = \beta[k_D C_{O3} C_B + k_{01} C_{O3}] \qquad (9.60)$$

Here the concentration of dissolved ozone can be found from Equation (8.24) with $k_T = k_D C_B + k_{01}$ [see Equation (8.26)]. Thus, solving Equation (9.60) at different conditions of I_0, C_B, etc., the contributions of the direct photolysis and the direct reaction to the ozone absorption rate can be estimated. Also, in this case, high values of I_0 can lead to a change in the kinetic regime of ozone photolysis for which Equation (9.60) is not valid. If in Equation (8.24), k_T is taken as $k_D C_B$ or k_{01}, the hypothetical ozone absorption rates for the individual ozone photolysis and ozone direct reactions can be calculated. In this way, comparison can be made with the overall combined ozone/UV process. Details of this comparison for two practical cases are given elsewhere.[47]

9.3.2.3 Weak UV Absorption

In a similar way as in the preceding cases, the ozone photolysis rate, given by Equation (9.51) is substituted in Equation (9.57) to obtain the ozone absorption rate for a slow kinetic regime:

$$N_{O3} = \beta[k_D C_{O3} C_B + k_{02} C_{O3}] \qquad (9.61)$$

Again, the ozone concentration is also calculated from Equation (8.24) with $k_T = k_D C_B + k_{02}$ for the combined process and with $k_T = k_D C_B$ or $k_T = k_{02}$ for the individual processes. Thus, comparisons can be made between the direct reaction and the direct photolysis of ozone (see also Reference 47 for examples of this case).

9.3.3 CONTRIBUTIONS OF THE DIRECT OZONE AND FREE RADICAL REACTIONS TO THE OXIDATION OF A GIVEN COMPOUND B*

The approach presented in the preceding section allows the estimation of both the ozone absorption rate and the importance of the direct and indirect ozonation pathways in the UV/O$_3$ oxidation system of a given compound. Here a similar approach is applied so far as the disappearance rate of B is concerned. This approach was also presented for the case of the ozone/hydrogen peroxide system in Section 8.4.3. The kinetic regime of the ozone absorption will also be considered slow. The overall rate of the disappearance of B due to the O$_3$/UV process presents three contributions, as shown below:

---

* Most of Section 9.3.3 is reprinted with permission from Beltrán, F.J., Theoretical aspects of the kinetics of competitive first-order reactions of ozone in the O$_3$/H$_2$O$_2$ and O$_3$/UV oxidation processes, *Ozone Sci. Eng.*, 19, 13–38, 1997. Copyright 1997, International Ozone Association.

Kinetics of the Ozone–UV Radiation System

$$-r_B = r_{UV/B} + r_D + r_{rad} \tag{9.62}$$

Determination of the contributions of the three possible reaction rates will indicate the importance of the direct or free radical ways of ozonation of a given compound in the presence of UV radiation. However, Equation (9.62) can be simplified according to the nature of solution regarding the absorption of UV radiation. Therefore, the three cases discussed previously for the ozone absorption rate (Section 9.3.2.) are also considered here.

1. Strong UV absorption exclusively due to dissolved ozone
2. Strong UV absorption due to the dissolved ozone and a given compound B
3. Weak UV absorption

9.3.3.1 Strong UV Absorption Exclusively due to Dissolved Ozone

The rate of B disappearance presents only two terms due to the direct ozonation and free radical oxidation. The rate equation is, then, similar to Equation (8.27) of the ozone/hydrogen peroxide oxidation system. Recall that for slow reactions the concentration of hydroxyl radicals is, in this case:

$$C_{HO\cdot} = \frac{r_i}{k_{HO}C_B + \sum k_S C_S} \tag{9.63}$$

with r_i the initiation rate of free radicals that for the UV/O_3 system has two possible contributions due to Reaction (8.1) and Reaction (9.2), although the latter contribution is negligible in most cases (see Section 9.3.4) due to the low molar absorptivity of hydrogen peroxide. With both contributions, however, the rate of free radical initiation is:

$$r_i = 2k_{i2}C_{HO_2^-}C_{O3} + \frac{2\Phi_H I_0 \varepsilon_H C_{H2O2T}}{\varepsilon_{O3} C_{O3}} \tag{9.64}$$

Equation (9.62) for the disappearance rate of B, taking into account Equation (9.64), reduces to

$$-r_{B0} = zk_D C_B C_{O3} + k_{HOB} C_B \frac{2k_{i2}C_{HO_2^-}C_{O3} + \dfrac{2\Phi_H I_0 \varepsilon_H C_{H2O2T}}{\varepsilon_{O3}C_{O3}}}{k_{HO}C_B + \sum k_S C_S} \tag{9.65}$$

where subindex 0 refers to initial conditions. In Equation (9.65), in addition to the concentration of dissolved ozone [which can be estimated from Equation (8.24)], the concentration of hydrogen peroxide is also needed. This problem can be solved from the molar balance of hydrogen peroxide applied at the start of ozonation. Under these conditions, the accumulation rate of hydrogen peroxide, dC_{H2O2t}/dt, is negligible and the concentration of hydrogen peroxide can be explicitly estimated from the corresponding mass balance:

$$-r_{O3/UV} = \left[k_{i2} C_{O3} \frac{10^{pH-pK}}{1+10^{pH-pK}} + \frac{I_0 \Phi_{H2O2} \varepsilon_H}{\varepsilon_{O3} C_{O3}} + k_H C_{HO\bullet} \right] C_{H2O2T} \qquad (9.66)$$

From Equation (9.66) and Equation (8.24) the concentrations of hydrogen peroxide and ozone, respectively, can be estimated at different conditions (C_B, k_D, I_0, etc.) and substituted in Equation (9.65) to determine the disappearance rate of B or the individual contributions of the direct reaction with ozone and free radical oxidation. In this way, the percentage of free radical oxidation can be estimated (see Reference 47 for more details).

9.3.3.2 Strong UV Absorption due to Dissolved Ozone and a Compound B

In this case, *a priori*, B is also removed through the three possible ways, and Equation (9.62) holds for the disappearance rate of this compound. Following the procedure applied to the preceding case, the contributions of direct ozone, direct photolysis of B, and free radical oxidation of B can be estimated and conclusions can be reached on whether or not the combined process is recommended.

9.3.3.2 Weak UV Absorption

This case is also similar to the previous one, although a possible simplification applies when B does not absorb radiation or, in other words, when the dissolved ozone absorbs most of the radiation. This, for example, is the case of the ozonation of organochlorine compounds such as trichloroethane. The procedure is similar to the previous cases and is not detailed here.

In any case, a previous work[47] gives examples of the three cases above considered with specific data on ozone absorption rate and contributions of direct and free radical processes to the disappearance rate of a compound B of known kinetics with ozone, hydroxyl radical, and direct photolysis. Note that all the cases discussed can be applied to a given surface or ground water polluted with a given compound B, provided the inhibiting character of the water, $\Sigma k_{HOS} C_S$, is known (see Section 7.3.1).

9.3.4 ESTIMATION OF THE RELATIVE IMPORTANCE OF THE RATES OF DIRECT PHOTOLYSIS/DIRECT OZONATION AND FREE RADICAL OXIDATION OF A COMPOUND B*

As in the ozone/hydrogen peroxide oxidation system, another way of estimating the relative importance of the direct or free radical oxidation of a compound B is through

---

* Most of Section 9.3.4 is reprinted with permission from Beltrán, F.J., Estimation of the relative importance of free radical oxidation and direct ozonation/UV radiation rates of micropollutants in water, *Ozone Sci. Eng.*, 21, 207–228, 1999. Copyright 1999, International Ozone Association.

Kinetics of the Ozone–UV Radiation System

the ratio of their corresponding oxidation/photolysis rates. In this case, however, some complication arises due to the two possible direct reactions (ozonation/photolysis) of compound B. Thus, the ratio of the free radical and direct types of oxidation is:

$$\frac{r_{Rad}}{r_D + r_{UV}} = \frac{k_{HOB} C_{HO} C_B}{z k_D C_{O3} C_B + I_0 \Phi_B F_B [1 - \exp(-\mu L)]} \quad (9.67)$$

However, comparison will be made between the free radical rate, r_{Rad}, and the direct ozonation or direct UV photolysis rates individually considered.

Also, r_{Rad} has two possible contributions due to the initiation steps of the direct photolysis of the hydrogen peroxide formed from the ozone photolysis, r_{i1}, and the ozone/hydroperoxide ion reaction, r_{i2}. Comparison is first made between these two initiation reaction rates.

9.3.4.1 Relative Importance of Free Radical Initiation Reactions in the UV/O₃ Oxidation System

Overall, the ratio between the free radical initiation reaction rates is:

$$\frac{r_{i2}}{r_{i1}} = \frac{2 k_{i2} 10^{pH-pK} C_{H2O2T} C_{O3}}{\left(1 + 10^{pH-pK}\right) 2 I_0 \Phi_{H2O2} F_H [1 - \exp(-\mu L)]} \quad (9.68)$$

Equation (9.68) can be simplified, once the variation of the hydrogen peroxide molar absorptivity with pH is considered:

$$\varepsilon_H = \frac{19 + 210 \times 10^{pH-pK}}{1 + 10^{pH-pK}} \quad (9.69)$$

and the amount of energy absorbed. This study is made for the two extreme cases where the photolysis rate is simplified: strong and weak UV absorption.

1. Strong UV Absorption ($\mu L > 2$)

This could be the usual case since ozone, hydrogen peroxide and B contribute to the absorption of energy; hence, it is likely that $\mu L > 2$. Equation (9.68) simplifies to yield:[46]

$$\frac{r_{i2}}{r_{i1}} = \frac{\mu k_{i2} 10^{pH-pK} C_{O3}}{\left(19 + 240 \times 10^{pH-pK}\right) 2.303 I_0 \Phi_{H2O2}} \quad (9.70)$$

which, in logarithmic form, becomes

$$\log \frac{r_{i2}}{r_{i1}} = \log \frac{\mu k_{i2} 10^{pH-pK} C_{O3}}{\left(19 + 240 \times 10^{pH-pK}\right) 2.303 \Phi_{H2O2}} - \log I_0 \quad (9.71)$$

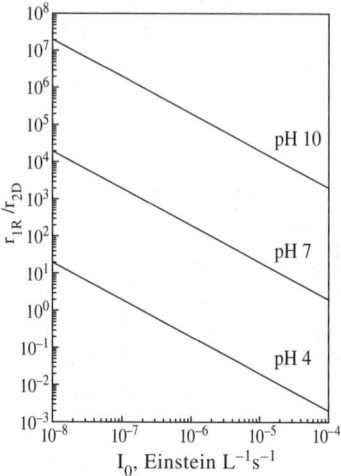

FIGURE 9.7 Comparison between rates of hydroxyl free radical initiation, due to hydrogen peroxide photolysis and direct ozonation of the hydroperoxide ion, of micropollutants in water as a function of the intensity of the incident radiation at different pH values in the combined ozonation with UV radiation. Conditions: 20°C, $\mu L > 2$, $\mu = 1$ cm^{-1}, $C_{O3} = 10^{-4}$ M, $L = 4$ cm. (From Beltrán, F.J., Estimation of the relative importance of free radical oxidation and direct ozonation/UV radiation rates of micropollutants in water, *Ozone Sci. Eng.*, 21, 207–228, 1999. Copyright 1999, International Ozone Association. With permission.)

Equation (9.71) presents two main difficulties because the concentration of ozone and the term μ have to be known. In any case, for specific values of C_{O3} and μ, a plot of the logarithm of the ratio between initiation rates against the logarithm of the intensity of incident radiation leads to parallel straight lines of slope equal to unity (in absolute value) and ordinates that depend on several parameters. Figure 9.7 shows a plot of this type. As can be seen from Figure 9.7, for pH > 4, the free radical initiation due to the reaction of ozone with the ionic form of the hydrogen peroxide (generated by ozone photolysis) is more important than the initiation due to direct hydrogen peroxide photolysis.

2. Weak UV absorption ($\mu L < 0.2$)

Now, the ratio of free radical initiation rates reduces to the following equation:

$$\frac{r_{i2}}{r_{i1}} = \frac{k_{i2}10^{\text{pH}-\text{pK}}C_{O3}}{\left(19+240\times 10^{\text{pH}-\text{pK}}\right)2.303 I_0 L\Phi_{H2O2}} \tag{9.72}$$

Or in logarithmic form:

$$\log\frac{r_{i2}}{r_{i1}} = \log\frac{k_{i2}10^{\text{pH}-\text{pK}}C_{O3}}{\left(19+240\times 10^{\text{pH}-\text{pK}}\right)2.303 L\Phi_{H2O2}} - \log I_0 \tag{9.73}$$

Kinetics of the Ozone–UV Radiation System

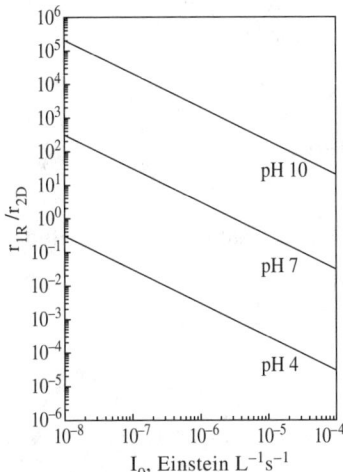

FIGURE 9.8 Comparison between rates of hydroxyl free radical initiation, due to hydrogen peroxide photolysis and direct ozonation of the hydroperoxide ion, of micropollutants in water as a function of the intensity of the incident radiation at different pH values in the combined ozonation with UV radiation. Conditions: 20°C, μL < 0.4, $C_{O3} = 5 \times 10^{-6}$ M, $L = 4$ cm. (From Beltrán, F.J., Estimation of the relative importance of free radical oxidation and direct ozonation/UV radiation rates of micropollutants in water, *Ozone Sci. Eng.*, 21, 207–228, 1999. Copyright 1999, International Ozone Association. With permission.)

The parallel straight lines that Equation (9.73) represents have been plotted in Figure 9.8 for specific values of C_{O3} and L.[46] As in the strong absorption case, for most practical cases, the free radical initiation reaction due to the ozone/hydrogen peroxide reaction is more important than the direct photolysis of hydrogen peroxide as can be deduced from Figure 9.8.

Note that in Equation (9.71) and Equation (9.73), I_0 should be limited to values lower than 10^{-4} Einstein L^{-1}s^{-1} to keep the kinetic regime of ozone absorption accompanied to the ozone photolytic reaction lower than 0.3, that is, to keep the slow kinetic regime of absorption. More information concerning this comparative test can be found elsewhere.[46]

9.3.4.2 Relative Importance of the Direct Reactions and Free Radical Oxidation Rates of Compound B

According to the preceding section, the reaction between ozone and hydrogen peroxide will be considered now as the initiation rate step. However, there are still two possibilities to consider that depend on the relative importance of the direct reactions: the ozone–compound B reaction and the photolytic reaction of B. If the first reaction is considered the main direct route to remove B, however, the comparison to be made is between the free radical rate and the direct reaction rate (the ozone direct reaction rate). This comparison has been discussed previously (see Section 8.4.3) because it reduces to the ozone/hydrogen peroxide oxidation system.

In this section, we discuss the relative importance of the free radical reaction rate and the direct photolysis rate for the removal of B. The ratio of these reactions is:[46]

$$\frac{r_{Rad}}{r_{UV}} = \frac{2k_{i2}k_{HOB}10^{pH-pK}C_{H2O2T}C_{O3}C_B}{\left(1+10^{pH-pK}\right)\left(k_{HOB}C_B + \sum k_S C_S\right)I_0\Phi_B F_B[1-\exp(-\mu L)]} \quad (9.74)$$

which can also be simplified according to the magnitude of absorbance.

1. Strong absorbance, $\mu L > 2$
Equation (9.74) reduces to:

$$\frac{r_{Rad}}{r_{UV}} = \frac{2k_{i2}k_{HOB}10^{pH-pK}C_{H2O2T}C_{O3}\mu}{2.303\left(1+10^{pH-pK}\right)I_0\Phi_B\varepsilon_B \sum k_S C_S} \quad (9.75)$$

if it is assumed that $k_{HOB}C_B \ll \sum k_S C_S$ as is the usual situation. In logarithmic form Equation (9.75) allows the ratio of reaction rates to be compared to the ratio of kinetic parameters of the free radical reaction, k_{HOB}, and direct photolysis, $\Phi_B \varepsilon_B$:

$$\log \frac{r_{Rad}}{r_{UV}} = \log \frac{k_{HOB}}{\Phi_B \varepsilon_B} + \log \frac{2k_{i2}10^{pH-pK}C_{H2O2T}C_{O3}\mu}{2.303\left(1+10^{pH-pK}\right)I_0 \sum k_S C_S} \quad (9.76)$$

Then a plot of the left side of Equation (9.76) against the logarithm of the ratio between $k_{HOB}/\Phi_B\varepsilon_B$ would allow both free radical and direct photolysis rates of B to be compared (see Figure 9.9). As seen in Equation (9.76), however, in addition to the concentration of ozone, the concentration of hydrogen peroxide has to be known. While a rough estimation of the concentration of ozone can be made by assuming the ozone solubility at the water interface, for the concentration of hydrogen peroxide the more rigorous procedure followed in Section 9.3.2 must be applied. However, the procedure presented here can be directly applied to the $O_3/H_2O_2/UV$ oxidation system where the concentration of hydrogen peroxide is known. Again, two extreme situations can be considered, depending on the magnitude of absorbance.

2. Weak absorbance solution, $\mu L < 0.2$
This is an unusual case because a low absorbance would mean that the direct ozone reaction of B could be even more important than the direct photolysis of B. Hence, comparison between the free radical rate and the photolytic rate would have no meaning. In any case, if these reaction rates are compared, Equation (9.74) will reduce to Equation (9.77):

$$\frac{r_{Rad}}{r_{UV}} = \frac{2k_{i2}k_{HOB}10^{pH-pK}C_{H2O2T}C_{O3}}{2.303\left(1+10^{pH-pK}\right)I_0\Phi_B\varepsilon_B L \sum k_S C_S} \quad (9.77)$$

Kinetics of the Ozone–UV Radiation System

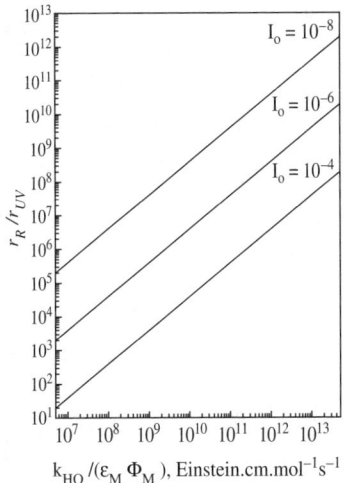

FIGURE 9.9 Comparison between rates of hydroxyl free radical oxidation and direct photolysis of micropollutants in water as a function of the ratio $k_{HOB}/(\Phi_B \varepsilon_B)$ at different intensity of the incident radiation in the combined ozonation with UV radiation. Conditions: 20°C, pH 7, $\mu L > 2$, $\mu = 1$ cm^{-1}, $C_{O3} = 10^{-4}$ M, $L = 4$ cm. (From Beltrán, F.J., Estimation of the relative importance of free radical oxidation and direct ozonation/UV radiation rates of micropollutants in water, *Ozone Sci. Eng.*, 21, 207–228, 1999. Copyright 1999, International Ozone Association. With permission.)

The logarithmic form of Equation (9.77) is:

$$\log \frac{r_{Rad}}{r_{UV}} = \log \frac{k_{HOB}}{\Phi_B \varepsilon_B} + \log \frac{2k_{i2} 10^{pH-pK} C_{H2O2T} C_{O3}}{2.303(1+10^{pH-pK}) I_0 L \sum k_s C_s} \quad (9.78)$$

In Figure 9.10, Equation (9.78) is plotted at different conditions. As observed, plots such as that of Figure 9.10 are very useful for the $O_3/H_2O_2/UV$ oxidation system because the concentration of hydrogen peroxide would be known.

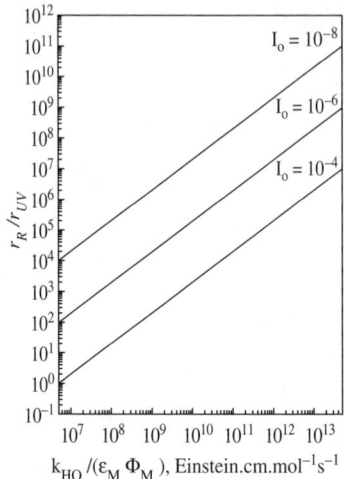

FIGURE 9.10 Comparison between rates of hydroxyl free radical oxidation and direct photolysis of micropollutants in water as a function of the ratio $k_{HOB}/(\Phi_B\varepsilon_B)$ at different intensities of the incident radiation in the combined ozonation with UV radiation. Conditions: 20°C, $\mu L < 0.4$, $\Sigma k_{HOS}C_S = 10^6$ s^{-1}, $C_{O3} = 5 \times 10^{-6}$ M, $C_{H2O2T} = 10^{-4}$ M, $L = 4$ cm. (From Beltrán, F.J., Estimation of the relative importance of free radical oxidation and direct ozonation/UV radiation rates of micropollutants in water, *Ozone Sci. Eng.*, 21, 207–228, 1999. Copyright 1999, International Ozone Association. With permission.)

References

1. Peyton, G.R. and Glaze, W.H., Destruction of pollutants in water by ozone in combination with ultraviolet radiation. 3. Photolysis of aqueous ozone, *Environ. Sci. Technol.*, 22, 761–767, 1988.
2. Friesen, K.J., Muir, D.C.G., and Webster, G.R.B., Evidence of sensitized photolysis of polychlorinated dibenzo-*p*-dioxins in natural waters under sunlight conditions, *Environ. Sci. Technol.*, 24, 1739–1744, 1990.
3. Owen, E.D., Principles of photochemical reactions in aqueous solutions, in *Organic Compounds in Aquatic Environments*, S.D. Faust and J.V. Hunter, Eds., Marcel Dekker, New York, 1971, 387–423.
4. Leifer, A., *The Kinetics of Environmental Aquatic Photochemistry: Theory and Practice*, American Chemical Society, New York, 1998.
5. Gal, E. et al., Photochemical degradation of parathion in aqueous solutions, *Water Res.*, 26, 911–915, 1992.
6. Beltrán, F.J. et al., Oxidation of polynuclear aromatic hydrocarbons in water. 2. UV radiation and ozonation in the presence of UV radiation, *Ind. Eng. Chem. Res.*, 34, 1607–1615, 1995.
7. Alfano, O.M., Romero, R.L., and Cassano, A.E., Radiation field modelling. 1. Homogeneous media, *Chem. Eng. Sci.*, 41, 421–444, 1986.
8. Jacob, S.M. and Dranoff, J.S., Radial scale-up of perfectly mixed photochemical reactors, *Chem. Eng. Prog. Symp. Ser.*, 62, 47–55, 1966.

9. Jacob, S.M. and Dranoff, J.S., Design and analysis of perfectly mixed photochemical reactors, *Chem. Eng. Prog. Symp. Ser.*, 64, 54–63, 1968.
10. De Laat, J., Tace, E., and Dore, M., Degradation of chloroethanes in dilute aqueous solution by H_2O_2/UV, *Water Res.*, 28, 2507–2519, 1994.
11. Glaze, W.H., Lay, Y., and Kang, J.W., Advanced oxidation processes. A kinetic model for the oxidation of 1,2-dibromo-3-chloropropane in water by the combination of hydrogen peroxide and UV radiation, *Ind. Eng. Chem. Res.*, 34, 2314–2323, 1995.
12. Jacob, S.M. and Dranoff, J.S., Light intensity profiles in a perfectly mixed photoreactor, *AIChE J.*, 16, 359–363, 1970.
13. Otake, T. et al., Light intensity profile in gas–liquid dispersion. Applicability of effective absorption coefficient, *Kagaku Kogaku Ronbunshu*, 7, 57–63, 1981.
14. Otake, T. et al., Light intensity profile in gas–liquid dispersion. Applicability of effective absorption coefficient, *Int. Chem. Eng.*, 23, 288–297, 1983.
15. Yokota, T. et al., Light absorption rate in a bubble column photochemical reactor, *Kagaku Kogaku Ronbunshu*, 7, 157–163, 1981.
16. Alfano, O.M., Romero, R.L., and Cassano, A.E., Radiation field modelling. II. Heterogeneous media, *Chem. Eng. Sci.*, 41, 1137–1153, 1986.
17. Beltrán, F.J. et al., Application of photochemical reactor models to UV radiation of trichloroethylene in water, *Chemosphere*, 31, 2873–2885, 1995.
18. Baxendale, J.H. and Wilson, J.A., Photolysis of hydrogen peroxide at high light intensities, *Trans. Faraday Soc.*, 53, 344–356, 1957.
19. Taube, H., Photochemical reactions of ozone in solution, *Trans. Faraday Soc.*, 53, 656–665, 1957.
20. Soley, J. et al., Kinetic study of 4-chloro-2-methylphenoxyacetic acid photodegradation, *Ind. Eng. Chem. Prod. Res. Dev.*, 25, 645–649, 1986.
21. Simmons, M.S. and Zepp, R.G., Influence of humic substances on photolysis of nitroaromatic compounds in aqueous systems, *Water Res.*, 7, 899–904, 1986.
22. Dulin, D., Drossman, H., and Mill, T., Products and quantum yields for photolysis of chloroaromatics in water, *Env. Sci. Technol.*, 20, 72–77, 1986.
23. Morooka, S. et al., Decomposition and utilization of ozone in water treatment reactor with ultraviolet radiation, *Ind. Eng. Chem. Res.*, 27, 2372–2377, 1988.
24. Gal, E. et al., Photochemical degradation of parathion in aqueous solutions, *Water Res.*, 26, 911–915, 1992.
25. Beltrán, F.J., Ovejero, G., and Acedo, B., Oxidation of atrazine in water by ultraviolet radiation combined with hydrogen peroxide, *Water Res.*, 27, 1013–1021, 1993.
26. Beltrán, F.J., Ovejero, G., and Rivas, F.J., Oxidation of polynuclear aromatic hydrocarbons in water. 3. UV radiation combined with hydrogen peroxide, *Ind. Eng. Chem. Res.*, 35, 883–890, 1996.
27. Beltrán, F.J. et al., Aqueous UV radiation and UV/H_2O_2 oxidation of atrazine first degradation products: deethylatrazine and deisopropylatrazine, *Env. Tox. Chem.*, 15, 868–872, 1996.
28. Beltrán, F.J. et al., Contribution of free radical oxidation to eliminate volatile organochlorine compounds in water by ultraviolet radiation and hydrogen peroxide, *Chemosphere*, 32, 1949–1961, 1996.
29. Beltrán, F.J., González, M., and González, J.F., Industrial wastewater advanced oxidation. 1. UV radiation in the presence and absence of hydrogen peroxide, *Chemosphere*, 31, 2405–2414, 1997.
30. De Laat, J. et al., Modeling of oxidation of atrazine by H_2O_2/UV. Estimation of kinetic parameters, *Ozone Sci. Eng.*, 19, 395–408, 1997.

31. Beltrán, F.J., Encinar, J.M., and Alonso, M.A., Nitroaromatic hydrocarbon ozonation in water. 2. Combined ozonation with hydrogen peroxide or UV radiation, *Ind. Eng. Chem. Res.*, 37, 32–40, 1998.
32. Bose, P., Glaze, W.H., and Maddox, D.S., Degradation of RDX by various advanced oxidation processes: I. Reaction rates, *Water Res.*, 997–1004, 1998.
33. Beltrán, F.J. et al., A kinetic model for advanced oxidation processes of aromatic hydrocarbons in water: application to phenanthrene and nitrobenzene, *Ind. Eng. Chem. Res.*, 38, 4189–4199, 1999.
34. Beltrán, F.J. et al., Determination of kinetic parameters of ozone during oxidations of alachlor in water, *Water Env. Res.*, 72, 689–697, 2000.
35. Beltrán, F.J., García-Araya, J.F., and Álvarez, P.M., Sodium dodecylbenzene sulfonate removal from water and wastewater. 1. Kinetics of decomposition by ozonation, *Ind. Eng. Chem. Res.*, 39, 2214–2220, 2000.
36. Chang, P.B.L. and Young, T., Kinetics of methyl tert-butyl ether degradation and by-product formation during UV/hydrogen peroxide water treatment, *Water Res.*, 34, 2233–2240, 2000.
37. Rivas, F.J., Beltrán, F.J., and Acedo, B., Chemical and photochemical degradation of acenaphthylene: Intermediate identification, *J. Haz. Mat.*, B75, 89–98, 2000.
38. Beltrán, F.J. et al., Kinetics of simazine advanced oxidation in water, *J. Environ. Sci. Health*, B35, 439–454, 2000.
39. Panadés, R., Ibartz, A., and Esplugas, S., Photodecomposition of carbendazim in aqueous solutions, *Water Res.*, 34, 2951–2954, 2000.
40. Miller, J.S. and Olejnik, D., Photolysis of polycyclic aromatic hydrocarbons in water, *Water Res.*, 35, 233–243, 2001.
41. Beltrán, F.J. et al., Treatment of wastewater from table olive industries: quantum yield of photolytic processes, *Bull. Environ. Contam. Toxicol.*, 67, 195–201, 2001.
42. Griggs, M., Absorption coefficients of ozone in the ultraviolet and visible regions, *J. Chem. Phys.*, 49, 857–859, 1968.
43. Kolthoff, I.M. and Belcher, R., *Volumetric Analysis*, Vol. III, Interscience, New York, 1957.
44. Bader, H. and Hoigné, J., Detemination of ozone in water by the indigo method, *Water Res.*, 15, 449–456, 1981.
45. Schechter, H., Spectrophotometric method for determination of ozone in aqueous solution, *Water Res.*, 7, 729–739, 1973.
46. Beltrán, F.J., Estimation of the relative importance of free radical oxidation and direct ozonation/UV radiation rates of micropollutants in water, *Ozone Sci. Eng.*, 21, 207–228, 1999.
47. Beltrán, F.J., Theoretical aspects of the kinetics of competitive first-order reactions of ozone in the O_3/H_2O_2 and O_3/UV oxidation processes, *Ozone Sci. Eng.*, 19, 13–38, 1997.
48. Beltrán, F.J. Theoretical aspects of the kinetics of competitive ozone reactions in water, *Ozone Sci. Eng.*, 1995, 17, 163–181.

10 Heterogeneous Catalytic Ozonation

Catalysts are substances used to accelerate the rate of different chemical reactions. These reactions are typically those encountered in the chemical industry where high temperatures and pressure are applied. These conditions, especially temperature, act on the surface of certain materials (metal oxides, activated carbon, zeolites, etc.) which, after adsorption of reactant molecules, accelerate the rate of numerous reactions. In water treatment, the high reactivity of ozone and the active surfaces of some materials can also be used to increase the ozonation rate. In an attempt to improve the performance of advanced oxidation of water contaminants with the use of ozone, numerous research studies in the mid-1990s focused on the combined application of ozone and solid catalysts. These systems constituted the catalytic ozonation of water contaminants.[1] At that time, however, catalytic ozonation was not a new process; the use of ozone and catalysts dates back to the 1970s.[2,3] In the first studies, however, attention focused mainly on the use of transition metal salts (such as nitrates, sulfates, etc.), which are soluble in water.[2] When dealing with catalytic ozonation, one should distinguish the homogeneous (HoCO) and the heterogeneous (HeCO) processes, depending on the water solubility of the catalyst.

One of the first studies on this matter was by Hill,[4,5] who observed the homogeneous decomposition of ozone catalyzed by Co^{2+} in acid media (acetic acid or perchloric acid). He proposed a mechanism of reactions with the formation and participation of hydroxyl free radicals formed in a first step from the direct oxidation of Co^{2+}:

$$O_3 + CO^{2+} + H_2O \xrightarrow{k=37 \text{ M}^{-1}\text{min}^{-1}} HO\bullet + CoOH^{2+} + O_2 + O_2 \qquad (10.1)$$

Formation of hydroxyl radicals signaled the beginning of a new advanced oxidation technology. Nonetheless, the first work on the catalytic ozonation of water pollutants appears to have been Hewes and Davidson[2] who, in 1972, reported data on the TOC elimination of a secondary municipal effluent with an average 18 ppm initial TOC. In this work, different transition metal salts (homogeneous catalytic ozonation, HoCO) were used with significant results at temperatures between 30 and 60°C and pH between 5 and 10. At the conditions investigated, total destruction of TOC was observed in less than 3 h. Some years later, Chen et al.[3] reported successful results on the heterogeneous catalytic ozonation (HeCO) of model compounds (i.e., phenol) and wastewaters (459 mgL^{-1} initial TOC) with a Fe_2O_3 type catalyst. The results were expressed as a function of COD and TOC removed per ozone consumed. Table 10.1 and Table 10.2 list studies on homogeneous and heterogeneous catalytic ozonation,

TABLE 10.1
Works on Homogeneous Catalytic Ozonation

Ozonation System	Reactor and Operating Conditions	Observations	Reference
Ozone decomposition study Hydrogen peroxide Catalyst: sulfates of Fe(II), Co(II), Ni(II), Cu(II), etc.	Batch reactors, acid pH with $0.1N$ H_2SO_4	Significant effect of Co(II), Ce(III), Ag(I), Cu(II), Tl(I), and Ce(IV) Fe(II) has no effect Proposed mechanism	6
Ozone decomposition study Catalyst: Co(II): 6×10^{-5} to $3.6 \times 10^{-4}\,M$	Batch reactors, pH: 1.6–3.5 with $HClO_4$, 0°C	Acetic acid acts as inhibitor Mechanism and kinetic study Values of rate constants	4
Ozone decomposition study Acetic acid Catalyst: Co(II): 4.5×10^{-5} to $7.1 \times 10^{-4}\,M$	Batch reactor, pH: 0.76 with $HClO_4$	Mechanism and kinetic study Values of rate constants	5
Secondary effluent (TOC: 10–35 mgL^{-1}) Phenol 7.8 mgL^{-1}, surfactant 18.92 mgL^{-1}, DDT: 2.54 mgL^{-1} Catalyst: Salts of Co(II), Ti(II), Mn(II), etc. (100 mgL^{-1})	Batch reactor with recirculated water 2 l, 30–60°C, pH: 5–10 buffered water	Improvements of COD removals First-order kinetics	2
Ozone–Fe(II) reaction study	Batch reactor, pH: 1.3–5.4	Stoichiometric determination: 1.84–2.53 mol Fe(II)/mol O_3 Influence of different free radical species Mechanism proposed	7
Removal of Fe(II), 0.84 mgL^{-1} and Mn(II), 0.16 mgL^{-1} Humic acids	Batch flasks, both prepared and natural water	Ozonation after sedimentation and before filtration Removal of Fe(II) and Mn(II) by direct means; on Mn(II) removal, positive influence of humics and negative influence of carbonates	8
Azoic dyes: 100 mgL^{-1} Catalyst: 20–100 mgL^{-1} Cu(II), Zn(II), Ag(I), and Cr_2O_3	Magnetically stirred semibatch bubble reactor	Reduction of decoloration time with respect to ozonation alone Highest reduction with Zn(II), 40% Cr_2O_3, worst catalyst	9
Ozone reaction with Fe(II) and Mn(II) in 1–2 mgL^{-1}	Batch flasks, 25°C, C_{O3}:1 mgL^{-1}	Stoichiometric determination Fast reaction with Fe(II) Positive effect of pH (5.5 to 7) with Mn(II) Rate constant data determined	10

TABLE 10.1 (continued)
Works on Homogeneous Catalytic Ozonation

Ozonation System	Reactor and Operating Conditions	Observations	Reference
Ozone reaction with Mn(II) (0.11 mgL^{-1}) River water Treatment plant water	Batch reactors, C_{O3} = 0.5–2 mgL^{-1}	Significant removal of Mn(II) Reaction: Mn(II) to Mn(VII), depending on ozone dose	11
Oxalic acid (1.4 × 10^{-3} M) Catalyst: Mn(II): 3.3 × 10^{-5} – 1.2 × 10^{-3} M	300-mL semibatch reactor, pH = 0 and 4.7 (phosphate buffer), also, batch reactor: spectrophotometer cell	Mechanism proposed Kinetic study Determination of rate constant	12
Ozone reaction with Fe(II)	Stopped flow spectrophotometer reactor (10^{-3}–400 s reaction time, pH = 0–2)	Proposed mechanism and kinetic study Rate constant determination Kinetic modeling fitting	13
Ozone–Mn(II) (0.6–1 mgL^{-1} reaction) Raw water: DOC: 7.3 mgL^{-1} pH7, and settled water: DOC: 3.5 mgL^{-1} pH6.6	Continuous bubble column (2 m high, 25.4 cm diameter) Presence of carbonates	95% Mn removal in settled water 20% removal in raw water 2.7 mg O$_3$ l^{-1} ozone transferred dose 15–25% DOC removed	14
Pesticides (B), Catalyst: VO(I), Fe(III), Co(II), Ni(II), Cu(II), and also heterogeneous catalysts	Semibatch reactors Pesticide/catalyst ratio: 10	Effects of pH, pesticide concentration Intermediate formed Good catalytic activity	15
Ozone–Mn(II) reaction Presence of Karmin Indigo	Batch reactor, pH: 2–8	Different Mn species appear Mechanism proposed Influence on Indigo elimination	16
Pyrazine: 4.3 × 10^{-3} M Oxalic acid: 2.8 × 10^{-4} M Catalyst: Mn(II): 2.5 × 10^{-5} M	Semibatch reactor pH: 4.2	Formation of Oxalic–Mn complex that acts as catalyst; effects of inhibitors and promoters	17
Humic acid (commercial): TOC: 11 mgL^{-1} Catalyst: Ag(I), Fe(II), Co(II), Fe(III), Cu(II), Mn(II), Zn(II), Cd(II), etc.: 6 × 10^{-5} M	Batch flasks pH 7 Phosphate buffer	TOC reduction: 33% O$_3$ alone, 63% O$_3$/catalyst Intermediate identification (GC/MS) Better catalysts: Ag(I), Mn(II)	18
Atrazine: 3 × 10^{-6} M Catalyst: Mn(II) (0–1.5 mgL^{-1})	3.65-L semibatch bubble column with recirculated water (85 Lh^{-1}), catalyst solution continuously fed; pH 7, 20°C, phosphate buffer	Nearly 90% removal ATZ with O$_3$/catalyst against 20% with O$_3$ alone	19

TABLE 10.1 (continued)
Works on Homogeneous Catalytic Ozonation

Ozonation System	Reactor and Operating Conditions	Observations	Reference
River water, TOC: 2.95 mgL^{-1}, UV absorbance: 0.057, Catalyst: Mn(II)	5-L semibatch bubble reactor; ozone dose: 1.7–2.8 mg/mgTOC, pH 7.8	TOC reduction: 16% O$_3$ alone, 22% O$_3$/Mn, UV reduction: 5% O$_3$ alone, 63% O$_3$/catalyst, intermediate identification; aldehyde production increased with O$_3$/catalyst	20
Pyruvic acid, 2.5 × 10^{-3} M Catalyst: Mn(II) (0.25 mgL^{-1}), also with MnO$_2$ (solid)	Semibatch stirred reactor, ozone condition: 36 Lh^{-1}, 3% O$_3$ volume, pH: 2–4, 25°C, phosphate buffer	No reaction with O$_3$ alone Significant oxidation with O3/Mn Negative effect of pH Mechanism and kinetic study	21
Atrazine: 3.6 × 10^{-6} M with humic acids (DOC: 1–6 mgL^{-1}) Catalyst: Mn(II): 1 mgL^{-1}	3.65-L semibatch bubble column (1.3 m high, 6 cm diameter) with recirculated water (85 Lh^{-1}); catalyst solution continuously fed; pH 7, 20°C, phosphate buffer	Presence of humics (<2 mgL^{-1}) increased ATZ removal in O$_3$/catalyst, a mechanism is proposed	22
Chlorobenzenes synthetic solution: 1–5 mgL^{-1} Catalysts: Fe(II), Fe(III), Mn(II): 6 × 10^{-5} M	3-L semibatch bubble reactor, pH 7 (pH 8.4 ozone alone)	% COD removal: 18% O$_3$ alone vs. 55% O$_3$/Fe(II) Slight increased removal of chlorobenzenes with O$_3$/catalyst	23
Glyoxalic acid: 0.27 mgL^{-1} Catalysts: Mn(II) (up to 2 × 10^{-5} M), MnO$_2$ filtrated solution and solid MnO$_2$ (see Table 10.2)	Semibatch stirred reactor, pH 2–5.4, phosphate buffer, ozone condition: 36 Lh^{-1}, 3% O$_3$ volume	Formic and oxalic acid identified Significant increase of ozonation with catalyst Mechanism proposed and kinetic study pH 4 optimum	24
Pyruvic acid: 4 × 10^{-3} M Catalyst: Mn(II) 12–24 mgL^{-1} MnO$_2$ filtrated solution (Mn(IV): 11 µM)	Semibatch stirred reactor pH 1–3, phosphate buffer, ozone condition: 36 Lh^{-1}, 3% O$_3$ volume	Formation of Mn(VII) Similar acid removal with Mn(II) and Mn(IV) Kinetic study Kinetic modeling	25
p-Nitrotoluene (PNT) in acetic anhydride: 3 × 10^{-4} M Catalyst: Mn(II) (1.4 × 10^{-4} M) in H$_2$SO$_4$ (1.27 M)	Semibatch bubble column, ozone condition: 10^{-2} Ls^{-1}, 4 × 10^{-4} ozone concentration	Mn(III) initiating species Mechanism and kinetic study Information on rate constant data and intermediates	26

TABLE 10.1 (continued)
Works on Homogeneous Catalytic Ozonation

Ozonation System	Reactor and Operating Conditions	Observations	Reference
$O_3/H_2O_2/Fe(II)$ system: fentozone Acid and disperse dyes: 80–120 mgL^{-1} Catalyst: Fe(II) 26 mgL^{-1}	1.5-L semibatch and batch bubble column (0.2 m high, 4 cm I.D.) pH 4.2–10.5, m_{O3gas} = 29.3 mg min^{-1} C_{H2O2} = 52 and 100 mgL^{-1}	Empirical kinetic study Two ozone demand periods: instantaneous and slow decay demands Formation of sludge, sludge lower with ozone processes, also used Ferral 2060 as catalyst (see also Table 10.2)	27
o-Chlorphenol: 100 mgL^{-1} Catalysts: Nitrates of Pb(I), Cu(II), Zn(II), Fe(II), and Mn(II): 1 mgL^{-1}	2.8-L semibatch ozone reactor, pH 3, 18 mgO$_3$min^{-1}	Removal efficiency in 60 min: 90% with Mn(II), 80% with Fe(II), 60% no catalyst TOC removal: 30% with Mn(II)/O$_3$	28

TABLE 10.2
Works on Heterogeneous Catalytic Ozonation

Ozonation System	Reactor and Operating Conditions	Observations	Reference
Phenol: COD: 1000 Ethyl acetoacetate: COD 210 TOC: 73, wastewater: TOC: 459 Catalyst: Fe$_2$O$_3$: up to 20000 Units in mgL^{-1}	Packed-bed reactor Continuous feed of water and gas pH: 4.3–6.3	LH mechanism for ozone–surface reaction proposed, 100% phenol conversion in 40 min Significant removals of COD	3
Peat water Catalyst: Activated carbon 150–700 µm	Three phase fluidized reactor, 1.25 m high, 4.2 cm diameter, 50 g catalyst, 150 cm^{3}min^{-1} with 51 mg O$_3$L^{-1}	Color removal 55% with O$_3$/catalyst 15% with O$_3$ alone	29
Phenol: 1.1×10^{-3} M Catalyst: CuO, MnO$_2$, Pd, supported on Al$_2$O$_3$	Semibatch bubble reactor, pH 6–9 20°C, 1.433 L min^{-1}, 3% O$_3$ w/w	No influence of catalysts Ozonation through direct means	30

TABLE 10.2 (continued)
Works on Heterogeneous Catalytic Ozonation

Ozonation System	Reactor and Operating Conditions	Observations	Reference
Anyline dye wastewater: COD 2.2, DBO: 1.4 Catalyst: CuO, Fe_2O_3, NiO, Cr_2O_3, Co_2O_3: 55 Size: 0.6–1.2 mm Other units in gL^{-1}	Semibatch bubble column, 12 cm high, 6 cm diameter, 12 Lh^{-1}, 35 mgL^{-1}, O_3	Good performance of catalysts; however, catalysts leached at pH 7 Better catalyst: NiO	31
Phenol, hydroquinone, carboxylic acids, and aldehydes (maleic, glyoxal, oxalic, etc.) 5×10^{-4} M, catalyst: Fe_2O_3/Al_2O_3 (0.45% Fe)	Semibatch fixed-bed bubble reactor, pH < 5, no buffers	Significant influence on TOC removal No effect on phenol or hydroquinone oxidation Better efficiency of O_3/catalyst	32
Tap water disinfection Catalyst: Ag/Al_2O_3	Fixed-bed cartridge 15 cm length, 10 cm diameter, 500 g catalyst	Killing bacteria and virus	33
Fulvic acid, protein, disaccharide TOC: 12 Catalyst: TiO_2, TiO_2/Al_2O_3, TiO_2/Clay: 4 mm Units in mgL^{-1}	Semibatch fixed-bed bubble reactor 30 g catalyst, pH 8	Removal of TOC with 20 g O_3/g TOC and catalyst: 86% for fulvic acid 81.4% for cellobiose 71% for albumine $O_3/TiO_2 > O_3/H_2O_2 > O_3$	34
Atrazine Volatile organochlorines Leachates Catalyst: Not given	Semibatch fixed-bed bubble reactor	Ecoclear process 99% Atrazine removal Mechanism through surface reactions No participation of HO radicals COD reduction in biotreated wastewater from 1000 to 250 mgL^{-1}	35
Oxalic acid: 2.1×10^{-4} M Natural organic matter Catalyst: TiO_2/Al_2O_3	Bubble column ozonation plus fixed-bed catalytic column 18–24°C, pH 7	No influence of carbonates No participation of HO radicals With 2 mgL^{-1} ozone dose: 40% TTHMFP removal with O_3/catalyst vs. 10% O_3 alone	36
Phenols: 455 mgL^{-1} Systems: O_3/UV, $O3/H_2O_2$, O_3/UV/TiO_2 Catalyst: TiO_2	Countercurrent falling film absorber C_{O3}: 15–17 mgL^{-1} pH: 6.4–12.1	Kinetic modeling No appreciable effect of TiO_2	37
Bromate control in surface water, Br^-: 80–100 μgL^{-1}, TOC: 3–4 mgL^{-1}	Bubble ozonation column and fixed-bed catalytic column	Bromate formation prevented at least 30% with O_3/catalyst compared to O_3 alone	38

TABLE 10.2 (continued)
Works on Heterogeneous Catalytic Ozonation

Ozonation System	Reactor and Operating Conditions	Observations	Reference
Wastewater: COD: 1575 mgL^{-1}, pH 7–12, K hydrogenphthalate Sulfuric acid (pH:–3) Leachate with carbonates: COD 500 mgL^{-1} Catalyst: not given	Fixed-bed catalytic reactor (30 cm high, 2 cm diameter), C_{O3}: 17 to 60 mgL^{-1}, also, pilot plant unit	Ecoclear process Reduce TOC, COD pH does not affect Not via HO radicals 76–88% COD conversion in leachates	39
Chlorobenzene: 10 mgL^{-1} Bicarbonates: 5×10^{-3}–0.01 M Catalyst: α-FeOOH (30–50 mesh), 0.05–0.2 gL^{-1}	Slurry semibatch bubble tank, 0.2 Lmin^{-1}, 3–22 mgO$_3$L^{-1}	Less ozone residual with catalyst CBZ removal: 83% with O$_3$/catalyst (1 gL^{-1}) vs. 60% with O$_3$ alone in 30 min	40
Oxalic acid: 0.27 g/L Catalyst: MnO$_2$ up 0.25 g/L 90–300 μm, S$_{BET}$: 6.9 m^{2}g^{-1} Different pH$_{zpc}$	Slurry semibatch bubble reactor, pH: 3.2–7, 36 Lh^{-1}, 3% O$_3$ volume	Removal rate increased with decreasing pH Surface reaction mechanism Better pH 3.2	41
Chloroaromatics (PCBs, PCDD, etc.) adsorbed on catalyst Catalyst: Wessalith DAY (hydrophobic zeolite)	Fixed-bed reactor, gas–solid catalytic reaction: 50–90°C, 60 g catalyst, 50 mgO$_3$ L^{-1}	85% removal of adsorbates after 3 h HO is a reactive species in the system	42
Phenols (Cl–phenol, methoxy phenol, pyrocatechol) Catalyst: activated carbon: 4 mm, S$_{BET}$: 1330 m^{2}g^{-1}	1.5-L slurry semibatch stirred reactor pH 2, 15–35°C	Activated carbon enhances O$_3$ selectivity; similar oxidation rates (O$_3$ and O$_3$/C) but lower O$_3$ consumption when C is present	43
Different pollutant and wastewater	Reference 35	Reference 35	44
Concentrated leachate from nanofiltration (COD: 8–9 gL^{-1}, AOX: 7–9 gL^{-1} TCA, TCE, BTX Catalyst: activated carbon	Full scale fixed-bed reactor 1 kgO$_3$h^{-1}	Ozone consumption: 0.8kg/kgCOD COD reduction to 2–3 gL^{-1} AOX reduction to 0.5–1 gL^{-1} Removal of TCA, BTX, TCE No influence of carbonates	45
Water contaminants Catalyst: different commercial catalysts	Fluidized-bed catalytic reactor with water recirculation and ozonation chamber	COD removal improvement with O$_3$/catalyst Patented work	46
Fulvic acids: DOC: 2.84, BDOC: 0.23 (units are mgL^{-1}) Catalyst: TiO$_2$/Al$_2$O$_3$ (1.5–2.5 mm)	1-L flask batch reactor 10 gL^{-1}, catalyst; 20°C, pH 7.5, phosphate buffer	O$_3$/catalyst leads to mineralization of by-products and better reduction of chlorine demand	47

TABLE 10.2 (continued)
Works on Heterogeneous Catalytic Ozonation

Ozonation System	Reactor and Operating Conditions	Observations	Reference
Nonionic and ionic surfactants TOC 8 mgL^{-1} Catalyst: Not given	Slurry batch reactor and semibatch bubble column for ozonation followed by fixed-bed reactor	Better TOC removals in O$_3$/catalyst experiments	48
Photocatalysis system Aniline, 10^{-3} M, TOC:78 mgL^{-1} TiO$_2$ (Degussa, P25), 27 nm	Pyrex cell photoreactor, 125 W med. pres Hg lamp: 9.24 × 10^{-4} Einstein L^{-1}min^{-1}, 2 gL^{-1} catalyst, pH 3	Better system: O$_3$/TiO$_2$/light, in 1 h, TOC removal: 95% with O$_3$/TiO$_2$/Light 55% with O$_3$ alone, 40% with light/TiO$_2$	49
Ozone decomposition O$_3$-p-chlorobezoate Acetate, methanol Natural water (lake water) (DOC:1.3 mgL^{-1}, pH 7.8) Catalyst: activated carbon (1300 m^{2}g^{-1}, carbon black (20–460 m^{2}g^{-1})	Slurry batch flasks 20 mgL^{-1} catalyst 10 mgL^{-1} O$_3$	Catalyst accelerates p-chlorobenzoate ozonation rate Process via HO radicals reacting in solution but generated in catalyst surface Effects of inhibitors and promoters of O$_3$ decomposition Stoichiometry HO/O$_3$ as in other advanced oxidation processes	50
p-chlorophenol, cyanuric acid: (0.4 gL^{-1}) Catalyst: mesoporous materials (aluminosilicates), S$_{BET}$ > 700 m^{2}g^{-1}	Slurry semibatch reactor	Oxidation goes through adsorption plus surface reaction	51
Chlorophenol, oxalic acid, chloroethanol (1 gL^{-1}) Catalyst: Al$_2$O$_3$ (285 m^{2}g^{-1}), Fe$_2$O$_3$/Al$_2$O$_3$, TiO$_2$/Al$_2$O$_3$	Slurry semibatch reactor 24 mgL^{-1}h^{-1} O$_3$	Significant improved of ozonation rate with O$_3$/catalyst In 300 min: conversion chlorophenol: 100% with O$_3$/catalyst vs. 38% O$_3$ alone	52
Salicylic acid, peptides, humic substances Catalyst: Me/Al$_2$O$_3$, Me/TiO$_2$, Me/clay Me: not given	Fixed-bed catalyst in recirculating loop to a semibatch bubble ozonation column	TOC$_o$ = 2.5–42 mgCL^{-1} Me/Al$_2$O$_3$: high stability Best removed with an attapulgite-based catalyst Mechanism proposed	53
Succinic acid: 10^{-3} M Catalyst:Ru/CeO$_2$ (40–200 m^{2}g^{-1})	Slurry semibatch reactor 0.8 gL^{-1} catalyst, pH 3.4 27.5 × 10^{-3} molO$_3$h^{-1}	No reaction with ozone alone No appreciable adsorption Better catalyst via impregnation and reduction Nearly 100% TOC removed in 60 min	54

TABLE 10.2 (continued)
Works on Heterogeneous Catalytic Ozonation

Ozonation System	Reactor and Operating Conditions	Observations	Reference
Commercial humic acid (TOC:5.34–7.3), river water (TOC 4.46) (units in mgL^{-1}) Catalyst: TiO$_2$/Al$_2$O$_3$ (2–6 mm, 2.5% TiO$_2$ in catalyst)	Slurry semibatch bubble reactor pH 7.2, phosphate buffer, 2.5–10 gL^{-1} catalyst 400 mgO$_3$h^{-1}	T = 30 min, TOC% removal improved if humic substance concent < 5.34 mgL^{-1} River water: TOC removal: 11.2% ozone alone vs. 16.4% O$_3$/catalyst Identification of by-products	55
Wet-air oxidation, O3/catalyst Formic acid: 0.3 M 20 catalysts tested: (Pt/Active carbon, Pt/Al2O3, etc.)	Slurry semibatch bubble reactor 9 mgL^{-1} O$_3$ in water	Reactions are not via HO radicals Best catalyst: Pt/Al$_2$O$_3$ Formic acid zero-order kinetics Activation energy: 5 kcalmol^{-1}	56
Glyoxalic acid: 0.27 mgL^{-1} Catalysts: Mn(II) (up to 2 × 10^{-5} M), MnO$_2$ filtrated solution and solid MnO$_2$	Semibatch stirred reactor pH 2–5.4, phosphate buffer ozone condition: 36 Lh^{-1}, 3% O$_3$ volume	Formic and oxalic acid identified Significant increase of ozonation with catalyst Mechanism proposed and kinetic study pH 4 optimum	24
UV/O$_3$/TiO2, UV/TiO2, UV/O$_3$/Fe(II) 2,4dichlorophenoxyacetic acid (2,4-D): 2 × 10^{-3} M Catalyst: TiO$_2$ (Degussa P-25), 59 m^{2}g^{-1}, 27 nm, Fe(II) 10^{-3} M	100 mL slurry semibatch photoreactor 2 gL^{-1} catalyst, pH 2.6 6 W Philips Black light 8.3 × 10^{-7} Einstein L^{-1}min^{-1} 1.4 gO$_3$h^{-1}	Data on 2,4-D, TOC, and Cl$^-$ concentration Better system: UV/O$_3$/Fe(II), in 15 min: 2,4D% removals were: 100% with UV/O$_3$/Fe(II), 96% with UV/O$_3$/TiO$_2$, 75% O$_3$, 15% UV/TIO$_2$ 70% TOC removal with UV/O$_3$/Fe(II)	57
Textile wastewater (4 dyes): COD: 250–1800 mgL^{-1} Catalyst: active carbon, 5 mm, 893 m^{2}g^{-1}	Fixed-bed semibatch column, 100 cm high, 6 cm diameter, 60–360 Lh^{-1}, up to 200 g catalyst	Total color removal, significant improved COD removal with O$_3$/C Kinetic study Effects of O$_3$/catalyst noted above 30 min with respect to adsorption effects Stability and activity of catalyst studied	58

TABLE 10.2 (continued)
Works on Heterogeneous Catalytic Ozonation

Ozonation System	Reactor and Operating Conditions	Observations	Reference
Succinic acid (up to 5×10^{-3} M) Catalyst: Ru/CeO$_2$ (up to 3.2 gL^{-1}), 40 m^{2}g^{-1}	500-mL slurry semibatch reactor 1.25 gO$_3$h^{-1}, pH 3.4	Impact of O$_3$ on catalytic surface Different forms of catalyst preparation Catalyst characterization Total acid conversion in less than 60 min with O$_3$/catalyst, nearly complete TOC removal at same conditions	59
1,2-dihydroxybenzene (1,2-DH) 1–5 mM Catalyst: activated carbon, 845 m^{2}g^{-1}, 1–2 mm	Fixed-bed reactor 6 mmolO$_3$min^{-1}, 5 Lh^{-1} residence time: 0.5–240 min	Gas ozone solid catalytic reaction, treatment of adsorbed 1,2-DH, near-destruction of 1,2-DH in 60 min, study of ozone effects on catalyst surface	60
2-chlorophenol Catalyst: γ-Al$_2$O$_3$: 120–190 m^{2}g^{-1}, 60–200 μm	Slurry semibatch stirred reactor 18 mgO$_3$h^{-1}, pH: 3–9 up to 2 gL^{-1} catalyst	Differences between O$_3$ and O$_3$/catalyst regarding TOC removal Kinetic study, toxicity effects, activity of catalyst	61
Sulfosalicylic acid, 2.5 × 10^{-3}M, TOC: 193.2 mgL^{-1} V-O/SiO$_2$ (92.6 m^{2}g^{-1}) V-O/TiO$_2$ (86.8 m^{2}g^{-1}) MnO$_2$	Slurry semibatch bubble reactor, 40 cm high, 3 cm diameter pH 3.2, 0.67 Lmin^{-1}, 37–50 mgO$_3$min^{-1}L^{-1}	Negligible effect of MnO$_2$ catalyst Significant improvement of TOC and ozonation rate with V-O catalysts Intermediate oxalic acid followed High HCO$_3^-$ concentration (>0.02 M) affects the ozonation rate	62
UV/O$_3$, UV/TiO$_2$, UV/O$_3$/TiO$_2$ Glyoxal (G), p-toluenesulfonic acid/pS), napthalene 1,5-sulfonic acid (NS), pyrrole-2-carboxylic acid (P): 1 mM Catalyst: TiO$_2$ (Degussa P-25) 0.5 gL^{-1}	Slurry semibatch bubble photoreactor, UVAHAND 250 W lamp with 360 nm cut-off filter, pH 3 or 7, 7 × 10^{-6} Einstein s^{-1}	Intermediates followed; UV/TiO2/O$_3$ better to remove these compounds except NS and DOC	63
Aromatic compounds (benzene, toluene, etc.): 1.7–2.4 mgL^{-1} Catalyst: perfluoro–octyl–alumine (PFOA)	200-mL slurry semibatch magnetically stirred reactor pH 6, 18°C, 2 gL^{-1} catalyst 13 mgO$_3$L^{-1} in water	Removal of aromatics with catalysts between 24–43% higher than without it	64

TABLE 10.2 (continued)
Works on Heterogeneous Catalytic Ozonation

Ozonation System	Reactor and Operating Conditions	Observations	Reference
Ozone decomposition Catalyst: activated carbon 800 m^{2}g^{-1}, 1–3.5 mm	Slurry batch-stirred reactor PH 2–9, 5–30°C, up to 2 gL^{-1} catalyst	Influence of variables, ozone decomposition rates higher with catalyst LH mechanism proposed, HO radical reactions included, and kinetic study	65
Oxalic acid: up to $8 \times 10^{-3} M$ Catalyst: TiO$_2$ in powder (Aldrich)	Slurry semibatch stirred reactor pH 2.5, 5–30°C, up to 5 gL^{-1} catalyst	TOC and concentration time data Total TOC elimination with O$_3$/TiO$_2$ LH mechanism proposed and kinetic study: oxalic acid zero order kinetics	66
O$_3$/H$_2$O$_2$/Ferral 2060 system Ferral 2060: natural clay with 2% Fe$_2$(SO$_4$)$_3$ + 6% Al$_2$(SO$_4$)$_3$, acid and disperse dyes: 80–120 mgL^{-1} Catalyst: 357 mgL^{-1}	1.5-L semibatch and batch bubble column (0.2 m high, 4 cm I.D.) pH 4.2–10.5, m_{O3gas} = 29.3 mgmin^{-1} C_{H2O2} = 52 and 100 mgL^{-1}	Empirical kinetic study Two ozone demand periods: instantaneous and slow decay demands Formation of sludge; sludge lower with ozone processes	27
Cl-acetic and Cl-succinic acids: 0.5–2 mM, Catalyst: Ru/CeO$_2$-TiO$_2$: 85 m^{2}g^{-1}, pH$_{pzc}$ = 5.7, 2% Ru	1-L magnetic-stirred semibatch slurry reactor pH = 2.6 and 3.6, m_{O3gas} = 21.25 mgmin^{-1}, 0.8 gL^{-1} catalyst	Complete mineralization of acids if pH < pK_a, adsorption of acids is a key feature for ozonation improvement; stability of catalyst also studied	67
Ozone decomposition on different metal oxides on Al$_2$O$_3$, SiO$_2$, TiO$_2$, SiO$_2$.Al$_2$O$_3$ 3–18% metal oxide by x-ray diffraction	Fixed-bed reactor through which aqueous ozone was passed	SiO$_2$ better support, although slightly better than Al$_2$O$_3$; efficient catalyst only noble metals: Pt, Pd, Ru, Rh, Ag, Os, Ir, with strong Me-O bond Catalyst active at least 24 h Mechanism as in ozone decomposition in gas phase catalysis (see Section 10.2)	68
Reactive and acid dyes Ferral 2060: natural clay with 2% Fe$_2$(SO$_4$)$_3$ + 6% Al$_2$(SO$_4$)$_3$ 0–487 mgL^{-1}	1.5-L semibatch and batch bubble column (0.2 m high, 4 cm I.D.) pH 4.2–10.5, m_{O3gas} = 29.3 mgmin^{-1} pH 3.08 to 9.01	Coagulation catalytic process In 1 min more than 91 and 80% removal of color and COD No flocculation at pH 3	69

TABLE 10.2 (continued)
Works on Heterogeneous Catalytic Ozonation

Ozonation System	Reactor and Operating Conditions	Observations	Reference
Ozone decomposition on soil and Fe(O) 5–50 gL^{-1} catalyst α-FeOOH also tested	Continuous magnetically stirred ozone cell C_{O3} = 3 mgL^{-1}	Two ozone demands: instantaneous and long decay demands Presence of metal oxides and natural organics are key factors of kinetics; p-CBA used as HO radical probe	70
Dichlorvos (DDVP): organophosphorous insecticide, 10^{-3} M Catalyst: a microporous silicate	Fixed-bed column (1.5 cm I.D, 5 cm high), 2 g catalyst, ozone dissolved fed to bed column: 5 mLmin^{-1}, pH 4, C_{O3} = 1 to 16 mgL^{-1}	Cl$^-$, PO$_S^{3-}$ formed, reduction of DOC, citotoxicity with O$_3$/catalyst High ozone adsorption noted	71
NTS (as above): 45 mgL^{-1} Catalyst: 7 different commercial activated carbons	2-L stirred ozone slurry reactor or fixed-bed reactor 76 mgO$_3$min^{-1}	Better catalytic activity with basic Acs (pH$_{PZC}$ high) and containing metals	72
3 different naphthalene sulfonic acids Catalyst: activated carbon (Filtrasorb 400 AC): 0.5 to 2 g 0.5–0.8 mm particle size, S_{BET} = 1000 m^{2}g^{-1}	As in Reference 72	Study of adsorption capacity of ozonated (at different times) activated carbon Increase of carboxylic acid groups, decrease of graphitic groups, better if pH ≪ pK$_{PZC}$, low time (10 min), ozonated carbon better for NS removal	73
As in Reference 73	As in Reference 72	Genotoxic study with three different tests Ozone alone by-products also genotoxic Decrease of genotoxicity with O$_3$/AC	74
Natural organic matter (NOM) of different type Catalyst: FeOOH, 147 m^{2}g^{-1}, 0.3–0.6 mm particle size	Semibatch mechanically stirred bubble column (50 cm high, 5 cm I.D.) H$_2$O$_2$/FeOOH also applied	Slightly higher removal of NOM fractions with O$_3$/FeOOH (33 without catalyst vs. 50% with catalyst) Mechanism goes through HO radicals; carbonates and pCBA used as probes	75

TABLE 10.2 (continued)
Works on Heterogeneous Catalytic Ozonation

Ozonation System	Reactor and Operating Conditions	Observations	Reference
1,3,6-naphthalenetrisulfonic acid (NTS): 45 mgL^{-1} Catalyst: activated carbon As in Reference 71	As in Reference 72	AC characterization with and without ozonation O_3 + virgin carbon better performance Identification of intermediates: carboxylic acids Catalytic activity diminishes with ozonated carbons	76

respectively, together with some of their main features. Here, attention is focused on the HeCO process, given that this form of oxidation has the cleanest, most economical, and most beneficial health properties, compared to the HoCO. Insights into the HeCO process are presented in a review[1] that examines aspects concerning catalyst preparation, ozonation performance, and possible mechanism of reactions.

Another heterogeneous catalytic ozonation system that is currently attracting the interest of different research groups involves the simultaneous application of ozone and UV light (A or B type) in the presence of n-type semiconductor catalysts such as TiO_2. This process has the aim of improving the oxidative (and reductive) capacity of the semiconductor photocatalyst where absorbed photons promote electrons from the semiconductor valence band to the empty and more energetic conduction band. In this way, a positive oxidant hole is created in the valence band, which is able to initiate an oxidative process leading to the appearance of hydroxyl radicals. The promoted electron of negative redox potential is able to reduce oxygen, leading to superoxide ion radicals that could initiate a new oxidizing process. Section 10.4 gives a more detailed description of the process together with some kinetic features of this system. Semiconductor photolysis has been applied successfully to numerous and different chemicals, and excellent reviews on this matter have been published.[77–82] Studies on photocatalytic ozonation, on the other hand, were also initiated by the decomposition kinetic study of ozone. This decomposition is significantly improved when irradiated with UV light in the presence of TiO_2.[83] Recent studies have been reported in the literature on the catalytic destruction of pollutants in the simultaneous presence of ozone, UV, and a semiconductor (TiO_2). Table 10.3 lists some of the most recent works on this matter and details about their main features. It should also be noted that homogeneous catalytic ozonation in the presence of UV light has also been the subject of research. In these cases, the synergism of Fenton-like systems (Fe(II) or Fe(III)/UVA light) with ozone was applied to improve the organic matter removal.[91] In this chapter, however, we have only mentioned heterogeneous photocatalytic ozonation. Section 10.4.3 presents details of the possible mechanism and kinetics of this ozonation process.

TABLE 10.3
Recent Works on Semiconductor Heterogeneous Photocatalytic Ozonation

Ozonation System	Reactor and Operating Conditions	Observations	Reference
Monochloracetic acid (MCA) and pyridine: 10^{-3} M O_3/UV/TiO_2 (Degussa P25)	0.4-L semibatch stirred photoreactor: UV lamp (360-nm cut-off filter), 7 μ Einstein $L^{-1}s^{-1}$, pH 3 0.5 gL^{-1} catalyst, 20 mgL^{-1} O_3	O_3/UV/TiO_2 leads to removals of 24 and 16 times higher than O_3 alone and 4 and 18 times higher than UV/TiO_2 for pyridine and MCA, respectively Mechanism proposed O_3/UV/TiO_2 lowest specific energy consumption	84
CN^-: 50 mgL^{-1} UV/O_3/TiO_2 (Degussa P25: 50 m^2g^{-1} and BDH: 10 m^2g^{-1})	1-L semibatch cylindrical photoreactor, medium pressure Hg lamp with 300-nm cut-off filter, pH 11.3	Adsorbed ozonide ion radical detected by ESR, O_3/UV/TiO best system: 100% CN^- removal in 20 min Formation of $CO_3^=$, CNO^-, and NO_3^-	85
Formic acid: 4.9 to 107 mgL^{-1} O_3/UV/TiO_2 (coated)	0.3-L annular flow photoreactor in series with a semibatch ozone bubble column, 6 W fluorescent lamp 10–50°C, pH 2–12, 0.96 mgO_3min^{-1}	With O_3/UV/TiO_2, formic acid oxidation rate increased 2 and 3 times with respect to UV/TiO_2 and O_3 alone, respectively, Langmuir–Hinshelwood kinetics applied, rate constant determination	86
Vinasse wastewater: COD: 110 gL^{-1}, O_3, O_3/H_2O_2, O_3/H_2O_2/UV, O_3/H_2O_2/UV/TiO_2	1-L semibatch ozone photoreactor, UV Hg vapor lamp (200–800 nm), TiO_2: 2 gL^{-1}, C_{H2O2} = 0.04 M C_{O3g} = 34 mgL^{-1}, 500 Lh^{-1}, pH 4.4	COD removal better with O_3/H_2O_2/UV/TiO_2 Effects of oxidation methods on anaerobic digestion Slight improvement in CH_4 production in anaerobic digestion of O_3/UV/TiO_2-treated wastewater	87
Cyanide wastewater (CN^-:100 mgL^{-1}, Cu(II): 20 mgL^{-1}, COD:20 mgL^{-1} TiO_2 Degussa P25	3-L photoreactor with 40 W low pressure Hg lamp, 140 quartz tube-coated TiO_2 film, C_{O3g} = 2,5 mgL^{-1}, pH 10, 2.4 Lh^{-1} plus three cation, anion, and mixed ion exchange columns	In 100 min: 100% CN^- removal, 87% COD removal, resins eliminates CNO^- and Cu(II)	88
Phenol: 100 mgL^{-1} MnO_2 or TiO_2/UV/O_3 30–138 m^2g^{-1}, 3 nm size	150 mL Pyrex glass semibatch photoreactor, General Electric UV Hg lamp (220–380 nm), 1700 μWcm^{-2} at 254 nm 2.15 mg$L^{-1}O_3$	Intermediates detected, studies with FT-IR, TPR, TPD, x-ray diffraction; oxidation rate increases about 10 times with O_3/UV/TiO_2 with respect to O_3/UV; mineralization increases about 3 times	89

TABLE 10.3 (continued)
Recent Works on Semiconductor Heterogeneous Photocatalytic Ozonation

Ozonation System	Reactor and Operating Conditions	Observations	Reference
TCE and PCE: 10 mgL^{-1} γ-Rays/TiO$_2$	TiO$_2$-coated glass tube reactor (7 cm length, 0.2 cm I.D.) with 2 mg TiO$_2$ Co^{60} source: 300 Gy, 20°C	80% and 99% organics removal and 35% and 40% TOC removal with O$_3$ or O$_3$/TiO$_2$ and O$_3$/TiO$_2$/γ-rays, respectively, removal of toxicity with O$_3$/TiO$_2$/γ-rays	90

10.1 FUNDAMENTALS OF GAS–LIQUID–SOLID CATALYTIC REACTION KINETICS

As in the classic ozonation process, one of the important points of HeCO deals with the kinetics of oxidation. HeCO is a gas (ozonated air or oxygen)/water solid (catalyst) system where mass transfer and chemical reaction steps must be considered for the appropriate formulation of the process rate. Before we describe the highlights of HeCO kinetics, we present the fundamentals of the kinetics of gas–liquid–solid catalytic reactions.

Gas–liquid–solid catalytic reactions are heterogeneous reacting systems that involve a series of consecutive–parallel steps of mass transfer and chemical reactions on the catalyst surface. The surface of the catalyst constitutes a key parameter for improving the reaction rate. Therefore, in most catalytic processes, the catalyst is usually supported in a porous material with internal surface areas varying from hundreds (alumina supported catalysts) to more than 1000 m^2/g (activated carbon) of catalyst.[92] The internal surface area is thus the zone where the reaction occurs. Figure 10.1 shows the steps of this process. If a general catalytic reaction

$$A + zB \xrightarrow{catalyst} 2P \qquad (10.2)$$

is assumed to develop on the catalyst surface, these steps are:

1. External diffusion of reactant gas molecules, A, from the bulk gas to the gas–liquid interface
2. External diffusion of reactant molecules, A, in the liquid, from the gas–liquid interface to the bulk liquid
3. External diffusion of reactant molecules A and B from the bulk liquid reaching up to the catalyst surface (pore mouth)
4. Internal diffusion of reactant molecules through the catalyst pores with simultaneous surface reaction on the internal catalyst surface

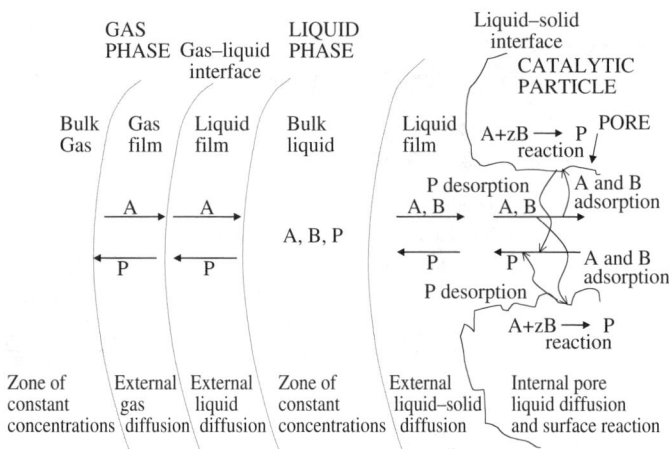

FIGURE 10.1 Mechanism steps of a gas–liquid–solid catalytic reaction.

5. Catalyst surface reaction that involves three consecutive steps:
   a. Adsorption of reactants molecules on the active centers of the catalyst surface
   b. Surface reaction of adsorbed molecules to yield the adsorbed products
   c. Desorption of adsorbed products

For reversible reactions, there are also these steps

6. Internal diffusion of product molecules through the catalyst pores from the internal catalyst surface to the pore mouth or external surface
7. External diffusion of product molecules from the catalyst surface reaching up to the bulk liquid

Here, we consider only irreversible reactions because diffusion of products does not influence the process rate.

Rate equations for catalytic reactions are established according to the kinetic regime, i.e., in accordance with the relative importance of mass transfer and (surface) chemical reaction steps. These kinetic regimes can be classified as follows:

- Slow kinetic regime
- Fast kinetic regime or external diffusion kinetic regime
- Internal diffusion kinetic regime

The rate equations for these kinetic regime systems are presented below.

10.1.1 SLOW KINETIC REGIME

This is the case when the mass-transfer resistances are negligible because external and internal diffusions are very fast steps compared to the surface reaction step. In

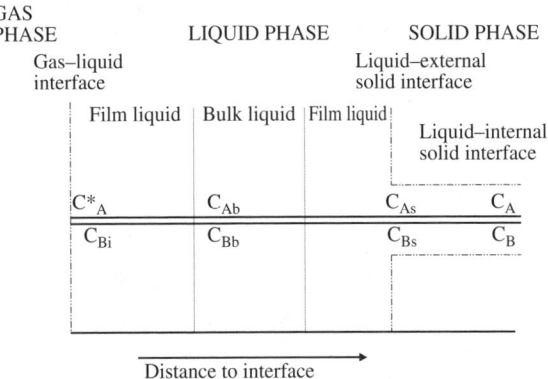

FIGURE 10.2 Concentration profiles of a gas–liquid–solid catalytic reaction in the slow kinetic regime.

the slow kinetic regime, the high diffusion rates make the concentration of reactants at any point in the liquid, both outside and inside the catalyst pores, equal to that in the bulk liquid. Figure 10.2 shows the concentration profile corresponding to this kinetic regime. The rate equation for the disappearance of reactants depends exclusively on the slowest step of the surface reaction, Step 3.1, 3.2, or 3.3. Hence, adsorption, chemical surface reaction, or desorption can control the process rate. The rate equation is established from a Langmuir–Hinshelwood (LH) mechanism involving these three consecutive steps.[93] For the case of Reaction (10.2), for $z = 1$, a possible mechanism would be:

- Adsorption of reactants:

$$A + S \underset{k_{A-}}{\overset{k_{A+}}{\rightleftarrows}} A \bullet S \qquad (10.3)$$

$$B + S \underset{k_{B-}}{\overset{k_{B+}}{\rightleftarrows}} B \bullet S \qquad (10.4)$$

- Surface chemical reaction:

$$A \bullet S + B \bullet S \underset{k_{s-}}{\overset{k_{s+}}{\rightleftarrows}} 2P \bullet S \qquad (10.5)$$

- Desorption:

$$P \bullet S \underset{k_{D-}}{\overset{k_{D+}}{\rightleftarrows}} P + S \qquad (10.6)$$

where S represents one free active center of the catalyst surface.

Once the mechanism is established, the slowest or controlling step is assumed from experimental results (catalytic experiments, FT-IR analysis, thermogravimetric analysis, etc.). For example, if the surface chemical Reaction (10.5) is assumed to be the controlling step, the process rate or kinetics of the catalytic reaction is:

$$-r' = -r'_S = k_{S+}\left[C_{A \bullet S}C_{B \bullet S} - \frac{C_{P \bullet S}^2}{K_S}\right] \qquad (10.7)$$

where the kinetics is expressed per mass of catalyst as a function of adsorbed species concentration, $C_{A \bullet S}$, $C_{B \bullet S}$, and $C_{P \bullet S}$, and K_S is the equilibrium constant of the surface chemical reaction, Step (10.5). These concentrations can be expressed as a function of the bulk concentrations if the other steps (adsorption and desorption steps) are considered at equilibrium, which is the logical consequence of their rapidity. For example, in the case studied, from the equilibrium equations of adsorption and desorption steps, concentrations of adsorbed species are:

From equilibrium (10.3):

$$C_{A \bullet S} = K_A C_A C_v \qquad (10.8)$$

From equilibrium (10.4):

$$C_{B \bullet S} = K_B C_B C_v \qquad (10.9)$$

From equilibrium (10.6):

$$C_{P \bullet S} = \frac{C_P C_v}{K_P} \qquad (10.10)$$

where C_v is the concentration of free active centers on the catalyst surface. When Equation (10.8) to Equation (10.10) are substituted in Equation (10.7), the kinetics becomes a function of bulk-species concentrations and free active-center concentration. Finally, this concentration can also be expressed as a function of the other concentrations if the total balance of active centers is considered:

$$C_t = C_v + C_{A \bullet S} + C_{B \bullet S} + C_{P \bullet S} \qquad (10.11)$$

where C_t is the total concentration of active centers, which according to the Langmuir–Hinshelwood theory, remains constant. If Equation (10.11) is considered, the process rate finally becomes, for this case, as follows:

$$-r' = -r'_S = k_{S+}C_t^2 \frac{\left(K_A K_B C_A C_B - \dfrac{C_P^2}{K_P^2 K_S}\right)}{\left(1 + K_A C_A + K_B C_B + \dfrac{C_P}{K_P}\right)^2} \qquad (10.12)$$

Heterogeneous Catalytic Ozonation

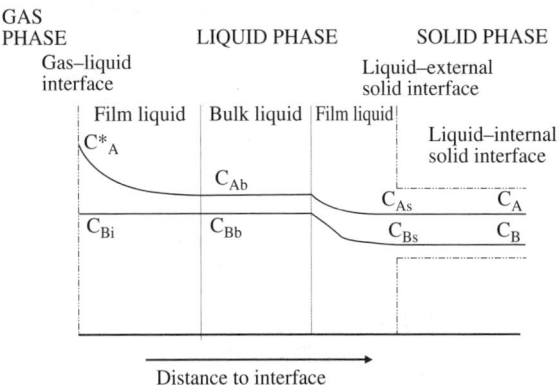

FIGURE 10.3 Concentration profiles of a gas–liquid–solid catalytic reaction in the fast kinetic regime.

Equation (10.12) should then be tested with experimental results. This complex type kinetic equation can often be simplified according to the experimental findings to yield, for example, a second-order kinetics, which results in easier mathematical treatment:

$$-r' = -r'_S = k_{S+}C_t^2 K_A K_B C_A C_B = k_T C_A C_B \qquad (10.13)$$

10.1.2 Fast Kinetic Regime or External Diffusion Kinetic Regime

In this kinetic regime, internal diffusion and surface chemical reactions are considered to be fast steps compared to external mass-transfer steps. Now, it is considered that the concentration of species varies within the films close to the gas–liquid and/or liquid–solid interfaces and that the concentration within the pores is the same as that at the external catalyst surface. Figure 10.3 illustrates this situation. There could be up to three consecutive external mass-transfer steps, according to Figure 10.3. Therefore, the process rate is the same in each of them. These step rates are expressed as the product of mass-transfer coefficient and the driving force:

For the gas phase, the mass-transfer step is

$$N_A = k_G(P_A - P_{Ai}) \qquad (10.14)$$

For the liquid (close to the gas–liquid interface), the mass transfer step is

$$N_A = k_L(C_A^* - C_A) \qquad (10.15)$$

For the liquid (close to the liquid–solid interface), the mass-transfer step is

$$N_A = k_c(C_A - C_{As}) \qquad (10.16)$$

where P_{Ai} and C_A^* are related through the Henry's law constant [see Equation (4.78)] and C_{As} is the concentration of A at the catalyst surface. Note that in this kinetic model, no homogeneous reaction between A and B is considered. If so, external mass transfer through the liquid film close to the liquid–solid interface will be simultaneously accompanied by a chemical reaction in the bulk water and Equation (10.16) would become, as an approximation:

$$N_A a = k_c a (C_A - C_{As}) + k_h C_A C_B \tag{10.17}$$

where the second term on the right side of Equation (10.17) represents the contribution of the homogeneous reaction between A and B, with k_h with the rate constant of this reaction that would take place within the bulk liquid and a the external surface of the catalyst per volume of solution. Note that a first-order reaction with respect to A and B is assumed in Equation (10.17). Equation rates (10.14) to (10.16) are also the rate of disappearance of A within the catalyst particle, that is:

$$N_A a = -r'w = k_T C_{As} C_{Bs} w \tag{10.18}$$

where w is the mass of catalyst per volume of solution. In most of the situations, external mass-transfer steps through both the gas and liquid films close to the gas–water interface are very fast so that the process rate is given by Equation (10.16). A combination of Equation (10.16) and Equation (10.18), together with Equation (10.19) (for liquid–solid diffusion of B)

$$N_A a = z k_c a (C_B - C_{Bs}) \tag{10.19}$$

leads to the final kinetic equation for this kinetic regime as a function of mass-transfer coefficients and bulk concentrations of A and B, with z the stoichiometric ratio of the catalytic reaction.

10.1.3 Internal Diffusion Kinetic Regime

In the absence of external mass-transfer limitations, the kinetics of gas–liquid–solid catalytic reactions is controlled by the internal diffusion of reactants through the pores of the catalyst. Diffusion of reactants through the pores develops simultaneously with the surface reaction steps mentioned previously, so that the final kinetics will depend on the relative importance of both diffusion and reaction steps. In this kinetic regime, there is a drop in reactant concentration through the pore from the external surface to the center of the pore, usually the center of the catalyst particle as far as kinetics is concerned. Figure 10.4 illustrates this situation. In some cases, the surface reactions are so fast that the internal diffusion exclusively controls the process rate. In these cases, surface reactions develop only in the outer sections of the pores. In other cases, when surface reactions are extremely low, internal diffusion does not affect the rate as was the case studied in Section 10.1.1. This represents,

Heterogeneous Catalytic Ozonation

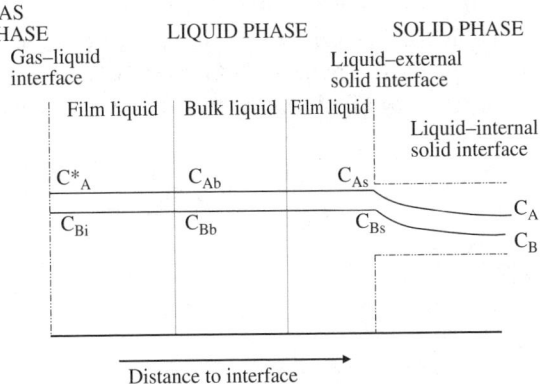

FIGURE 10.4 Concentration profiles of a gas–liquid–solid catalytic reaction in the internal diffusion kinetic regime.

of course, the ideal situation, since surface reactions would take place in all the internal surface of the pores.

Because the internal diffusion kinetic regime represents a simultaneous mass transfer and chemical reaction process, the steps followed to obtain the kinetic rate equation are similar to those presented in Chapter 4 for the case of fast gas–liquid reactions. Here a spherical catalytic particle is considered. The process rate is given by Fick's law applied to the external surface of the catalytic particle:

$$N_A\big|_{r=R} = -D_{eA} \frac{\partial C_A}{\partial r}\bigg|_{r=R} \quad (10.20)$$

where D_{eA} is the effective diffusivity of A, which in the case of liquids is calculated from the molecular diffusivity (see Section 5.1.1) once the porosity, ε_p, and tortuosity factor, τ_p, of the particle have been accounted for:

$$D_{eA} = D_A \frac{\varepsilon_p}{\tau_p} \quad (10.21)$$

As can be deduced from Equation (10.20), the concentration profile of reactant A through the pore is needed to determine the process rate. The concentration profile of A is obtained from the solution of the microscopic mass-transfer equation of A applied to the catalyst particle. This equation, for constant effective diffusivity, diffusion in the radial direction, and steady-state situation, is:[93]

$$-D_{eA} \frac{d\left(r^2 \frac{dC_A}{dr}\right)}{dr} + r^2 S_g \rho_p r''_A = 0 \quad (10.22)$$

where r_A'' represents the surface chemical reaction rate per internal surface of the catalyst, S_g is the internal surface per mass of the catalyst (commonly known as the BET surface), and ρ_p is the apparent density of the catalyst.[94] Equation (10.22), once developed, should be expressed in dimensionless form with the changes:

$$\varphi_A = \frac{C_A}{C_{As}} \quad \text{and} \quad \lambda = \frac{r}{R} \tag{10.23}$$

Boundary conditions are:

$$\begin{array}{ll} \lambda = 1 & \varphi = 1 \\ \lambda = 0 & \dfrac{d\varphi}{d\lambda} = 0 \end{array} \tag{10.24}$$

There is no analytical solution of Equation (10.22) for a second-order kinetics ($-r_A'' = k_T C_A C_B$), but for first-order kinetics it is:

$$\varphi = \frac{\sinh(\phi_1 \lambda)}{\lambda \sinh \phi_1} \tag{10.25}$$

where ϕ_1 is the Thiele number for a first-order reaction defined as:

$$\phi_1 = R \sqrt{\frac{k_T S_g \rho_p}{D_{eA}}} \tag{10.26}$$

The square of this number represents the ratio between the maximum rates of surface reaction and internal diffusion. It is evident that for high values of this number the process rate will be preferentially controlled by the internal diffusion while the opposite situation holds at low values of the Thiele number, i.e., control of the surface reaction.

Knowing the concentration profile of A, applying Fick's law [Equation (10.20)] leads to the reaction kinetic equation. For first-order reactions, the kinetics is:

$$N_A\big|_{r=R} = -D_{eA} \frac{\partial C_A}{\partial r}\bigg|_{r=R} = -\frac{D_{eA} C_{As}}{R} \frac{d\varphi_A}{d\lambda}\bigg|_{\lambda=1} = -\frac{D_{eA} C_{As}}{R} \left(\phi_1 \cosh \phi_1 - 1\right) \tag{10.27}$$

The process rate, however, is usually defined as a function of another parameter, the effectiveness factor, η, which represents the number of times the maximum surface reaction rate (evaluated at the external catalyst surface concentration: C_{As}) is diminished due to internal mass-transfer effects. Then, the kinetic equation is:

Heterogeneous Catalytic Ozonation

$$-r''_A = \eta(-r''_A)_{C_{As}} \tag{10.28}$$

with

$$\eta = \frac{3}{\phi_1^2}(\phi_1 \cosh\phi_1 - 1) \tag{10.29}$$

and

$$(-r''_A)_{C_{As}} = k_T C_{As} S_g \tag{10.30}$$

It can easily be deduced from Equation (10.29) that the lower the values of ϕ_1, the higher the values of η, i.e., the actual reaction rate is closer to the maximum one given by Equation (10.30). Then internal diffusion effects should be limited as much as possible.

10.1.4 General Kinetic Equation for Gas–Liquid–Solid Catalytic Reactions

In previous sections, rate equations for gas–liquid–solid catalytic reactions have been presented for different kinetic regimes. However, if all the steps (external and internal diffusion, and surface reaction) that constitute the mechanism of these reactions proceed at similar rates, all of them will have an effect on the process rate. For these cases, a global effectiveness factor is defined as follows:[93]

$$\Gamma = \frac{-r'_A}{(-r'_A)_{C_{Ab}}} = \frac{-r''_A}{(-r''_A)_{C_{Ab}}} \tag{10.31}$$

This new parameter represents the number of times the maximum surface reaction rate (evaluated at the bulk-liquid concentration, C_{Ab}) diminishes due to the internal and external mass transfer effects. The global effectiveness factor is obtained from the rate equations for external mass transfer [Equation (10.16)] and internal diffusion plus surface reaction [Equation (10.28)] once the external and internal surface areas have been accounted for. Thus, since the external mass transfer and internal diffusion with surface reaction are consecutive steps, the process rate can be expressed as:

$$k_{cA}(C_{Ab} - C_{As})a_c = \eta(-r''_A)_{C_{As}} S_g = \eta k_T C_{As} S_g \tag{10.32}$$

where a_c is the external surface area of the catalyst particle per mass of catalyst. From these equations the general process rate is:

$$-r''_A = \frac{\eta}{1 + \dfrac{\eta k_T S_g}{k_{cA} a_c}} k_T C_{Ab} \tag{10.33}$$

where the expression for the global effectiveness factor can be deduced once Equation (10.31) is accounted for. In Equation (10.33) parameters for all steps of the mechanism influence the process rate. Note that Equation (10.32) and Equation (10.33) consider that external mass transfer through gas and liquid films close to the gas–liquid interface is negligible. For more details on this matter the reader should refer to any specialized book.[93–95]

10.1.5 Criteria for Kinetic Regimes

The conditions that allow the kinetic regime to be established depend mainly on the values of external mass-transfer coefficient, effective diffusivity, and surface reaction rate constant. There are two main criteria usually followed in gas–liquid–solid catalytic reactions to distinguish the right kinetic regime. The first is the criterion of Weisz–Prater,[96] which distinguishes between the internal diffusion and surface chemical reaction. This criterion depends on the product of the effectiveness factor and the square of the Thiele number:

$$E = \eta \phi_1^2 = \frac{(-r'_A)_{exp} \rho_p R^2}{D_{eA} C_{As}} \qquad (10.34)$$

where E is the ratio between the actual or experimental process rate and the maximum internal diffusion rate, respectively, i.e., the same definition as the reaction factor in gas–liquid reactions (see Chapter 4). When $E \ll 1$ the process rate is surface–chemical controlled. In the opposite situation, when $(E \gg 1)$ internal diffusion is the controlling step of the catalytic reaction rate. Note that the Weisz–Prater criterion is applied when external mass transfer to solid catalyst surface is negligible. To confirm that the external mass-transfer regime does not control the process rate, the criterion of Mears should be used.[97] According to this criterion the external mass-transfer resistance is negligible when Equation (10.35) holds:

$$\frac{(-r'_A)_{exp} \rho_b R n}{k_{cA} C_{Ab}} < 0.15 \qquad (10.35)$$

where n is the reaction order and ρ_b, is the density of the catalytic bed in the reactor used.

We should emphasize that the kinetics of many gas–liquid–solid catalytic reactions is strongly dependent on temperature, so that kinetic equations for nonisothermal catalytic processes are deduced from the simultaneous solution of microscopic mass and energy balance equations. These equations then lead to important deviations from the corresponding isothermal ones. However, here only the fundamentals of isothermal reaction kinetics are given, since catalytic ozone reactions do not present significant variations of temperature in water.

As far as the kinetics of heterogeneous catalytic ozonation is concerned, two aspects should be considered: ozone decomposition kinetics and the catalytic ozonation of compounds.

10.2 KINETICS OF HETEROGENEOUS CATALYTIC OZONE DECOMPOSITION IN WATER*

As established in Chapter 2, because of its high reactivity, ozone decomposes in water to yield free radicals. The instability of the ozone molecule is both an advantage and a drawback. The advantage is that, when ozone decomposes, it yields hydroxyl free radicals and thus ozonation becomes an advanced oxidation process by itself. The drawback is that ozone cannot be used as the final disinfectant in drinking water treatment. For these reasons, the literature reports many studies on the kinetics of ozone decomposition in water (see Section 2.5 and Section 11.6). However, with a few exceptions where the ozone decomposition was studied in the presence of transition metal salts,[4,5,7,10] most of the studies are for buffered systems that lack the presence of true metal catalysts. The literature also presents studies on heterogeneous catalytic ozonation of different compounds and wastewater, as indicated previously (see Table 10.2).

In contrast, many studies on ozone decomposition (mechanism and kinetics included) over a catalyst surface in the gas phase have been conducted. These works are related to the destruction of gas ozone at the contactor outlet in water treatment plants due to the hazardous character of ozone in the surrounding atmosphere. Table 10.4 lists some of the main works published on this matter and their chief characteristics. Here detailed studies on the mechanism of the chemisorption of ozone on the catalyst surface are presented. Different types of catalysts have been used, from transition metal to activated carbon catalysts. A common mechanism to explain the ozone gas catalytic decomposition considers the adsorption of ozone on the catalyst surface and the formation of active oxygen adsorbed species (ozonide, superoxide, atomic oxygen), which finally react with another ozone molecule:[109]

$$O_3 + S \longrightarrow O_3 - S \quad (10.36)$$

$$O_3 - S \longrightarrow O - S + O_2 \quad (10.37)$$

$$O - S + O_3 \longrightarrow 2O_2 + S \quad (10.38)$$

The works are sometimes completed with infrared studies that show the wavelengths where ozone or oxygen adsorbed species absorb infrared radiation and allow their identification. Thus, Bulanin et al.[110,111] reported that peaks at 1034 and 1108 cm^{-1} of FT-IR analysis of catalyst samples treated with ozone correspond to ozone adsorbed species.

---

* Part of Section 10.2 is reprinted with permission from Beltrán, F.J. et al., Kinetics of heterogeneous decomposition of ozone in water on an activated carbon, *Ozone Sci. Eng.*, 24, 227–237, 2002. Copyright 2002, International Ozone Association.

TABLE 10.4
Works on Ozone Gas Decomposition on Catalyst Surfaces

Catalyst	Observations	Reference
Fe_2O_3, NiO_2, Co_2O_3	LH mechanism and kinetic study, first-order kinetics at 20–60°C	98
γ-Al_2O_3	LH mechanism proposed	99
14 different oxides of transition metals	Fixed-bed reactor: 25–200°C. Proposed formation of adsorbed oxygen species: O_2^-, O, O_3^-, O^-, etc. Better catalyst: Ag_2O (ozone decomposes at <25°C)	100
α-Fe_2O_3	23–65°C, LH mechanism; nitrogen oxides accompanying ozone from the ozonator act as inhibitors of ozone decomposition	101
TiO_2 powders of 2 to 186 m^2g^{-1} and 10 to more than 200 nm size. TiO_2 was coated on the inner surface of the Pyrex tube	Flow-type cylindrical Pyrex ozone reactor (390 cm^3, 5 cm I.D., 21 cm length); it was also supplied with 8 W low pressure Hg lamp ($\lambda > 200$ nm), 20–25°C, C_{O3} in air: 80 mgL^{-1}. Hydroxyl groups determination on the TiO_2 surface. Ozone decomposition depends on the hydroxyl groups sites of TiO_2 surface; decomposition increases with TiO_2 and TiO_2/UV	102
MnO_2, 64 m^2g^{-1}	Fixed-bed reactor, 10–80°C, $C_{O3} = 5.8 - 29 \times 10^{-5}$ M. LH mechanism; benzene oxidation is also studied	103
Activated carbon: 110–900 m^2g^{-1} 0.75 mm	6-mm high fixed-bed reactor, 25°C, C_{O3} in gas: 6×10^{-7} M. Type of activated carbon affects ozone decomposition	104
Activated carbon/MeO/ cordierite foam ($2MgO.2Al_2O_3.5SiO_2$), Me = Fe, Co, Ni, Mn, Ag, etc.	Fixed-bed reactor, gas lineal velocity: 0.7 ms^{-1}. Laser Raman spectroscopy study. MnO_2–Fe_2O_3 best catalyst	105
Co(II)/SiO_2, 0.7–3 mm, 360 m^2g^{-1}	Fixed-bed reactor: 2.6 cm diameter; gas residence time: up to 300 min; gas ozone: $5–50 \times 10^{-6}$ M. Mechanism and kinetic study; HO radical formation proposed. Very good yields of ozone destruction	106
Charred cherry stones	Fixed-bed reactor, charring temperature: 450–900°C. Ozone decomposition: 25–150°C, 10–60 Lh^{-1}, up to 2.5% O_3 volume. LH mechanism and kinetic study, activation energy: 41.6 $kcalmol^{-1}$	107
Cu(110)	Theoretical study, models of ozone adsorption, and decomposition are given	108

In contrast to the number of studies on catalytic ozone decomposition kinetics in the gas thus far few works deal with this subject in water.[50,65] These works also report the ozone decomposition on activated carbon surfaces.

The positive effect of the simultaneous use of ozone and activated carbon for removing pollutants from water has been the subject of various works, as reported later (see Section 10.3). A few attempts have been made to clarify the mechanism and kinetics of the ozone decomposition on activated carbon in water.

There are two ways to study the kinetics of ozone decomposition in water in the presence of solid catalysts. Both methods differ in the way ozone is fed to the reactor: as an ozone aqueous solution or continuously fed as an ozone–oxygen or ozone–air mixture. In the first case, a batch well-agitated reactor is usually used, while in the second case the reactor works as a semibatch well-agitated reactor where the water containing the catalyst has been previously charged. The semibatch way, however, is not recommended since ozone rapidly (in a few minutes) reaches a stationary concentration in water. Hence, only a few samples can be drawn from the reactor to prepare an ozone concentration profile with time, which is needed to study the kinetics. Another difficulty is the influence of the mass-transfer rate, which makes it necessary to know the mass-transfer coefficients of ozone from the water interface to the bulk water and from the bulk water to the water–solid interface. Accordingly, batch ozone decomposition studies are recommended. In this way, the system is a liquid–solid catalytic reaction because external mass transfer due to the diffusion of ozone through the film layer of water close to the gas–liquid interface is eliminated.

Once the reactor has been chosen, experiments should be developed according to certain rules. Because the main goal of the kinetics is to determine the rate constants of the catalytic reaction, it is convenient that the experiments be carried out in the slow kinetic regime. This would allow an LH mechanism[112] to be proposed and a rate equation to be deduced without the problems related to mass transfer. First, a series of experiments must be completed to establish the conditions of agitation and catalyst particle size where external and internal diffusion are fast steps and do not control the process rate. Thus, according to the preceding sections, in the external and internal diffusion kinetic regimes, the ozone decomposition rate will depend on the mass-transfer coefficient, k_c, and the particle size, R, respectively. Values of mass-transfer coefficients are proportional to the agitation speed.[113] The usual technique to apply in order to avoid external mass-transfer limitations is to conduct experiments at increasing agitation speeds. Once this variable does not affect the process rate, it can be concluded that external mass-transfer effects are negligible. The internal diffusion kinetic regime is avoided by completing experiments at decreasing catalyst particle size, R. When this variable does not affect the process rate, it can be said that internal diffusion is also negligible. When both mass-transfer effects are negligible, the process rate is in the chemical or slow kinetic regime. In the case of ozone decomposition kinetics in a batch reactor, it includes a series of experiments at different agitation speeds to fix the lowest values of this variable above which the process rate is constant. A second series of experiments at different particle size should then be completed to fix the maximum particle size below which the process rate is unaffected when this variable is changed. Finally, experiments at different ozone concentrations, catalyst mass, and temperature, and/or pH are developed to confirm the mechanism and theoretical rate equation proposed. This sequence of steps was described in a previous paper as discussed later in this chapter.[65]

The first work on the kinetics of ozone decomposition in the presence of a solid catalyst was by Jans and Hoigné,[50] who used an activated carbon and a carbon black as solid catalysts and small flasks as batch well-agitated reactors to study the kinetics. With these types of reactors, gas–water mass transfer was avoided, but the slow kinetic regime was confirmed. Experimental conditions applied included the presence of substances known to act as true initiators, inhibitors, or promoters of the ozone decomposition (see Chapter 2), although no experiments with agitation and different particle size of catalyst were carried out. A priori, the rate constants obtained should not be considered as due exclusively to the true surface reaction kinetics since no attempt was made to eliminate external and internal diffusion. However, from the results Jans and Hoigné obtained, it could be argued that the experiments were carried out in the slow kinetic regime. In any case, the main aim of their work was to clarify that the catalytic process was an advanced oxidation system, i.e., a system leading to the formation of hydroxyl radicals. The authors reported that the ozone decomposition rate increases in the presence of carbon (activated or black) compared to noncatalytic experiments under similar conditions.[50] They also observed that the decay rate of ozone followed apparent first-order kinetics, regardless of the presence or absence of the substances indicated above. For example, it was reported that at pH 5, in the presence of glucose (promoter) and acetate (inhibitor), the apparent rate constant increased from 8.1×10^{-4} s^{-1} in the absence of activated carbon to 2.6×10^{-3} s^{-1} in its presence. The apparent pseudo first-order rate constant also increased with pH (e.g., in the presence of 4 mgL^{-1} activated carbon, the rate constant increased from 2.6×10^{-3} s^{-1} at pH 5 to 15×10^{-3} s^{-1} at pH 7.4). The best fits in experimental data resulted by assuming a first ozone adsorption step at the carbon surface followed by one initiation step for free radicals. Hence, they concluded that ozone decomposition on the carbon surface acts as an initiating step of a radical mechanism leading to HO radicals in water. To confirm this, some decomposition experiments were also carried out in the presence of p-chlorobenzoic acid, a compound that does not adsorb onto the carbon surface (at the conditions investigated) and does not significantly react with ozone, but does react with hydroxyl radicals. The authors observed that although the presence of activated carbon accelerated the decomposition of ozone, the oxidizing efficiency or substrate selectivity did not change at all. In other words, they found the same stoichiometric ratio, 1 mole of hydroxyl radicals formed per 2 ozone moles decomposed, in the O_3/activated carbon system as in other advanced ozone oxidation systems (O_3/H_2O_2, O_3/UV). In view of these findings they concluded that activated carbon initiates the ozone decomposition in free radicals that subsequently react in water.

The study of the catalytic decomposition kinetics of ozone in water as a liquid–solid catalytic system following the rules above indicating avoidance of mass-transfer limitations has also been the subject of publications. Beltrán et al.[65] conducted such a study based on experimental results obtained in a perfectly mixed basket reactor (see Figure 10.5). The experiments were similar to those of Jans and Hoigné[50] with an aqueous solution of ozone charged into the basket reactor and put

Heterogeneous Catalytic Ozonation

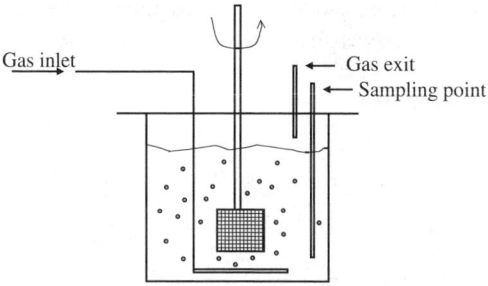

FIGURE 10.5 Perfectly mixed basket reactor.

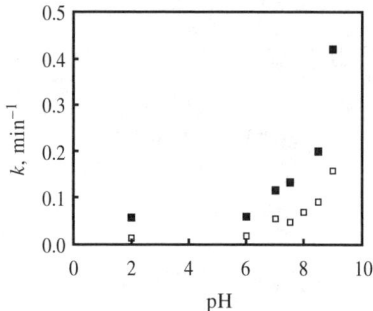

FIGURE 10.6 Variation of the homogeneous (□) and heterogeneous (■) apparent first-order rate constant of ozone decomposition in water with pH. (Reprinted with permission from Beltrán, F.J. et al., Kinetics of heterogeneous decomposition of ozone in water on an activated carbon, *Ozone Sci. Eng.*, 24, 227–237, 2002. Copyright 2002, International Ozone Association.)

into contact with a known mass of activated carbon situated in the basket. The activated carbon used was a porous material with 789 m^2g^{-1} BET surface area of spherical particles. Experimental conditions were established so that the process rate (ozone decomposition rate) developed in the slow or chemical kinetic regime (see Section 10.1.1). This was achieved after performing experiments at different agitation speed and particle size. Experiments at different mass of catalyst, temperature, and concentration of reactants (ozone in this case) were then carried out. It should be highlighted that the ozone decomposition in water in the presence of activated carbon was due to two mechanisms: the homogeneous reaction (see Section 3.1) and the catalytic reaction. Beltrán et al.[65] observed, as Jans and Hoigné[30] did previously, that, regardless of the presence and absence of activated carbon, the process rate followed a first-order kinetics with respect to ozone. They also observed that the catalytic decomposition rate of ozone increased with increasing pH. Figure 10.6, shows a plot of the apparent first-order rate constant of ozone decomposition in water in the presence and absence of activated carbon. As can be observed from Figure 10.6, the rate constant of ozone decomposition in the presence of activated carbon was always higher than in its absence. We can also see that the rate constant remained virtually unchanged in the 2 to 6 pH range while it increased significantly for pH > 6. In view of these results and information in the literature on catalytic decomposition of ozone in the gas phase, the following LH mechanism was proposed:[65]

Homogeneous decomposition:

$$O_3 + OH^- \xrightarrow{k_1 = 70 M^{-1}s^{-1}} HO_2^- + O_2 \qquad (10.39)$$

$$O_3 + HO_2^- \xrightarrow{k_{i1} = 2.8 \times 10^6 M^{-1}s^{-1}} HO_2\bullet + O_3^-\bullet \qquad (10.40)$$

$$O_3 + In \xrightarrow{k_{i2}} O_3^-\bullet + In^+ \qquad (10.41)$$

$$HO_2\bullet \underset{}{\overset{K_4 = 1.6 \times 10^{-5} M}{\rightleftharpoons}} O_2^-\bullet + H^+ \qquad (10.42)$$

Heterogeneous decomposition surface reaction:

For pH 2 to 6:

$$O_3 + S \underset{}{\overset{K_{C1} = \frac{k_{C1D}}{k_{C1I}}}{\rightleftharpoons}} O_3 - S \qquad (10.43)$$

$$O_3 - S \underset{}{\overset{K_{C2} = \frac{k_{C2D}}{k_{C2I}}}{\rightleftharpoons}} O - S + O_2 \qquad (10.44)$$

$$O_3 + O - S \underset{}{\overset{K_{C3} = \frac{k_{C3D}}{k_{C3I}}}{\rightleftharpoons}} 2O_2 + S \qquad (10.45)$$

In addition, for pH > 6:

$$OH^- + S \underset{}{\overset{K_{C4} = \frac{k_{C4D}}{k_{C4I}}}{\rightleftharpoons}} OH - S \qquad (10.46)$$

$$O_3 + OH - S \underset{}{\overset{K_{C5} = \frac{k_{C5D}}{k_{C5I}}}{\rightleftharpoons}} \bullet O_3 - S + HO\bullet \qquad (10.47)$$

$$\bullet O_3 - S \underset{}{\overset{K_{C6} = \frac{k_{C6D}}{k_{C6I}}}{\rightleftharpoons}} \bullet O - S + O_2 \qquad (10.48)$$

$$O_3 + \bullet O - S \underset{}{\overset{K_{C7} = \frac{k_{C7D}}{k_{C7I}}}{\rightleftharpoons}} O_2^- \bullet + S + O_2 \qquad (10.49)$$

Heterogeneous Catalytic Ozonation

Homogeneous propagation and termination reactions:

$$O_3 + O_2^{-}\bullet \xrightarrow{k_2=1.6\times 10^9 M^{-1}s^{-1}} O_3^{-}\bullet + O_2 \tag{10.50}$$

$$O_3^{-}\bullet + H^{+} \xrightarrow{k_3=5\times 10^{10} M^{-1}s^{-1}} HO_3\bullet \tag{10.51}$$

$$HO_3\bullet \xrightarrow{k_4=1.4\times 10^5 s^{-1}} HO\bullet + O_2 \tag{10.52}$$

$$HO\bullet + P \xrightarrow{k_t} \text{End products} \tag{10.53}$$

This mechanism supports the conclusion of Jans and Hoigné[50] about the character of activated carbon for initiating the radical chains leading to hydroxyl radicals that would react in the aqueous solution [Step (10.47) and Step (10.53)].

Thus, two kinetics of the activated carbon catalytic ozone decomposition in water were proposed depending on the pH range:

- For pH 2 to 6, Step (10.39) to Step (10.42) and Step (10.50) to Step (10.53) (homogeneous decomposition) and Step (10.43) to Step (10.45) catalytic decomposition. In this case, ozone adsorption [Step (10.46)] was considered as the controlling or slow step.
- For pH > 6, the same steps for the homogeneous decomposition and Step (10.46) to Step (10.49) for the catalytic decomposition. The authors assumed Step (10.47) was the slower one.[65]

From the mechanism proposed, first-order rate equations for the decomposition of ozone were deduced, thus confirming the experimental findings.[65] The rate equation deduced at pH > 6 was particularly interesting since it allowed the equilibrium constant of the hydroxyl ion adsorption Step (10.46) to be determined. Thus, the mechanism proposed led to the following rate equation:[65]

For pH > 6

$$-r_{O3}|_{Het} = -\frac{dC_{O3}}{dt}\bigg|_{het} = \frac{k_{C5D}K_{C4}C_t C_{OH^-} C_{O3}}{1+K_{C4}C_{OH^-}} = k_{Het2}C_{O3} \tag{10.54}$$

where

$$k_{Het2} = \frac{k'_{C5}C_{OH^-}}{1+K_{C4}C_{OH^-}} \tag{10.55}$$

and

$$k'_{C5} = k_{C5D}K_{C4}C_t \tag{10.56}$$

Since experiments were carried out in a perfectly mixed batch reactor, the rate of ozone decomposition was:

$$-r_{O3} = -\frac{dC_{O3}}{dt} = r_{O3}|_{Hom} + r_{O3}|_{Het} = \left(k_{Hom} + k_{Het1} + k_{Het2}\right)C_{O3} = k_{T2}C_{O3} \quad (10.57)$$

Determination of different rate constants, k_{Hom} (homogeneous process), k_{Het1} (catalytic process) for Step (10.39) to Step (10.42) and k_{Het2} was made by plotting the experimental results of $\ln C_{O3}$ against time to obtain straight lines (for more details see Reference 65). Values of k_{Het2} were calculated as the difference $k_{T2} - (k_{Hom} + k_{Het1})$ corresponding to each pH and temperature. The catalytic rate constant k_{Het2} was not a true chemical rate constant but a function of hydroxyl ion concentration as shown in Equation (10.55), which is a Langmuir isotherm type equation. This is a typical equation commonly found from LH mechanisms. The inverse of this equation shows a linear relationship between $1/k_{T2}$ and $1/C_{OH^-}$:

$$\frac{1}{k_{Het2}} = \frac{K_{C4}}{k'_{C5}} + \frac{1}{k'_{C5}C_{OH^-}} \quad (10.58)$$

so that a plot of its left side against the inverse of the hydroxyl ion concentration leads to a straight line of slope $1/k'_{C5}$. From the ordinate, the equilibrium constant, K_{C4}, could be determined. As a final conclusion of this work,[65] Arrhenius type equations were found for both the homogeneous and heterogeneous ozone decomposition kinetics. Regardless of the pH, the activation energy of the heterogeneous process was always lower than that of the homogeneous process.

10.3 KINETICS OF HETEROGENEOUS CATALYTIC OZONATION OF COMPOUNDS IN WATER

A few studies on the catalytic ozonation of compounds in water deal with the corresponding kinetics. Nonetheless, the kinetics, as in other ozonation processes, is a basic tool to establish the reactivity of ozone. Kinetic studies of catalytic ozonation are usually carried out in semibatch reactors where ozone is continuously fed as an ozone–oxygen or ozone–air mixture. The batch reactor with a charged aqueous ozone–compound solution is not recommended in this case because ozone is likely to be consumed in such a short period of time that the small amount of concentration data collected is not appropriate for a rigorous study. The semibatch reactor is then charged with the compound aqueous solution and catalyst. The catalyst can be in powder form or small pellets. When a powder catalyst is fed, the reactor works as a well-agitated slurry reactor. A basket placed along the agitating bar is also used to contain the catalyst pellets in the second case (see Figure 10.5).

Kinetic studies are developed in the different kinetic regimes (see Section 10.1) that allow the determination of rate constant, mass transfer, and effective diffusivity

Heterogeneous Catalytic Ozonation

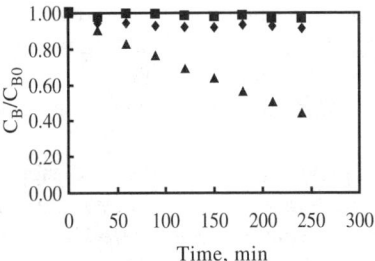

FIGURE 10.7 Evolution of dimensionless remaining oxalic acid concentration with time corresponding to experiments of single adsorption (♦), single ozonation (■) and catalytic ozonation with activated carbon (▲) of oxalic acid in water. Experimental conditions: initial $C_{oxalic} = 8 \times 10^{-3} M$, $C_{O3}(gas) = 30$ mg/L, mass of carbon = 1.25 g/L, carbon size = 1.0 to 1.6 mm, gas flow rate = 15 Lh^{-1}, agitation = 200 rpm, $T = 20°C$, pH = 2.5. (Reprinted with permission from Beltrán, F.J. et al., Kinetics of catalytic ozonation of oxalic acid in water with activated carbon, *Ind. Eng. Chem. Res.*, 41, 6510–6517, 2002. Copyright 2002, American Chemical Society.)

data that could later be used for kinetic modeling. In the treatment that follows, a catalytic reaction between ozone and a compound B will be considered:

$$O_3(gas) + zB(water) \xrightarrow{solid\ catalyst} Products \qquad (10.59)$$

For an easier mathematical treatment, an ozone first-order and a compound zero-order reaction are considered. These kinetic orders have been observed in a few catalytic ozonation reactions.[66]

10.3.1 THE SLOW KINETIC REGIME

Determination of the rate constant of the catalytic reaction is the main objective in this kinetic regime. In order that the catalytic reactions proceed in the specified kinetic regime, experimental conditions are established to avoid mass transfer limitations. However, the first experiments are aimed at confirming that the catalytic ozonation process is recommended compared to the noncatalytic process and the ozone-free adsorption process. Therefore, experiments on adsorption kinetics are necessary to compare if the adsorption rate competes with the catalytic ozonation rate. Catalyst-free ozonations are needed as second blank experiments for similar reasons. For example, Figure 10.7 shows that oxalic acid can be significantly removed with ozone in the presence of an activated carbon and that catalyst-free ozonation and ozone-free adsorption barely remove oxalic acid from water.[114] Then the catalytic process is a recommended option to follow. Next, the sequence of experiments at different agitation speeds and catalyst particle sizes (when the catalysts are in pellet form) is recommended. In this case, however, a series of experiments, at different gas flow rates of the inlet ozone containing gas, should first be carried out together with experiments for different agitation speeds. High values of both variables make the process rate unaffected by external mass transfer. Determination of critical or minimum gas flow rate and agitation speed is the objective of these experiments. When the catalyst is in pellet form, experiments for different particle sizes should be carried out to determine the maximum size that minimizes the internal diffusion effects.

As explained in the previous section, the next step includes experiments at different concentrations and temperatures and/or pH at the fixed minimum gas flow rate, agitation speed, and maximum particle size (in the case of pellet catalysts). Effects of these variables on the compound and ozone concentration, together with other complementary data on adsorption kinetics, FT-IR, and/or thermogravimetric analysis, should help propose an LH mechanism. For instance, Beltrán et al.[66] proposed the following mechanism of adsorption, surface, and desorption steps for the catalytic ozonation of oxalic acid in a powered TiO_2 slurry well-agitated semi-batch reactor.

Adsorption of ozone:

$$O_3 + S \underset{}{\overset{K_1}{\rightleftarrows}} O=O-O-S \tag{10.60}$$

$$O=O-O-S \xrightarrow{k_{12}} O-S+O_2 \tag{10.61}$$

Adsorption of oxalic acid:

$$B + S \underset{}{\overset{K_3}{\rightleftarrows}} B-S \tag{10.62}$$

Surface reaction plus desorption of products:

$$O-S + B-S \underset{}{\overset{K_4}{\rightleftarrows}} 2CO_2 + H_2O + 2S \tag{10.63}$$

In this work,[66] in addition to adsorption data for ozone and oxalic acid, the proposed mechanism was also based on the fact that the catalytic ozonation of oxalic acid led to total mineralization. In this particular case, in addition, catalyst-free ozonation was also considered. Step (10.61) was taken as the limiting one and the oxalic acid reaction rate was given by Equation (10.64):

$$-\frac{dC_B}{dt} = k_{i1} C^*_{O3} + w k_{i2} C_{O_3-S} \tag{10.64}$$

where the first term represented the contribution of the noncatalytic process, w was the mass of catalyst per unit volume, and C_{O3-S} was the concentration of ozone

Heterogeneous Catalytic Ozonation

adsorbed species. Equation (10.64) could further be reduced to the following zero-order oxalic acid kinetics, which supported the experimental results:

$$-\frac{dC_B}{dt} = k_{i1}C_{O3}^* + wk_{i2}K_1C_{O3}^* \frac{C_T}{1+K_1C_{O3}^*} = k_T \quad (10.65)$$

where K_1 is the equilibrium constant of Step (10.60). The zero-order rate constant was then obtained from the slope of the straight lines of the oxalic acid concentration profiles with time. Thus, the actual apparent zero-order rate constant of the catalytic reaction, k_{het} was:[66]

$$k_{het} = k_{i2}K_1C_{O3}^* \frac{C_T}{1+K_1C_{O3}^*} \quad (10.66)$$

with

$$k_{het} = \frac{k_T - k_{i1}C_{O3}^*}{w} \quad (10.67)$$

Here, Equation (10.66) is also a Langmuir type expression. The inverse form of Equation (10.66)

$$\frac{1}{k_{het}} = \frac{1}{k_{i2}C_T K_1 C_{O3}^*} + \frac{1}{k_{i2}C_T} \quad (10.68)$$

allows the rate constant ($k_{i2}C_T$) and equilibrium constant, K_1, to be determined from experiments at different ozone gas concentrations. Note that gas and water phases are perfectly mixed in this kind of reactor so that C_{O3}^*, the dissolved ozone concentration, is obtained from the ozone gas pressure at the reactor outlet by application of Henry's law. In Beltrán[66] an activation energy of 8±3 kcal mol^{-1} (32 ± 12 kJ mol^{-1}) was found for the catalytic reaction, much lower than that corresponding to the noncatalytic process.

10.3.2 External Mass Transfer Kinetic Regime

There are three forms to express the external mass transfer rate in gas–liquid–solid catalytic reactions, as shown in Section 10.1.2. For ozonation systems, however, these reduce to two since gas resistance to ozone transfer is negligible.[115] The external mass-transfer steps can even be reduced to only one, which corresponds to the ozone mass transfer through the water film close to the gas–water interface when the catalyst is used in powder form. Here both possibilities due to the nature of the catalyst are discussed for purposes of kinetic study.

10.3.2.1 Catalyst in Powder Form

When catalysts are in powder form, the external mass-transfer resistance across the water film surrounding the particle is negligible. With very small particle size, the mass-transfer coefficient, k_c, is very high[93] and the concentration profile through the indicated water film is uniform ($C_{O3s} = C_{O3}$). The external mass-transfer resistance is reduced to that corresponding to the water film close to the gas–water interface. The rate equation for this case is similar to Equation (10.16) and it was seen in the slow kinetic regime of catalyst-free ozone reactions [Equation (5.71)]:

$$N_{O3} = k_L a \left(C_{O3}^* - C_{O3} \right) \tag{10.69}$$

Thus, determination of mass-transfer coefficients is carried out as indicated in Section 5.3.6. However, another way to determine the mass-transfer coefficient is based on the knowledge of the rate constant of the catalytic reaction. The catalytic rate given by Equation (10.69) is also given by Equation (10.28) because both the external mass-transfer step considered and the internal diffusion plus the simultaneous surface reaction step are consecutive. In addition, the catalytic rate expressed by Equation (10.28) reduces to the maximum surface chemical reaction rate because in powdered catalysts there is no internal diffusion. Therefore, the internal effectiveness factor is unity. For a first-order ozone catalytic reaction, Equation (10.70) holds:

$$\left(-r'_{O3} \right)_s w = k C_{O3s} w = k C_{O3} w \tag{10.70}$$

where w represents the mass of catalyst per volume of slurry. From Equation (10.69) and Equation (10.70) the catalytic rate for an ozone first-order reaction becomes:

$$N_{O3} = \frac{C_{O3}^*}{\dfrac{1}{k_L a} + \dfrac{1}{wk}} \tag{10.71}$$

The ozone disappearance rate, N_{O3}, is not usually applied to determine $k_L a$ with Equation (10.71). This equation is used to determine the catalytic ozonation rate of compound B. In a semibatch reactor, a mass balance of B is given by Equation (10.72), once Equation (10.71) has been considered:

$$-\frac{dC_B}{dt} = z \frac{C_{O3}^*}{\dfrac{1}{k_L a} + \dfrac{1}{wk}} = k_o \tag{10.72}$$

Integration of Equation (10.72) with the initial condition:

$$t = t_i \quad C_B = C_{Bs} \tag{10.73}$$

leads to the following equation:

$$C_B = C_{Bs} - k_o(t - t_i) \tag{10.74}$$

where t_i is the time from which concentration of ozone in the gas at the reactor outlet becomes constant ($=C_{O3gs}$) and C_{Bs} is the concentration of compound B at time t_i. From Equation (10.74) a plot of the concentration of B against $(t - t_i)$ should lead to a straight line of slope $-k_o$. Then the rate constant of the catalytic process, k, or the mass transfer coefficient, k_La, can be calculated. This procedure was applied to experimental results of the catalytic ozonation of oxalic acid in the presence of powdered TiO_2 when conditions for significant external mass-transfer resistance held.[66]

10.3.2.2 Catalyst in Pellet Form

External mass-transfer resistance through the water films close to both the gas–water and water–solid interfaces can be responsible for external diffusion limitations. In the absence of homogeneous ozonation reaction, the catalytic ozonation is a single three-steps-in-series process, and the ozonation rate can be expressed through each one of the rates of these steps, which is:

$$N_{O3} = k_L a (C_{O3}^* - C_{O3}) = k_c a_p (C_{O3} - C_{O3s}) = \eta(-r'_{O3})_s w \tag{10.75}$$

where a_p is the external surface area of the catalyst per volume of solution. If experimental conditions are such that internal diffusion effects are negligible (small size of catalyst particles), the internal effectiveness factor is unity, and for a first-order reaction, $(-r'_{O3})_s = kC_{O3s}$, combination of the three rate equations in Equation (10.75) leads to:

$$N_{O3} = \frac{C_{O3}^*}{\dfrac{1}{k_L a} + \dfrac{1}{k_c a} + \dfrac{1}{wk}} \tag{10.76}$$

Equation (10.76) coupled with the mass-balance equation of B in the semibatch system, once integrated with initial Condition (10.73), allows the determination of one of the two mass-transfer coefficients, $k_c a$ or $k_L a$, provided k is known.

10.3.3 INTERNAL DIFFUSION KINETIC REGIME

If internal diffusion of ozone through catalyst pores is a slow step, provided the external mass-transfer resistances are negligible, the catalytic ozonation rate is given by Equation (10.28) with the internal effectiveness factor, for the case of a first-order reaction, given by Equation (10.29). Then the simultaneous use of the mass balance of B and Equation (10.28), Equation (10.29), and Equation (10.26) (equation for the Thiele number) allow the determination of the rate constant of the catalytic reaction, k, or the effective diffusivity of ozone in the catalyst pores, as shown below.

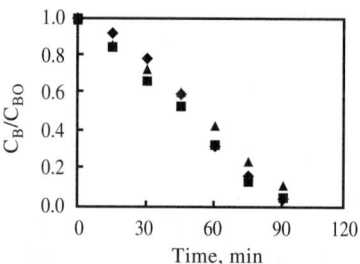

FIGURE 10.8 Evolution of dimensionless remaining oxalic acid concentration with time corresponding to experiments of catalytic ozonation with a TiO_2/Al_2O_3 catalyst and different catalyst particle size. Experimental conditions: initial $C_{oxalic} = 8 \times 10^{-3} M$, $C_{O3}(gas) = 40$ mg/L, mass of catalyst = g/L, gas flow rate = 15 Lh^{-1}, agitation = 300 rpm, $T = 20°C$, pH 2.5; particle size, mm: ▲ = 2 to 2.5, ♦ = 1.0 to 1.6, ■ = 1.6 to 0.5.

10.3.3.1 Determination of the Effective Diffusivity and Tortuosity Factor of the Porous Catalyst

This procedure requires the experimental results of catalytic ozonation at different catalyst particle sizes, once critical or minimum gas flow rate and agitation speed have been determined, to avoid external mass-transfer limitations. Figure 10.8 shows, as an example, the experimental results of oxalic acid concentration with time in catalytic ozonation in the presence of a TiO_2/Al_2O_3 catalyst with different particle sizes. It is seen that there exists a critical value of the particle size (1 mm) below which the oxalic acid rate is not affected. At conditions of particle size greater than the critical one, internal diffusion influences the kinetics. For a given first-order catalytic ozone reaction and zero-order for B (the usual case in heterogeneous catalytic ozonation[66]), and according to Equation (10.28) and Equation (5.48), from the slopes of the straight lines resulting from plots of the concentration of B against time, the product ηk can be obtained. The internal effectiveness factor, by definition, is also the ratio between the actual ozonation rate when internal diffusion is significant and the maximum surface ozone reaction rate when internal diffusion effects are negligible. These two ozonation rates can be obtained from the ratio of slopes of the B concentration profiles with time corresponding to experiments completed in the presence of large particle size and critical particle size (3 and 1 mm, respectively, in the example in Figure 10.8). Knowing η, the Thiele number ϕ_1 is obtained from Equation (10.29) and the effective diffusivity of ozone with Equation (10.26). With this procedure, D_e for ozone in the pellet catalyst can be determined. The tortuosity factor, τ_p, can then be obtained from Equation (10.21).

10.3.3.2 Determination of the Rate Constant of the Catalytic Reaction

With experimental results in the internal diffusion kinetic regime, the rate constant of the catalytic reaction can also be obtained provided other parameters such as the effective diffusivity are known. The method implies a trial-and-error method as follows:

- Assume a value for the rate constant, k.
- With Equation (10.28) and the mass balance of B applied [Equation (5.48)] to experimental results in the indicated kinetic regime, calculate the internal effectiveness factor, η.
- With Equation (10.29) and η, determine the Thiele number, ϕ_1.
- With Equation (10.26) and ϕ_1, S_g, R, D_e, etc., determine k.
- Compare the assumed and calculated values of k and decide whether to restart the process or finish.

The above kinetic studies can be applied to particular cases of the catalytic ozonation of model compounds and can also be used in moderately contaminated wastewater with low COD. In these cases, TOC can be used as a surrogate parameter to follow the mineralization of the wastewater. Thus far, however, there have been no publications on this topic.

10.4 KINETICS OF SEMICONDUCTOR PHOTOCATALYTIC PROCESSES

When certain materials, called semiconductors, are present in irradiated aqueous systems, fractions of the incident radiation of appropriate wavelength (or energy) are absorbed by the surface of these materials, thus triggering photochemical reactions. In fact, there are two possible mechanisms that photocatalysts undergo when irradiated. These possibilities depend on where the initial excitation occurs. An adsorbate molecule can be excited (absorbs radiation) and then interacts with the catalyst in a process called catalyzed photoreaction. The second possibility is the absorption of radiation by the photocatalyst, which then transfers an electron or energy to one adsorbed or bulk molecule in a process called sensitized photoreaction. It is this last type of photocatalytic reaction that semiconductor catalysts undergo. As different metals that present a continuum of electronic states, semiconductors have a void energy region called the band gap that extends from the highest energy electron-filled band or valence band to the lowest energy empty band or conduction band. Photons of at least equal energy (E_{hv}) to the band gap energy (E_{bg}) are absorbed, promoting one electron (e^-) from the valence band to the conduction band, as shown in Figure 10.9. This process, in addition, gives rise to the appearance of an oxidant

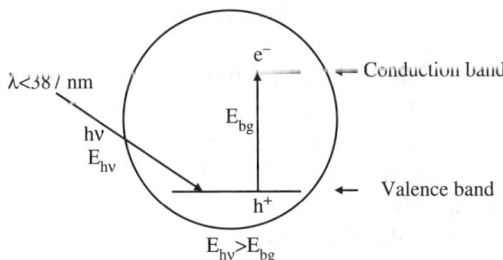

FIGURE 10.9 Formation of charge carriers during photon absorption by semiconductor valence band.

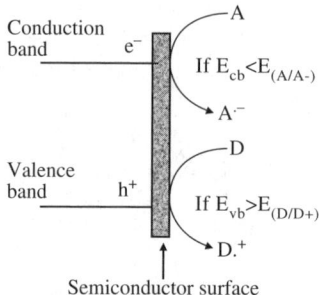

FIGURE 10.10 Photoinduced oxidation and reduction by charge carriers in semiconductor photocatalysis. D = electron donor, A = electron acceptor.

positive hole (h$^+$) in the valence band. These charge carriers can recombine, liberating heat, or be involved in oxidative and reductive pathways that can further trigger thermal or photocatalytic reactions to mineralize the matter present in water. Both charge carriers can be trapped by surface sites on the catalyst surface to generate more active sites. The criteria for photocatalyst reactions of a given adsorbed molecule or bulk molecule depend on two factors: the relative value of its redox potential compared to those of the semiconductor valence and conduction bands and the kinetic competition of these photoreactions with the charge carrier recombination. Thus, regardless of kinetic considerations, photoinduced oxidation will occur when the target molecule presents an oxidant potential less positive than the semiconductor valence band (E_{vb}). In these conditions, interfacial electron transfer is thermodynamically possible. On the other hand, photoinduced reduction will take place when the reduction potential of the target molecule is less negative than the semiconductor conduction band (E_{cb}). Figure 10.10 shows this situation when the target compound is an acceptor of electrons (A) or a donor of electrons (D). In aqueous semiconductor photocatalyst systems, the highest available electron donor is water, so that for the photoinduced oxidation reaction to occur, $E_{vb} > E_{(H2O/HO•)} = 2.8$ (at pH 0). On the other hand, the highest available electron acceptor could be oxygen, so that conduction band electron trapping or photoinduced reduction will take place if $E_{cb} < E_{(O2/HO2•)} = 0.2$. Accordingly, a good semiconductor is one with band gap energy higher than 3 eV, provided the previous conditions are fulfilled. Among semiconductors, in practice, only TiO$_2$ (band gap energy = 3.2 eV) and some other compounds such as SrTiO$_3$ fulfilled these conditions.[77] In addition, TiO$_2$ is cheap, nontoxic, and has high stability and photocatalytic ability. Among the crystalline forms of TiO$_2$, anatase and rutile, it has been experimentally demonstrated that the anatase form is the most active one although redox potentials of the valence and conduction bands of both forms are quite similar.[116] As reported,[116] the way of preparation rather than the crystalline form could be the main reason for the different activity observed when using anatase or rutile TiO$_2$, since this is intimately related to the porosity and surface area. The material prepared by Degussa and called P25 is the most used anatase, TiO$_2$, a semiconductor for water purification photocatalysis. In the following sections, the particular mechanism and kinetics of photocatalysis with this material are discussed.

TABLE 10.5
Mechanism of the TiO_2 Semiconductor Photocatalysis of Organics in Water[117]

Light absorption:

$$TiO_2 \xrightarrow{hv \geq 3.2eV (\lambda \leq 387nm)} e^- + h^+ \quad (10.77)$$

Charge carrier recombination:

$$e^- + h^+ \longrightarrow heat \quad (10.78)$$

Adsorption steps:

$$O_L^=(*) + Ti(IV) + H_2O \rightleftharpoons O_L H^- + Ti(IV) - OH^- \quad (10.79)$$

$$Ti(IV) + H_2O \rightleftharpoons Ti(IV) - H_2O \quad (10.80)$$

$$B + S \rightleftharpoons B - S \quad (10.81)$$

Charge carrier trapping:

$$Ti(IV) + e^- \rightleftharpoons Ti(III) \quad (10.82)$$

$$Ti(III) + O_2 \rightleftharpoons Ti(IV) - O_2^- \bullet \quad (10.83)$$

$$Ti(IV) - H_2O + h^+ \rightleftharpoons Ti(IV) - HO \bullet + H^+ \quad (10.84)$$

$$Ti(IV) - OH^- + h^+ \rightleftharpoons Ti(IV) - HO \bullet \quad (10.85)$$

$$B - S + h^+ \rightleftharpoons B - S^+ \quad (10.86)$$

Hydroxyl radical reactions:

$$Ti(IV) - HO \bullet + B \longrightarrow Ti(IV) + P \quad (10.87)$$

$$Ti(IV) - HO \bullet + B - S \longrightarrow Ti(IV) + P - S \quad (10.88)$$

$$Ti(IV) - HO \bullet \rightleftharpoons Ti(IV) + HO \bullet \quad (10.89)$$

$$HO \bullet + B \longrightarrow P \quad (10.90)$$

$$HO \bullet + B - S \longrightarrow P - S \quad (10.91)$$

Other charge carrier and free radical recombination reactions:

$$e^- + Ti(IV) - O_2^- \bullet + 2H^+ \rightleftharpoons Ti(IV) - H_2O_2 \quad (10.92)$$

$$Ti(IV) - O_2^- \bullet + H^+ \rightleftharpoons Ti(IV) - HO_2 \bullet \quad (10.93)$$

$$Ti(IV) - H_2O_2 + Ti(IV) - HO \bullet \rightleftharpoons Ti(IV) - HO_2 \bullet + Ti(IV) - H_2O \quad (10.94)$$

(*) = Lattice oxygen.

10.4.1 MECHANISM OF TiO_2 SEMICONDUCTOR PHOTOCATALYSIS

In addition to the diffusion steps mentioned in Section 10.1 for heterogeneous catalysis, the main adsorption–reaction–desorption steps of the mechanism of TiO_2 photocatalytic degradation of organics in water have been reported by Turchi and Ollis[117] (see Table 10.5). Of particular interest is that oxidation and reduction pathways are usually initiated from the oxygen and water adsorbed molecules on TiO_2 surface active sites as shown in Step (10.79), Step (10.80), Step (10.82), and Step (10.83) rather than from the direct interaction of holes with bulk molecules of the target compound B, although in many cases this is also thermodynamically favorable. According to results obtained in organic-free water, only partial mineralization of organics was achieved that supports the interaction between holes and adsorbed water for the photoinduced oxidation to take place.[117] Reaction (10.86) then seems to be of negligible contribution. As can be deduced from the mechanism of Table 10.5, one main drawback of these systems is the recombination of charge carriers that can be accomplished in a few nanoseconds.[81] Since the main photoinduced reaction generally involves the direct or indirect action of the positive hole of the valence band, the presence of electron trapping species is necessary to avoid recombination. Thus, oxygen is usually the reducible species that takes the conduction band electrons as shown in Step (10.82) and Step (10.83). Once charge carriers have been trapped to yield free radical adsorbed species, they are desorbed or react immediately with the target compound. Given the very rapid nature of these steps it is likely that these species react before they can be desorbed from the catalyst surface to participate in Step (10.90) or Step (10.91). Turchi and Ollis,[117] however, determined that the rate of diffusion of hydroxyl radicals from their adsorbed site to the bulk solution could be a few orders of magnitude higher than the rate of the photogenerated HO• at the semiconductor surface. Reaction (10.90) or Reaction (10.91) is also plausible.

10.4.2 Langmuir–Hinshelwood Kinetics of Semiconductor Photocatalysis

As has been observed experimentally[77] and deduced from the mechanism presented in Table 10.5, semiconductor photocatalysis usually follows a Langmuir–Hinshelwood kinetics, so that the rate of removal of a given target compound is given by Equation (10.95):

$$r_B = \frac{k_{exp} K_{exp} C_B}{1 + K_{exp} C_B} \quad (10.95)$$

where k_{exp} and K_{exp} represent the observed chemical rate constant of the photocatalytic reaction and the equilibrium constant of the adsorption of B, respectively. Equation (10.95) can be completed if competition of intermediates is accounted for:

$$r_B = \frac{k_{exp} K_{exp} C_B}{1 + K_{exp} C_B + \sum_{j=2}^{n} K_j C_j} \quad (10.96)$$

Equations of this type were deduced by Turchi and Ollis and checked through experimental results of the TiO_2 photocatalytic removal of benzene and some organochlorine volatile compounds.[117]

Another important fact in kinetic studies is the influence of the intensity of absorbed radiation, I_a, and oxygen concentration. For the former, experimental and mechanistic observations lead to a first-order and half-order dependence between the experimental rate constant k_{exp} and I_a so that:

For low I_a:
$$k_{exp} = kI_a \qquad (10.97)$$

For high I_a:
$$k_{exp} = kI_a^{0.5} \qquad (10.98)$$

Since most of the experimental studies work with intensities higher than 10^{-7} Einstein $m^{-2}s^{-1}$ the half-order is usually found in these kinetic studies.

Concentration of oxygen is another important variable, not only because oxygen traps electrons in the conduction band but also because it influences the reaction rate given by Equation (10.95) or Equation (10.96). Also, a Langmuir–Hinshelwood influence has been observed for oxygen concentrations, as Equation (10.99) shows:

$$k_{exp} = k' \frac{K_{O2}C_{O2}}{1 + K_{O2}C_{O2}} \qquad (10.99)$$

where K_{O2} is the Langmuir adsorption equilibrium constant for oxygen in Ti(III) sites. Mills and Morris[118] checked Equation (10.99) from experimental results of the p-chlorophenol anatase TiO_2 photodegradation after plotting the reciprocal of carbon dioxide generation rate against the reciprocal of oxygen concentration.

Finally, other variables that influence the kinetics of semiconductor photocatalysis systems are the concentration of the target compound, the pH of water, the presence of anions, metal doping on the catalyst surface, etc. Detailed explanations of these effects are outside the scope of this work but interested readers can refer to various review papers.[77,79–82]

10.4.3 MECHANISM AND KINETICS OF PHOTOCATALYTIC OZONATION

As can be deduced from the mechanisms of ozone in water (see Table 2.4) and TiO_2 semiconductor photocatalysis (Table 10.5), there is a clear possibility for ozone to improve the semiconductor photocatalytic rate or vice versa. This improvement has been observed in different photocatalytic ozonation studies as depicted in Table 10.3. In these studies, although no rigorous kinetic studies were conducted, some possible link reactions between both ozonation and semiconductor photocatalytic oxidation mechanisms are presented. Thus, the fact that the potential of the conduction band is more negative than that of the pair ($O_3/O_3^-{}^\bullet$) makes plausible the ozone trapping of conduction band electrons according to the following sequence:[102]

$$Ti(IV) + e^- \rightleftarrows Ti(III) \qquad (10.82)$$

$$Ti(III) + O_3 \rightleftarrows Ti(IV) - O_3^- \bullet \qquad (10.100)$$

In fact the ozonide ion radical, $O_3^-\bullet$, has been detected during the ozone decomposition on TiO_2 surfaces in the gas phase[102] and by ESR measurements during the free cyanide decomposition in water with ozone, UV light, and P25 Degussa and BDH TiO_2 catalysts.[85] Once the ozonide ion radical is formed, a mechanism similar to that shown in Table 2.4 for the ozone decomposition can take place to eventually yield another way of hydroxyl radical generation. Furthermore, hydroxyl radicals can also be formed from the reaction between photogenerated and adsorbed hydrogen peroxide and ozone in a way similar to the one that occurs in the ozone/hydrogen peroxide and ozone/UV systems (see Chapter 8 or Chapter 9):

$$Ti(IV)-O_3\bullet + Ti(IV)-H_2O_2 \rightleftarrows Ti(IV)-O_3^-\bullet + Ti(IV)-HO_2^-\bullet + H^+ \quad (10.101)$$

As far as the kinetics is concerned, in previous studies, Langmuir–Hinshelwood photocatalytic ozonation kinetics was observed in the removal of monochloroacetic acid and pyridine,[84] cyanides,[85] and formic acid.[86] The studies were carried out in semibatch photocatalytic ozonation reactors, so that for a water phase of batchwise operation, the equation usually checked is

$$-\frac{dC_B}{dt} = k_{exp} C_{O_3-S} C_{B-S} \quad (10.102)$$

with k_{exp} the second-order rate constant of the photocatalytic ozonation process, and C_{O3-S} and C_{B-S} the concentration of adsorbed ozone and B species on TiO_2 active sites, which can be expressed by the corresponding Langmuir adsorption isotherms:

$$C_{O3-S} = \frac{K_{O3} C_{O3}}{1 + K_{O3} C_{O3}} \quad (10.103)$$

and

$$C_{B-S} = \frac{K_B C_B}{1 + K_B C_B} \quad (10.104)$$

with K_{O3} and K_B the adsorption equilibrium constants for ozone and B, respectively. Equations of this type have been checked by Hernández–Alonso et al.[85] in the TiO_2 photocatalytic ozonation of free cyanides.

In view of the experimental results obtained in this field, a promising future for the possible application of these systems is expected, as preliminary cost studies have also shown.[84] Obviously, before this occurs, considerable research effort should be devoted to elucidate many aspects such as those concerning the improvement of semiconductor catalyst preparation and the mechanism and kinetics of the process.

References

1. Legube, B. and Karpel Vel Leitner, N., Catalytic ozonation: a promising advanced oxidation technology for water treatment, *Catal. Today*, 53, 61–72, 1999.
2. Hewes, C.G. and Davison, R.R., Renovation of wastewater by ozonation, *AIChE Symp. Series*, 129, 71–80, 1972.
3. Chen, J.W. et al., Catalytic ozonation in aqueous system, *AIChE Symp. Series*, 166, 206–212, 1976.
4. Hill, G.R., Kinetics, mechanism, and activation energy of the cobaltous ion catalyzed decomposition of ozone, *J. Am. Chem. Soc.*, 70, 1306–1307, 1948.
5. Hill, G.R., The kinetics of oxidation of cobaltous ion by ozone, *J. Am. Chem. Soc.*, 71, 2434–2435, 1949.
6. Taube, H. and Bray, W.C., Chain reactions in aqueous solutions containing ozone, hydrogen peroxide, and acid, *J. Am. Chem. Soc.*, 62, 3357–3373, 1940.
7. Yang, T.C. and Neely, W.C., Relative stoichiometry of the oxidation of ferrous ion by ozone in aqueous solution, *Anal. Chem.*, 58, 1551–1555, 1986.
8. Paillard, H. et al., Iron and manganese removal with ozonation in the presence of humic substances, *Proc. 8th Ozone World Congress*, Vol. 1, C71–C96, Zurich, September 1987.
9. Abdo, M.S.E., Shaban, H., and Bader, M.S.H., Decolorization by ozone of direct dyes in presence of some catalysts, *J. Env. Sci. Health*, A23, 697–710, 1986.
10. Reckhow, D.A. et al., Oxidation of iron and manganese by ozone, *Ozone Sci. Eng.*, 13, 675–695, 1991.
11. Toui, S., The oxidation of manganese and disinfection by ozonation in water purification processing, *Ozone Sci. Eng.*, 13, 623–637, 1991.
12. Andreozzi, R. et al., The kinetics of Mn(II)-catalysed ozonation of oxalic acid in aqueous solution, *Water Res.*, 26, 917–921, 1992.
13. Logager, T. et al., Oxidation of ferrous ion by ozone in acidic solutions, *Inorg. Chem.*, 31, 3523–3529, 1992.
14. McKnight, K.F. et al., Comparison of ozone efficiency for manganese oxidation between raw and settled water, *Ozone Sci. Eng.*, 15, 331–341, 1993.
15. Goncharuk, V.V., Study of the processes of catalytic ozonation of organic matter in water, *Proc. 11th Ozone World Congress*, Vol. 2, S-16-13-S16-30, San Francisco, CA, August–September 1993.
16. Séby, F. et al., Study of the ozone–manganese reaction and the interactions of disulfonate Indigo Carmin/oxidized manganese forms as a function of pH, *Ozone Sci. Eng.*, 17, 135–147, 1995.
17. Andreozzi, R. et al., Manganese catalysis in water pollutants abatement with ozone, *Environ. Technol.*, 16, 885–891, 1995.
18. Gracia, R., Aragües, J.L., and Ovelleiro, J.L., Study of the catalytic ozonation of humic substances in water and their ozonation by products, *Ozone Sci. Eng.*, 18, 195–208, 1996.
19. Ma, J. and Graham, N.J.D., Preliminary investigation of manganese catalyzed ozonation for the destruction of atrazine, *Ozone Sci. Eng.*, 19, 227–240, 1997.
20. Gracia, R., Aragües, J.L., and Ovelleiro, J.L., Mn(II)-catalysed ozonation of raw Ebro River water and its ozonation by-products, *Ozone Sci. Eng.*, 32, 57–62, 1998.
21. Andreozzi, R. et al., The ozonation of pyruvic acid in aqueous solutions catalyzed by suspended and dissolved manganese, *Water Res.*, 32, 1492–1496, 1998.
22. Ma, J. and Graham, N.J.D., Degradation of atrazine by manganese catalysed ozonation: influence of humic substances, *Water Res.*, 33, 785–793, 1999.

23. Cortés, S. et al., Comparative efficiency of the systems O_3/high pH and O_3/catalyst for the oxidation of chlorobenzenes in water, *Ozone Sci. Eng.*, 22, 415–426, 2000.
24. Andreozzi, R., Marotta, R., and Sanchirico, R., Manganese catalysed ozonation of glyoxalic acid in aqueous solutions, *J. Chem. Technol. Biotechnol.*, 75, 59–65, 2000.
25. Andreozzi, R. et al., Kinetic modeling of pyruvic acid ozonation in aqueous solutions catalyzed by Mn(II) and Mn(IV) ions, *Water Res.*, 35, 109–120, 2001.
26. Potapenko, E.V. et al., Oxidation of 4-nitrotoluene by ozone in acetic anhydride in the presence of manganese sulfate, *Kinet. Catal.*, 42, 796–799, 2001.
27. Hassan, M.M. and Hawkyard, C.J., Decolourisation of aqueous dyes by sequential oxidation treatment with ozone and Fenton's reagent, *J. Chem. Technol. Biotechnol.*, 77, 834–841, 2002.
28. Ni, C.H., Chen, J.N., and Yang, P.Y., Catalytic ozonation of 2-dichlorophenol by metallic ions, *Water Sci. Technol.*, 47, 77–82, 2002.
29. McKay, G. and McAleavy, G., Ozonation and carbon adsorption in a three phase fluidized bed for colour removal from peat water, *Chem. Eng. Res. Des.*, 66, 531–536, 1988.
30. Johsi, M.G. and Shambaugh, R.L., The kinetics of ozone–phenol reaction in aqueous solution, *Water Res.*, 16, 933–938, 1982.
31. Munter, R.R. et al., Catalytic treatment of wastewater with ozone, *Khim. Tekhnol. Vody*, 7, 17–19, 1985.
32. Al-Hayek, N., Legube, B., and Doré, M., Catalytic ozonation (Fe(III)/Al_2O_3) of phenol and its ozonation by-products, *Environ. Technol. Let.*, 10, 416–426, 1989.
33. Heinig, C.F., O_3 or O_2 and Ag: a new catalyst technology for aqueous phase sanitation, *Ozone Sci. Eng.*, 15, 533–546, 1993.
34. Allemane, H. et al., Comparative efficiency of three systems (O_3, O_3/H_2O_2, and O_3/TiO_2) for the oxidation natural organic matter in water, *Ozone Sci. Eng.*, 15, 419–432, 1993.
35. Campen, J.P., Ecoclear: a catalytic ozonation process which is applicable to a variety of mildy polluted wastewater streams, *Proc. 11th Ozone World Congress*, IOA, Vol. 1, S-16-1-S-16-12, San Francisco, CA, 1993.
36. Pines, D. et al., A catalytic oxidation process for removal of ozone by-products, *Proc. 1994, Water Quality Technology Conference, Am. Water Works Assoc.*, Part II, 1483–1495, San Francisco, CA, November 1994.
37. Kallas, J. et al., Ozonation and AOP parameter estimation from countercurrent film absorber data, *Ozone Sci. Eng.*, 17, 527–550, 1995.
38. Ciba, N. et al., The impact of catalytic ozonation on bromate formation, *Proc. 11th Ozone World Congress*, IOA, 585–594, Lille, France, April 1995.
39. Kaptijn, J.P., Plugge, M.F.C., and Annee, J.H.J., Water treatment with a fixed bed catalytic ozonation process, *Proc. 11th Ozone World Congress*, IOA, 407–417, Lille, France, April 1995.
40. Bhat, N.N. and Gurol, M.D., Oxidation of chlorobenzene by ozone and heterogeneous catalytic ozonation, *Proc. Industrial Mid-Atlantic Conference*, 371–382, Bethlehem, PA, July 1995.
41. Andreozzi, R. et al., The use of manganese dioxide as heterogeneous catalyst for oxalic acid ozonation in aqueous solution, *Appl. Catal.*, 138, 75–81, 1996.
42. Leichsenring, S. et al., Catalytic oxidation of chloroaromatic trace contaminants adsorbed on Wessalith Day catalyst, *Chemosphere*, 33, 343–352, 1996.
43. Zaror, C.A., Enhanced oxidation of toxic effluents using simultaneous ozonation and activated carbon adsorption, *J. Chem. Technol. Biotechnol.*, 70, 21–28, 1997.

44. Logemann, F.P. and Annee, J.H.J., Water treatment with a fixed bed catalytic ozonation process, *Water Sci. Technol.*, 35, 353–360, 1997.
45. Kaptijn, J.P., The Ecoclear process: results from full scale installations, *Ozone Sci. Eng.*, 19, 297–305, 1997.
46. Barrat, P. et al., Advanced oxidation of water using catalytic ozonation, GB Patent, No. PCT/GB96/02525, 1997.
47. Volk, C. et al., Comparison of the effect of ozone, ozone–hydrogen peroxide system, and catalytic ozone of the biodegradable organic matter of a fulvic acid solution, *Water Res.*, 31, 650–656, 1997.
48. Karpel Vel Leitner, N. et al., Impact of catalytic ozonation on the removal of a chelating agent and surfactants in aqueous solution, *Water Sci. Technol.*, 38, 203–209, 1998.
49. Sánchez, L., Peral, J., and Domenech, X., Aniline degradation by combined photocatalysis and ozonation, *Appl. Catal. B Environ.*, 19, 59–65, 1998.
50. Jans, U. and Hoigné, J., Activated carbon and carbon black catalyzed transformation of aqueous ozone into OH-radicals, *Ozone Sci. Eng.*, 21, 67–89, 1999.
51. Cooper, C. and Burch, R., Mesoporous materials for water treatment processes, *Water Res.*, 33, 3689–3694, 1999.
52. Cooper, C. and Burch, R., An investigation of catalytic ozonation for the oxidation of halocarbons in drinking water preparation, *Water Res.*, 33, 3695–3700, 1999.
53. Karpel Vel Leitner, N. et al., Effects of catalysts during ozonation of salycilic acid, peptides, and humic substances in aqueous solutions, *Ozone Sci. Eng.*, 21, 261–276, 1999.
54. Karpel Vel Leitner, N. et al., Reactivity of various Ru/CeO_2 catalysts during ozonation of acid aqueous solutions, *New J. Chem.*, 24, 229–233, 2000.
55. Gracia, R. et al., Heterogeneous catalytic ozonation with supported titanium dioxide in model and natural waters, *Ozone Sci. Eng.*, 22, 461–471, 2000.
56. Lin, J. et al., Effective catalysts for wet oxidation of formic acid by oxygen and ozone, *Ozone Sci. Eng.*, 22, 241–247, 2000.
57. Piera, E. et al., 2,4-Dichlorophenoxyacetic acid degradation by catalyzed ozonation: $TiO_2/UVA/O_3$ and $Fe(II)/UVA/O_3$ systems, *Appl. Catal. B Environ.*, 27, 169–177, 2000.
58. Lin, S.H. and Lai, C.L., Kinetic characteristics of textile wastewater ozonation in fluidized and fixed bed activated carbon bed, *Water Res.*, 34, 763–772, 2000.
59. Delanoe, F. et al., Relationship between the structure of Ru/CeO_2 catalysts and their activity in the catalytic ozonation of succinic acid aqueous solutions, *Appl. Catal. B Environ.*, 29, 315–325, 2001.
60. Zaror, C. et al., Ozonation of 1,2-dihydroxybenzene in the presence of activated carbon, *Water Sci. Technol.*, 44, 125–130, 2001.
61. Ni, C.H. and Chen, J.N., Heterogeneous catalytic ozonation of 2-chlorophenol aqueous solution with alumna as catalyst, *Water Sci. Technol.*, 43, 213–220, 2001.
62. Ping, T.S. et al., Catalytic ozonation of sulfosalycilic acid, *Ozone Sci. Eng.*, 24, 117–122, 2002.
63. Gilbert, E., Influence of ozone on the photocatalytic oxidation of organic compounds, *Ozone Sci. Eng.*, 24, 75–82, 2002.
64. Kasprzyk, B. and Nawrocki, J., Preliminary results on ozonation enhancement by perfluorinated binded alumina phase, *Ozone Sci. Eng.*, 24, 63–68, 2002.
65. Beltrán, F.J. et al., Kinetics of heterogeneous decomposition of ozone in water on an activated carbon, *Ozone Sci. Eng.*, 24, 227–237, 2002.

66. Beltrán, F.J., Rivas, F.J., and Montero-de-Espinosa, R., Catalytic ozonation of oxalic acid in an aqueous TiO_2 slurry reactor, *Appl. Catal. B Environ.*, 29, 221–231, 2002.
67. Fu, H., Karpel vel Leitner, N., and Legube, B., Catalytic ozonation of chlorinated carboxylic acids with Ru/CeO_2-TiO_2 catalyst in aqueous system, *New J. Chem.*, 26, 1662–1666, 2002.
68. Lin, J., Kawai, A., and Nakajima, T., Effective catalysts for decomposition of aqueous ozone, *Appl. Catal. B Environ.*, 39, 157–165, 2002.
69. Hassan, M.M. and Hawkyard, C.J., Ferral-catalyzed ozonation of aqueous dyes in a bubble-column reactor, *Catal. Comm.*, 3, 281–286, 2002.
70. Lim, H.N. et al., Characterization of ozone decomposition in a soil slurry: kinetics and mechanism, *Water Res.*, 36, 219–229, 2002.
71. Kim, B.S. et al., Catalytic ozonation of an organophosphorous pesticide using microporous silicate and its effect on total toxicity reduction, *Water Sci. Technol.*, 46, 35–41, 2002.
72. Rivera-Utrilla, J. and Sánchez-Polo, M., Ozonation of 1,3,6-naphthalenesulphonic acid catalysed by activated carbon in aqueous phase, *Appl. Catal. B Environ.*, 39, 319–329, 2002.
73. Rivera-Utrilla, J. and Sánchez-Polo, M., The role of dispersive and electrostatic interactions in the aqueous phase adsorption of naphthalenesulphonic acids on ozone-treated activated carbons, *Carbon*, 40, 2685–2691, 2002.
74. Rivera-Utrilla, J. et al., Effect of ozone and ozone/activated carbon treatments on genotoxic activity of naphthalenesulfonic acids, *J. Chem. Technol. Biotechnol.*, 77, 883–890, 2002.
75. Park, J.S., Choi, H., and Ahn, K.H., The reaction mechanism of catalytic oxidation with hydrogen peroxide and ozone in aqueous solution, *Water Sci. Technol.*, 47, 179–184, 2002.
76. Sánchez-Polo, M. and Rivera-Utrilla, J., Effect of the ozone–carbon reaction on the catalytic activity of activated carbon during the degradation of 1,3,6-naphthalene-trisulphonic acid with ozone, *Carbon*, 41, 303–307, 2003.
77. Mills, A., Davies, R.H., and Worsley, D., Water purification by semiconductor photocatalysis, *Chem. Soc. Rev.*, 22, 417–425, 1993.
78. Kamat, P., Photochemistry of nonreactive and reactive (semiconductor) surfaces, *Chem. Rev.*, 93, 267–300, 1993.
79. Fox, M.A. and Dulay, M.T., Heterogeneous photocatalysis, *Chem. Rev.*, 93, 341–357, 1993.
80. Linsebigler, A.L., Lu, G., and Yates, J.T., Jr., Photocatalysis on TiO_2 surfaces: principles, mechanisms, and selected results, *Chem. Rev.*, 95, 735–758, 1995.
81. Hoffmann, M.R. et al., Environmental applications of semiconductor photocatalysis, *Chem. Rev.*, 95, 69–96, 1995.
82. Bhatkhande, D.S., Pangarkar, V.G., and Beenackers, A.C.M., Photocatalytic degradation for environmental applications — a review, *J. Chem. Technol. Biotechnol.*, 77, 102–117, 2001.
83. Sierka, R.A. and Hendricks, K.L., Radical species production during ozonation of illuminated titanium dioxide suspensions, *Proc. 11th Ozone World Congress*, International Ozone Association, S-15-8-S-15-24, San Francisco, CA, 1993.
84. Kopf, P., Gilbert, E., and Eberle, S.H., TiO_2 photocatalytic oxidation of monocholoroacetic acid and pyridine: influence of ozone, *J. Photochem. Photobiol. A Chem.*, 136, 163–168, 2000.
85. Hernández-Alonso, M.D. et al., Ozone enhanced activity of aqueous titanium dioxide suspensions for photocatalytic oxidation of free cyanide ions, *Appl. Catal. B Environ.*, 39, 257–267, 2002.

86. Wang, S., Shiraishi, F., and Nakano, K., A synergistic effect of photocatalysis and ozonation on decomposition of formic acid in an aqueous solution, *Chem. Eng. J.*, 87, 261–271, 2002.
87. Martín, M.A. et al., Kinetic study of the anaerobic digestion of vinasse pretreated with ozone, ozone plus ultraviolet light, and ozone plus ultraviolet light in the presence of titanium dioxide, *Proc. Biochem.*, 37, 699–706, 2002.
88. Wada, H., Murayama, T., and Kuroda, Y., Recycling of cyanide wastewater applying combined UV-ozone oxidation with a titanium dioxide catalyst and ion exchange resin method, *Bull. Chem. Soc. Jpn.*, 75, 1399–1405, 2002.
89. Villaseñor, J., Reyes, P., and Pecchi, G., Catalytic and photocatalytic ozonation of phenol on MnO_2 supported catalysts, *Catal. Today*, 76, 121–131, 2002.
90. Jung, J. et al., TCE and PCE decomposition by a combination of gamma-rays, ozone and titanium dioxide, *J. Radioanal. Nuclear Chem.*, 252, 451–454, 2002.
91. Canton, C., Esplugas, S., and Casado, J., Mineralization of phenol in aqueous solution by ozonation using iron or copper salts and light, *Appl. Catal. B Environ.*, 1320, 1–11, 2002.
92. Satterfield, C.N., *Heterogeneous Catalysis in Practice*, McGraw-Hill, New York, 1980.
93. Fogler, H.S., *Elements of Chemical Reaction Engineering*, 3rd ed., Prentice-Hall, Englewood Cliffs, NJ, 1999.
94. Smith, J.M., *Chemical Engineering Kinetics*, 2nd ed., McGraw-Hill, New York, 1970.
95. Shah, Y.T., *Gas–Liquid–Solid Reactor Design*, McGraw-Hill, New York, 1979.
96. Weisz, P.B. and Prater, C.D., Interpretation of measurements in experimental catalysis, *Adv. Catal.*, 6, 143–196, 1954.
97. Mears, D.E., Test for transport limitations in experimental catalytic reactors, *Ind. Eng. Chem. Process Des. Develop.*, 10, 541–547, 1971.
98. Calderbank, P.H. and Lewis, J.M.O., Ozone decomposition catalysis, *Chem. Eng. Sci.*, 31, 1216–1216, 1976.
99. Klimovskii, A.O. et al., Interaction of ozone with γ-Al_2O_3 surface, *React. Kinet. Catal. Lett.*, 23, 95–98, 1983.
100. Imamura, S. et al., Decomposition of ozone on a silver catalyst, *Ind. Eng. Chem. Res.*, 30, 217–221, 1991.
101. Mehandjiev, D. and Naidenov, A., Ozone decomposition on α-Fe_2O_3 catalyst, *Ozone Sci. Eng.*, 14, 277–282, 1992.
102. Ohtani, B. et al., Catalytic and photocatalytic decomposition of ozone at room temperature over titanium (IV) dioxide, *J. Chem. Soc. Faraday Trans.*, 88, 1049–1053, 1992.
103. Naidenov, A. and Mehandjiev, D., Complete oxidation of benzene on manganese dioxide by ozone, *Appl. Catal.*, 97, 17–22, 1993.
104. Rakitskaya, T.L. et al., Kinetics of ozone decomposition on activated carbons, *Kinet. Catal.*, 35, 90–92, 1994.
105. Heisig, C., Zhang, W., and Oyama, T., Decomposition of ozone using carbon-supported metal oxide catalysts, *Appl. Catal. B Environ.*, 14, 117–129, 1997.
106. Rakitskaya, T.L. et al., Kinetics and mechanism of low temperature ozone decomposition by Co-ions adsorbed on silica, *Catal. Today*, 53, 715–723, 1999.
107. Gómez-Serrano, V. et al., Formation of oxygen structures by ozonation of carbonaceous materials prepared from cherry stones. II. Kinetic study, *Carbon*, 40, 523–529, 2002.
108. Lin, J. and Nakayima, T., An AM1 study of decomposition of ozone on a Cu(110) surface, *Ozone Sci. Eng.*, 24, 39–47, 2002.
109. Dhandapani, B. and Oyama, S.T., Gas phase ozone decomposition catalysts, *Appl. Catal. B Environ.*, 11, 129–166, 1997.

110. Bulanin, K.M., Lavalley, J.C., and Tsiganenko, A.A., Infrared study of ozone adsorption on TiO_2 (Anatase), *J. Phys. Chem.*, 99, 10294–10298, 1995.
111. Bulanin, K.M. et al., Infrared study of ozone adsorption on CeO_2, *J. Phys. Chem. B*, 102, 6809–6816, 1998.
112. Hougen, O.A. and Watson, K.M., Solid catalysts and reaction rates: general principles, *Ind. Eng. Chem.*, 35, 529–541, 1943.
113. Chaudhari, R.V. and Ramachandran, P.A., Three phase slurry reactors, *AIChE J.*, 26, 177–201, 1980.
114. Beltrán, F.J., Rivas, F.J., Fernández, L.A., Álvarez, P., and Montero-de-Espinosa, R., Kinetics of catalytic ozonation of oxalic acid in water with activated carbon, *Ind. Eng. Chem. Res.*, 41, 6510–6517, 2002.
115. Sotelo, J.L., Beltrán, F.J., and González, M., Kinetic regimes changes in the ozonation of 1,3-cyclohexanedione in aqueous solutions, *Ozone Sci. Eng.*, 13, 397–419, 1991.
116. Mills, A., Morris, S., and Davies, R., Photomineralization of 4-chlorphenol sensitized by titanium dioxide: a study of the intermediates, *J. Photochem. Photobiol. A Chem.*, 70, 183–191, 1993.
117. Turchi, C.S. and Ollis, D.F., Photocatalytic degradation of organic water contaminants: mechanisms involving hydroxyl radical attack, *J. Catal.*, 122, 178–192, 1990.
118. Mills, A. and Morris, S., Photomineralization of 4-chlorphenol sensitized by titanium dioxide: a study of the initial kinetics of carbon dioxide photogeneration, *J. Photochem. Photobiol. A Chem.*, 71, 75–83, 1993.

11 Kinetic Modeling of Ozone Processes

The last step of a kinetic study is building a kinetic model in which all the information obtained from some of the methods presented thus far is applied. As in any system that involves chemical reactions and mass transfer, the kinetic model for ozonation processes consists of the mass-balance equations of the species present (ozone, reacting compounds, hydrogen peroxide, etc.) in the system, which is the reactor volume. In addition, for the particular case of gas–liquid reacting systems, depending on the kinetic regime of ozone absorption, the mathematical model can also include microscopic mass-balance equations applied to the film layer close to the gas–water interface, which are needed to determine the mass flux of species to or from the liquid or gas phases through the interface. In the first case (slow regime), the mathematical model is usually a set of nonlinear ordinary or partial differential or algebraic equations of different mathematical complexity. In the second case (fast regime) the mathematical complexity is even higher since the solution implies trial-and-error methods, together with numerical solution techniques for both the bulk mass-balance equations and microscopic differential equations.[1] In any case, solution of this model will allow the concentrations of the different species to be known at the reactor outlet or at any time, depending on the regime of ozonation (batch, semibatch, or continuous).

The kinetic model is built from the application of mass balances to an increment volume of reaction, ΔV, where the concentration of any species can be considered uniform and constant in space. Thus, for a species i, the mass-balance equation applied to a small fraction of reactor volume, ΔV, is:

$$F_{i0} - F_i + G_i \beta \Delta V = \frac{\Delta n_i}{\Delta t} \qquad (11.1)$$

where F_{i0} and F_i represent the molar rates of species i, at the entrance and exit of the reaction volume, ΔV, respectively; G_i is the generation rate that represents the i mole rate per unit of volume that is formed or removed; $\Delta n_i/\Delta t$ is the accumulation rate of i in that volume; and β is the liquid holdup or liquid fraction of the reaction volume. Since water ozonation systems do not involve variations in temperature, ozonation can be considered an isothermic system, and an energy balance is not required.

The small volume considered, ΔV, is divided into liquid and gas fractions that can be measured through the liquid holdup, β, defined in Equation (5.43) or Equation (5.44). Also, volume variations in ozonation systems are negligible, especially for the water phase (with density being constant). In the case of the gas phase, some variation due to the drop in pressure should be taken into account, especially in real ozone contactors several meters high. For laboratory or pilot plant ozone reactors, variation in total gas pressure can be neglected. According to this, for a system of constant volume or constant volumetric flow rates through the ozone reactor, Equation (11.1) reduces to Equation (11.2) and Equation (11.3), for any gas or liquid component, respectively.

For a gas component:

$$v_g \frac{\Delta C_i}{(1-\beta)\Delta V} + \frac{\beta}{1-\beta} G_i = \frac{\Delta C_i}{\Delta t} \qquad (11.2)$$

For a liquid component:

$$v_L \frac{\Delta C_i}{\beta \Delta V} + G_i = \frac{\Delta C_i}{\Delta t} \qquad (11.3)$$

with C_i the concentration of species i in this volume

$$C_i = \frac{F_i}{v_x} \qquad (11.4)$$

where the subindex x can be L or g, the liquid or gas phase, respectively. Consequently, v_L and v_g are the actual liquid and gas volumetric flow rates through the reactor, respectively.

For the liquid phase:

$$v_L = v_{L0}\beta \qquad (11.5)$$

and for the gas phase:

$$v_g = v_{g0}(1-\beta) \qquad (11.6)$$

where v_{L0} and v_{g0} are the corresponding volumetric flow rates for the water and gas phases at empty reactor conditions. For continuous systems, the volumetric flow rates can also be expressed as a function of hydraulic residence times, τ, since:

$$\tau = \frac{V}{v_0} \qquad (11.7)$$

The generation term in Equation (11.1) is a very important function that in gas–liquid reaction systems such as ozonation has different algebraic forms depending on the kinetic regime of absorption and the nature of i species. Here also, one should determine the balance equation in the gas or liquid phase for the form of the generation rate term.

Kinetic Modeling of Ozone Processes

The forms of the generation rate term in the most common cases are as indicated in the following sections.

11.1 CASE OF SLOW KINETIC REGIME OF OZONE ABSORPTION

When reactions of ozone develop in the bulk water (see Chapter 5) the kinetic regime is slow or the ozone reactions are slow. This is the typical kinetic regime for drinking water ozonation systems. In these cases the generation rate term of Equation (11.1) is as follows:

For the water phase:

1. For any nonvolatile species i:

$$G_i = r_i \qquad (11.8)$$

where r_i is the reaction rate of i due to chemical reactions

$$r_i = \sum r_{ij} \qquad (11.9)$$

where subindex j refers to any j reaction that species i undergoes in water. For example, in an ozonation system, r_i of a given compound will involve at least two reactions: the direct reaction with ozone and the free radical reaction with the hydroxyl radical. For a general case where UV radiation is also applied, another possible contribution is due to the direct photolysis. Then, the reaction rate of the compound i is expressed as the sum of the rates due to these three contributions:

$$r_i = r_D + r_{UV} + r_{Rad} \qquad (11.10)$$

Note that these contributions, once substituted in Equation (11.2), have negative values because the stoichiometric coefficients of their corresponding reactions are negative (see Section 3.1). Also, the exact form of Equation (11.10) will depend on the expression for the concentration of hydroxyl radicals, which is usually defined as in Equation (7.12), Equation (8.5), or Equation (9.39).

2. For any volatile species i:

$$G_i = N_{vi} + r_i = (k_L a)_i (C_i - C_i^*) + r_i \qquad (11.11)$$

where N_{vi} represents the desorption rate of i. Here, C_i^* is the concentration of i at the water interface, which can be expressed as a function of the partial pressure of i or the gas concentration of i, C_{gi}, with the corresponding Henry's law:

$$P_i = He_{vi} C_i^* = C_{gi} RT \qquad (11.12)$$

In this equation it is assumed that the gas phase resistance to mass transfer is negligible.

3. For ozone in the water phase:

$$G_i = N_{O3} + r_{O3} \qquad (11.13)$$

where r_{O3} is the reaction rate term that involves all the chemical reactions ozone undergoes in water and N_{O3} is the ozone transfer rate from the gas to the water phase:

$$N_{O3} = k_L a \left(C_{O3}^* - C_{O3} \right) \qquad (11.14)$$

and

$$P_{O3} = He C_{O3}^* = C_{O3g} RT \qquad (11.15)$$

where C_{O3g} is the concentration of ozone in the gas phase. In ozonation systems, Equation (11.15) always holds because the gas resistance to ozone transfer is negligible (ozone is barely soluble in water;[2] see also Section 4.2.3). The exact form of the reaction rate term, r_{O3}, is deduced from the mechanism of the reactions proposed. For example, the concentration of ozone is a function of the concentration of hydroxyl radicals that depends on the oxidizing system used, as observed in Chapter 7 to Chapter 9.

For the gas phase:

1. For any i volatile species:

$$G_i = N_{vgi} = -N_{vi} = (k_L a)_i (C_i - C_i^*) \qquad (11.16)$$

2. For ozone:

$$G_i = N_{O3g} = -N_g = -k_L a \left(C_{O3}^* - C_{O3} \right) \qquad (11.17)$$

where the minus sign means that ozone is transferred from the gas into the water phase.

11.2 CASE OF FAST KINETIC REGIME OF OZONE ABSORPTION

This is a rather unusual case in drinking water treatment because the fast kinetic regime mainly predominates when the concentration of compounds that react with ozone in water is high enough so that the Hatta number of ozone reactions is higher than 3 (see Table 5.5). Another possibility when the Hatta number is higher than 3 arises when the rate constants of the reactions of ozone and compounds present in water are also very high, although the usual case is the former one. As a consequence, the fast kinetic regime develops mainly in the ozonation of wastewater as presented in Chapter 6 and in a few other specific cases. In the following text a few examples of the absence or presence of the fast regime are given. Thus, let us assume that the water contains some herbicide such as mecoprop. The direct rate constant of the ozone–mecoprop reaction is 100 $M^{-1}s^{-1}$.[3] For the fast kinetic regime condition to be applied (see Table 5.5), the concentration of mecoprop in water should be higher than 0.8 M. This is an unrealistic value for the concentration of the herbicide because the kinetic regime in an actual case would likely be slower and values of G_i would correspond to equations in Section 11.1. However, if the compound present in water is a phenol (present, for example, in wastewater), the situation could change because the ozone–phenol reaction rate constant, let us say at pH 7, would be about 2×10^6 $M^{-1}s^{-1}$. In this case, the kinetic regime would be fast if the concentration of phenol is at least 5×10^{-5} M, which is a possible situation. Another possible case of fast regime arises when a phenol compound is treated at high pH. Because of the dissociating character of phenols, the increase in pH leads to an increase in the concentration of the phenolate species, which reacts with ozone faster than the nondissociating phenol species (see Chapter 2). Then, an increase in the rate constant yields an increase in the Hatta number and the conditions for the fast regime hold. For example, the literature reports studies about the kinetic modeling of certain chlorophenol compounds in alkaline conditions where the fast kinetic regime holds.[4-7] However, these cases are more likely specific to wastewater where the concentration can be followed with the COD that will simplify the mathematical model, as will be shown later (see also Chapter 6).

Generally, when the kinetic regime is fast, the parameter G_i is difficult to determine, except in the case of ozone, when it undergoes a simple irreversible reaction. In fact, G_{O3} (in absolute value) has the same expression for the gas and water phases:

$$G_i = N_{O3} = k_l a C_{O3}^* E \qquad (11.18)$$

where E, the reaction factor, depends on the fast kinetic regime type (moderate, fast, of pseudo first order, instantaneous, etc.) to take one of the forms discussed in Chapter 4. However, N_{O3} can be used in Equation (11.1) only for ozone in the gas phase. In the water phase, Equation (11.1) for ozone is not used since the concentration of ozone, C_{O3}, is zero when the kinetic regime is fast.

A different situation is presented when Equation (11.1) is applied to any other species reacting with ozone. The generation term, G_i, will depend on the concentrations of such species (including ozone), as indicated in Equation (11.8) or Equation (11.11). But if C_{O3} is zero, how can this situation be dealt with? In the fast kinetic regime, the concentration of ozone is not zero only within the liquid film layer, as already shown in Figure 4.10 to Figure 4.12. In fact, the concentration of ozone varies from C_{O3}^* at the gas–water interface to zero at a given point within the film layer (between the interface and bulk water). The concentration of the reacting species also changes within the film layer. In these cases, the maximum value of C_i is in bulk water. If concentrations are not constant within the film layer, how can G_i be calculated? There are several possible ways to solve this problem, all of which involve the solution of the microscopic mass-balance Equation (4.34) and Equation (4.35). One of these possibilities follows the complicated steps shown below:

- Calculate the concentration profiles of reacting species, including that of ozone, with the position in the film layer (depth of penetration). This requires the solution of the microscopic mass-balance equations of species [Equation (4.13) or Equation (4.34) and Equation (4.35) if film theory is applied] using numerical methods.
- Determine the generation rate terms from the mean values of the reaction rate terms once the concentrations of reactants are known at different positions within the film layer. This can be accomplished as follows:

$$G_i = \frac{1}{\delta} \int_0^\delta r_i dx \qquad (11.19)$$

where r_i is given by Equation (11.10).
- Solve the system of macroscopic mass-balance Equation (11.1) with the known values of G_i.

The second possibility is the determination of the mass flux of reacting species and ozone gas through the edge of the liquid film layer in contact with the bulk liquid and through the gas–film layer in contact with the bulk gas, N_{ib}^l and N_{O3b}^g, respectively.

For any reacting species, i:

$$G_i = N_{ib}^l = -D_i \left[\frac{\partial C_i}{\partial x}\right]_{x=\delta} \qquad (11.20)$$

and for ozone gas:

$$G_{O3} = N_{O3b}^g = -D_i \left[\frac{\partial C_i}{\partial x}\right]_{x=0} = k_L C_{O3}^* E \qquad (11.21)$$

Note that in Equation (11.21) the flux of ozone through the gas–film layer is the same as through the interface because of the absence of gas resistance to mass

Kinetic Modeling of Ozone Processes

transfer.[1] As also seen in Equation (11.21), the ozone flux is finally expressed as a function of the reaction factor, E. Values of E and the bulk mass flux of compounds, N_{ib}^l, can be calculated from the solution of the continuity Equation (4.13) or Equation (4.34) and Equation (4.35) as film theory is applied. For example, in the case of an irreversible second-order reaction between ozone and B [Reaction (4.32)], values of E can be known from the equations deduced in Section 4.2.1.2 (see also Table 5.5). E and the bulk mass flux of compounds through the liquid film layer–bulk water are then used in the bulk mass balances of species [Equation (11.2) and Equation (11.3) applied to the whole reactor volume (see Section 11.6.3)] to obtain the concentration profiles with time or position, depending on the type of flow of the gas and water phases through the reactor and the time regime (stationary or nonstationary) of ozonation. For example, Hautaniemi et al.[4] used this approach to predict the concentration profiles of some chlorophenol compounds and ozone, when ozonation was carried out at basic conditions in a semibatch, perfectly mixed tank.

It is evident that the resulting complex mathematical model results are very difficult to solve, especially for multiple series-parallel ozone reactions, which would be the usual case. Nonetheless, there is one possible case that could even lead to one analytical solution, i.e., when ozone, while being absorbed in water, undergoes a unique irreversible reaction with the compound B already present in water. This can either be the typical case of wastewater ozonation where COD can represent the concentration of the matter present in water that reacts with ozone [Reaction (6.5)], or just the case of one irreversible reaction between ozone and a compound B with a high rate constant (i.e., a phenol compound). Two methods can be applied, depending on the time regime conditions. In both cases, however, the only generation term needed is that of ozone, $G_{O3} = N_{O3}$. At nonsteady-state conditions the method needs the mass balance of B in bulk water, and at steady-state conditions a total balance is the recommended option, so that the corresponding generation rate term of B or COD is not needed in this second approach. In this chapter, the procedure based on the total balance will be followed to present the different solutions except in those cases where the use of the bulk mass balance of B is already applied (see Section 11.6.2.1). Section 11.8 presents an example of the kinetic model for the ozonation of industrial wastewater in the fast kinetic regime. Section 6.6.3.1 discussed a kinetic study to determine the rate coefficient of the reaction of ozone and wastewater of high reactivity.

11.3 CASE OF INTERMEDIATE OR MODERATE KINETIC REGIME OF OZONE ABSORPTION

When ozone reactions develop both in the film close to the gas–water interface and in bulk water, the kinetic regime is called intermediate or moderate. In this case, there is a need to quantify the fraction of ozone reactions in both zones of water. The problem is similar to that presented for fast reactions in Section 11.2, but it includes the difficulty of reaction in bulk water as well. Again, the solution to the problem implies the simultaneous solution of microscopic equations in the film layer and macroscopic equations in the bulk water. This complex problem has been recently treated by Debellefontaine and Benbelkacer,[8] who introduced the concept of the depletion factor, DF, previously defined by Schlüter and Schulzke.[9] In a way similar to the reaction or enhancement factor, this dimensionless number, E, compares the ozone absorption (in this case) at the edge of the film in contact with bulk water $(N_{O3})_{x=\lambda}$ with the physical absorption of ozone. Definition of the depletion factor is:[8]

$$DF = \frac{N_{O3}|_{x=\lambda}}{k_L(C^*_{O3} - C_{O3})} = \frac{-D_{O3}\frac{dC_{O3}}{dx}|_{x=\lambda}}{k_L(C^*_{O3} - C_{O3})} \qquad (11.22)$$

Note that the depletion factor is defined as the number of times the ozone physical absorption rate is increased due to chemical reactions in the bulk water, while the reaction factor is defined as the number of times the maximum physical ozone absorption rate ($k_L C^*_{O3}$) is increased due to chemical reactions in the film layer. If a moderate regime is considered, chemical reactions develop both in the film and in the bulk water (see Figure 4.9) so that in most cases the bulk ozone concentration is different from zero ($C_{O3} \neq 0$). Hence, in this situation, the reaction factor can also be defined as follows:

$$E = \frac{N_{O3}|_{x=0}}{k_L(C^*_{O3} - C_{O3})} = \frac{-D_{O3}\frac{dC_{O3}}{dx}|_{x=0}}{k_L(C^*_{O3} - C_{O3})} \qquad (11.23)$$

According to definitions of E and DF, it is evident that the ratio between the two dimensionless numbers (DF/E) represents the fraction of unconverted ozone that leaves the film and enters the bulk water. Applying Equation (11.22) and Equation (11.23) allows the generation rate terms of ozone and reacting species in the bulk water and the film layer, respectively, to be known separately. These terms are as follows:

For the generation rate of ozone (reacting) in the film:

$$G_{O3\,film} = (E - DF)k_L a(C^*_{O3} - C_{O3}) \qquad (11.24)$$

For the generation rate of ozone (reacting) in the bulk water:

$$G_{O3\,bulk} = r_{O3} \qquad (11.25)$$

In a similar manner, for any compound B reacting with ozone the generation rate terms in both the film and bulk water will be similar to those of Equation (11.24) and Equation (11.25) once the stoichiometric coefficients are accounted for. For example, for a compound i that reacts with ozone according to the stoichiometry given by Reaction (3.5), the generation rate terms would be:

In the film layer:

$$G_{i-film} = \frac{1}{z_i}\left[(E-DF)k_L a\left(C_{O3}^* - C_{O3}\right)\right] \quad (11.26)$$

In the bulk water:

$$G_{i-bulk} = r_i \quad (11.27)$$

With this approach, Debellefontaine and Benbelkacen prepared the kinetic model for the ozonation of maleic and fumaric acids.[10,11] More details of the use of Equation (11.24) to Equation (11.27) are given in Section 11.6.3.

11.4 TIME REGIMES IN OZONATION

Once the generation rate terms have been specified, Equation (11.1) and Equation (11.2) can be simplified further according to the effect of time on the performance of the system. Thus, although the gas phase is continuously fed to the ozone contactor, the water phase could be initially charged (batch system) or continuously fed (continuous system). Either way, the time regime is directly related to the size of the ozone contactor, which depends on the volume of treated water. Usually, in laboratory contactors, a semibatch system (continuous for the gas phase and batch for the water phase) is used to carry out the ozone reactions. In some pilot plant contactors, both the semibatch and continuous systems are possible, while in actual ozone contactors in water or wastewater treatment plants, the continuous system is the means of operation. The time regime (batch or continuous) is, thus, an important aspect in reactor design since Equation (11.1) can significantly be simplified depending on the time regime type. For example, in semibatch systems, for the water phase, there is no mass flow rates at the inlet and outlet of the reaction volume, and F_{i0} and F_i are not present in Equation (11.1), which then becomes

$$G_i = \frac{\Delta C_i}{\Delta t} \quad (11.28)$$

In fact, for the water phase, this is the equation that has been used for kinetic studies (see Chapter 5). Laboratory ozonation systems are examples of where these equations are applied since they usually are nonstationary processes where concentrations in water vary with time.

For continuous systems (some pilot plants and commercial contactors), although convection flow rates, F_{i0} and F_i, cannot be removed from Equation (11.1), the accumulation rate terms, $\Delta n_i/\Delta t$, are not present since these are steady-state processes. In a steady-state process, Equation (11.1) reduces to:

$$\frac{\Delta C_i}{\Delta \tau} + G_i = 0 \quad (11.29)$$

It is evident that, in a practical case, there will be a period of time at the start of the process when ozonation is a nonsteady-state operation and Equation (11.1) cannot be simplified. This represents the most difficult case to treat mathematically. Similarly, also for practical applications, ozone contactors are designed for the steady-state operation so that Equation (11.1) is solved starting from Equation (11.29). In fact, solving Equation (11.1) without any simplification is a rather academic exercise, although it allows the process time to reach the steady-state operation.

11.5 INFLUENCE OF THE TYPE OF WATER AND GAS FLOWS

Once the time regime has been established (semibatch or continuous systems, stationary or nonstationary operation), Equation (11.1) or Equation (11.2) and Equation (11.3) have to be applied to the whole reaction volume to proceed with its solution. This means the type of phase flow must be known. There are two main ideal flows for which Equation (11.1) can be expanded to the whole reaction volume: the perfectly mixed flow (PMF) and the plug flow (PF), which are based on the hypothesis given in Appendix A1. It is also necessary to remember that the G_i values in Equation (11.1) can involve the solution of microscopic differential mass-balance Equation (4.34) and Equation (4.35) within the liquid–film layer, in cases where the kinetic regime of ozonation is fast or moderate.

For the cases of PMF and PF, Equation (11.1) applies as follows:

- Perfectly mixed flow (PMF)

$$\frac{1}{\tau}(C_{i0} - C_i) + G_i = \frac{dC_i}{dt} \qquad (11.30)$$

where C_{i0} and C_i refer to the concentrations of i at the reactor inlet and outlet, respectively. The hydraulic residence time, τ, coincides with the mean residence time obtained from the residence time distribution function (see Appendix A3).

Note, however, that some authors consider the whole reactor volume divided into three zones of perfect mixing conditions: the water phase with volume V_L, the bubble phase with volume V_B, and the free board or space above the free surface of water with volume V_F.[12] Thus, in some kinetic modeling works, Equation (11.30) is applied to yield a system with three mass-balance equations[12,13] (see Section 11.6.1.1) because a different ozone concentration is assumed in each phase.

- Plug flow (PF)
  In this case, Equation (11.31) applies:

$$-\frac{\partial C_i}{\partial \tau} + G_i = \frac{\partial C_i}{\partial t} \qquad (11.31)$$

This equation can be integrated from the start of the process ($t = 0$) and for the whole reaction volume ($\tau = 0$ to $\tau = V/v_o$).

One important difference observed between Equation (11.30) and Equation (11.31) is that when the systems are at the steady state, the model with PMF is a set of algebraic nonlinear equations, while models with PF consist of a set of first-order partial differential equations.

In actual contactors (even of laboratory size), however, the type of gas and water flows can deviate from the ideal cases. Hence, tracer studies have to be carried out to determine the residence time distribution function, RTDF, as shown in Appendix A3. The RTDF can allow the real flow to be simulated as a combination of ideal flows or as another ideal flow model of specific characteristics. These are called models for nonideal flow.[14] The most commonly applied nonideal flow models are the N perfectly mixed tanks in series model and the axial dispersion model described in Appendix A3. When the flow is simulated with N perfectly mixed tanks in series, Equation (11.30) also applies but it has to be solved N times. This is so because the concentration of any species at the outlet of the last N-th reactor would represent the concentration of the treated species at the actual contactor outlet. The dispersion model represents a more complicated picture because it assumes that the flow is due to both convection and axial diffusion.[14] As a consequence, the mass flow rates [F terms in Equation (11.1)] are due not only to the convection flow contribution (volumetric flow rate times the concentration) but also to the axial diffusion transport, which is given by Fick's law:

$$N_{ad} = -D_i \frac{1}{U} \frac{\partial C_i}{\partial \tau} \tag{11.32}$$

where U represents the superficial velocity of the phase through the reactor. Then the total flow rate, F_i, in this model is:

$$F_i = v_0 C_i + S N_{ad} = v_0 C_i - D_i \frac{S}{U} \frac{\partial C_i}{\partial \tau} \tag{11.33}$$

where D_i is the axial dispersion coefficient of the i species in the phase.

For an element dV, the mass balance (11.1) is:

$$\partial F_i + G_i \beta \partial V - \frac{\partial n_i}{\partial t} \tag{11.34}$$

which becomes Equation (11.35), once Equation (11.7) and Equation (11.33) have been taken into account:

$$-\frac{\partial C_i}{\partial \tau} + \frac{D_i}{U^2} \frac{\partial^2 C_i}{\partial \tau^2} + G_i = \frac{\partial C_i}{\partial t} \tag{11.35}$$

or as a function of the contactor height, h:

$$-U\frac{\partial C_i}{\partial z} + D_i\frac{\partial^2 C_i}{\partial z^2} + G_i = \frac{\partial C_i}{\partial t} \qquad (11.36)$$

Equation (11.36) has to be integrated from the start of the process ($t = 0$), and for the whole reaction volume ($\tau = 0$ to $\tau = V/vo$ or better for $h = 0$ to $h = H$), which usually requires numerical methods.[15,16]

In addition to the classic or ideal models described above, the literature also reports several more sophisticated models that represent modifications of the N perfectly mixed tanks in series and axial dispersion models. For example, El-Din and Smith[17] proposed the nonisobaric steady-state one-phase axial dispersion model (1P-ADM), which consists of the nonlinear second-order ordinary differential equations representing the mass balance of species in the water phase. These equations are the same as those in the axial dispersion model [Equation (11.35)] with the concentration of ozone in the gas phase at any point in the column, h, which is present in the ozone mass-transfer rate term, G_i, expressed as an exponential function of position:

$$C_{O3g} = C_{O3g0}\exp(-\varsigma h) \qquad (11.37)$$

where C_{O3g0} is the concentration of ozone at the column entrance. Of course, the coefficient ς is an empirical parameter that has to be determined experimentally. The use of Equation (11.37) allows the omission of the ozone mass balance in the gas phase. This model can be useful in the case of kinetic models of ozone absorption and decomposition in water because balance equations for reacting compounds in water are not needed. For detailed information on this model see Reference 17.

Another kinetic model reported in the literature that presents a modification of the ideal N perfectly mixed tanks in series model is called the transient back flow cell model (BFCM).[18] As in the N tanks in series model, both the gas phase and the water phase are simulated with N tanks or cells in series. In this model, it is assumed that back flow exists between consecutive liquid cells, while no back flow is considered between gas cells (the gas phase is assumed to be in PF). The model has been tested with tracer studies and compared to the classic N tanks in series and axial dispersion models. Although it offers some advantages in accounting for variable backmixing and cross-sectional area along the column length, its mathematical solution seems difficult, especially applied to ozonation systems where generation rate terms are present. For more details see Reference 18, the original work.

11.6 MATHEMATICAL MODELS

In this section, the kinetic models are first applied to the case of the slow kinetic regime, which is the most common case for drinking water ozonation systems. The fast kinetic regime is then reviewed for the case of wastewater ozonation. Some details are also given for the moderate kinetic regime models.

Kinetic Modeling of Ozone Processes

Regardless of the kinetic regime of ozonation, different possibilities can be considered depending on the flow of the gas and water phases through the contactor and on the time regime of ozonation (semibatch, continuous, etc.).

11.6.1 Slow Kinetic Regime

Six cases are presented here:

- Both gas and water phases in perfect mixing flow
- Both gas and water phases in plug flow
- The water phase in perfect mixing flow and the gas phase in plug flow
- The water phase as N perfectly mixed tanks in series and the gas phase in plug flow
- Both the gas and water phases as N and N' perfectly mixed tanks in series
- Both gas and water phases with axial dispersion flow

11.6.1.1 Both Gas and Water Phases in Perfect Mixing Flow

This is the most common case presented in the literature. Ozonation in laboratory standard agitated tanks usually follows this model. The mathematical model consists of equations of the Equation (11.30) type, with the characteristics of G_i given according to the species i. Thus, the mathematical model is reduced to the following set of equations:

1. For ozone in the gas phase:

$$\frac{1}{\tau_g}\left(C_{O3g0} - C_{O3g}\right) + N_g \frac{\beta}{1-\beta} = \frac{dC_{O3g}}{dt} \quad (11.38)$$

   where subindex g represents ozone in the gas and N_g is given by Equation (11.17).

2. For ozone in the water phase:

$$\frac{1}{\tau_L}\left(C_{O30} - C_{O3}\right) + N_{O3} + r_{O3} = \frac{dC_{O3}}{dt} \quad (11.39)$$

   where r_{O3} and N_{O3} are as given in Equation (11.10) and Equation (11.14), respectively. In Equation (11.14) the term C_{O3}^*, the concentration of ozone at the water interface, can be expressed as a function of the concentration of ozone in the gas at the reactor outlet once the Henry and perfect gas laws are accounted for [Equation (11.15)].

3. For any reacting nonvolatile species i in the water phase:

$$\frac{1}{\tau_L}\left(C_{i0} - C_i\right) + r_i = \frac{dC_i}{dt} \quad (11.40)$$

   where r_i is defined in Equation (11.10).

4. A special case is ozonated water that contains volatile species, *vi*. For this species the mass balances are:

In the water phase:

$$\frac{1}{\tau_L}(C_{vi0} - C_{vi}) + N_{vi} + r_{vi} = \frac{dC_{vi}}{dt} \tag{11.41}$$

with N_{vi} as given in Equation (11.16) and r_{vi} as in Equation (11.10).

In the gas phase:

$$\frac{1}{\tau_g}(C_{vgi0} - C_{vgi}) + \frac{\beta}{1-\beta} N_{vgi} = \frac{dC_{vgi}}{dt} \tag{11.42}$$

with $N_{vgi} = -N_{vi}$.

In a general case, the system of Equation (11.38) to Equation (11.42) is solved numerically, for example with the fourth-order Runge–Kutta method (see Appendix A5), with the initial condition:

$$T = 0 \quad C_{O3g} = 0 \quad C_{vig} = 0 \quad C_{O3} = 0 \quad C_i = C_{i0} \quad C_{vi} = C_{vi0} \tag{11.43}$$

However, two possible simplifications apply:

1. For steady-state continuous operation, all accumulation rates are zero ($dC/dt = 0$) and the mathematical model reduces to a set of nonlinear algebraic equations that can be solved with Newton's method (see Appendix A5).
2. For semibatch operation (continuous system for the gas phase), convection water flow terms are removed from mass-balance equations ($F_i = 0$). In this case, C_{i0} and C_{vi0} are the initial concentrations of nonvolatile and volatile species in the water charged to the reactor, respectively. The solution is obtained in a way similar to the general case.

Recall that in studies where the reactor volume is divided into three volume fractions,[12] there are also three ozone mass-balance equations, one for each volume zone. In such cases, Equation (11.38) for the ozone mass balance in the gas phase is called the ozone mass balance in the bubble phase:

$$\frac{1}{\tau_g}(C_{O3g0} - C_{O3B}) + \frac{V_L}{V_B} N_g = \frac{dC_{O3B}}{dt} \tag{11.44}$$

Kinetic Modeling of Ozone Processes

where C_{O3B} is the ozone concentration in the bubble gas and N_g is defined as in Equation (11.17) but C_{O3}^* represents the ozone equilibrium concentration with the ozone bubble gas:

$$C_{O3}^* = \frac{C_{O3_B} RT}{He} \qquad (11.45)$$

Also, the ozone mass balance in the water phase remains as in Equation (11.39) with the difference in the ozone mass transfer rate, N_{O3}, where C_{O3}^* is expressed by Equation (11.45). The third and additional equation refers to the ozone mass balance in the free board of reactor:

$$V_g \left(C_{O3B} - C_{O3ge} \right) = V_F \frac{dC_{O3ge}}{dt} \qquad (11.46)$$

where V_g is the gas flow rate and C_{O3ge} is the concentration of ozone in the exiting gas. Note that for volatile compounds there are also three mass-balance equations as in the case of ozone. In this chapter, however, unless otherwise indicated, only systems with the reactor volume divided in gas and water phases will be considered. Table 11.1 gives a few examples of ozone works following this model.

11.6.1.2 Both Gas and Water Phases in Plug Flow

This is another possible practical case when, for example, ozonation is carried out in bubble columns. The mathematical model consists of the mass-balance equations as a set of nonlinear partial differential equations where the concentrations of ozone and reacting species vary with time and position, z, in the bubble column. This corresponds to Equation (11.31). The mathematical model is solved by numerical methods. The exact form of these equations also depends on the relative direction of gas and water flows through the column, i.e., countercurrent or parallel flow operation. For example, here we present the equations for countercurrent operation when the mathematical system is solved from the top of the column ($z = 0$).

1. For the ozone in the gas phase:

$$U_{g0} \frac{\partial C_{O3g}}{\partial h} + N_{O3g} \frac{\beta}{1-\beta} = \frac{\partial C_{O3g}}{\partial t} \qquad (11.47)$$

where U_{g0} is the actual gas phase velocity at empty column conditions:

$$U_{g0} = \frac{U_g}{1-\beta} \qquad (11.48)$$

TABLE 11.1
Works on Kinetic Modeling of Ozonation Systems

Ozonation System	Reactor System	Kinetic Regime and Phase Flow Type	Reference No. (Year)
Ozone–phenol	Semibatch stirred reactor pH acid, lab scale	Slow regime, water and gas perfectly mixed; intermediates considered	19 (1983)
Ozone/H_2O_2/volatile organochlorine compounds	70-L semicontinuous sparged stirred tank; continuous hydrogen peroxide feed, lab scale	Slow–fast regimes, gas and water phase perfectly mixed	20, 21 (1989)
Ozone–toluene	Continuous packed column, 1.24 m, 5 cm I.D., 6 mm Raschig ring packing	Slow regime, water and gas in plug flow	22 (1990)
Ozone decomposition	Simulation; application of SBH and TFG mechanisms	Homogeneous aqueous system, water phase in perfect mixing	23 (1992)
Ozone transfer to water	75-L continuous bubble column, 4.2 m, 15 cm I.D., pilot scale	Slow regime, column divided in three parts according to tracer studies: perfect mixing at the top and bottom and plug flow in the middle	24 (1992)
Ozone/UV/volatile organochlorine compounds	Simulation of a continuous-bubble photoreactor column	Slow regime, gas phase in plug flow, water phase perfectly mixed	25 (1993)
Ozone transfer to water	Simulation of a continuous bubble column	Slow regime; gas phase always plug flow; water phase flow as perfect mixing, plug flow, 3 perfect mixing reactors of different size (dispersion)	26 (1993)
Ozone/H_2O_2/atrazine	Ozone contactors at water treatment plants: simulation	Homogeneous aqueous system, water as a series of perfectly mixed reactors of equal size	27 (1994)
Ozone–bromide	Batch reactor; influence of pH, ammonia, and bromide	Homogeneous aqueous system, water perfectly mixed	28 (1994)
Ozone transfer to water	Simulation applied to a countercurrent bubble column and a countercurrent flow chamber (absorption with five subsequent flow chambers)	Slow regime, water with axial dispersion flow, and gas in plug flow	29 (1994)
Ozone/distillery and tomato wastewater	Laboratory and pilot plant bubble columns of different height	Fast pseudo first-order and slow regimes for distillery and tomato wastewater, respectively; COD, ozone partial pressure, and dissolved ozone	30 (1995)

TABLE 11.1 (continued)
Works on Kinetic Modeling of Ozonation Systems

Ozonation System	Reactor System	Kinetic Regime and Phase Flow Type	Reference No. (Year)
Ozone/H_2O_2 natural water	Continuous bubble columns; simulation of water treatment plant ozone contactors	Slow regime, reactor divided in zones that behave as a series of perfectly mixed tanks; total ozone mass balance used instead of gas balance	31 (1995)
Ozone transfer to natural water	Continuous bubble column; pilot scale: 2.5 m, 15 cm I.D.	Slow regime, water and gas as a series of equal size perfectly mixed reactors	32 (1996)
Ozone decomposition with UV radiation	Batch photoreactor, 254-nm UV lamps	Homogeneous aqueous system, water perfectly mixed	33 (1996)
Ozone/H_2O_2/volatile organochlorine compounds	Continuous tubular reactor; pilot scale: 14.8 m, 1.8 cm I.D.	Slow regime, homogeneous aqueous system, water in plug flow	34 (1997)
Ozone mass transfer	Cocurrent down flow jet pump contactor, lab scale	Slow regime, water phase in plug flow, total ozone mass balance used instead of ozone gas mass balance	35 (1997)
Ozone decomposition in the presence of NOM	Homogeneous batch reactor, NOM up to 0.25 mM as organic carbon	Homogeneous aqueous system; use and comparison of SBH and THG mechanisms of ozone decomposition; influence of NOM	36 (1997)
Ozone/H_2O_2 general model applied to TCE and PCE	Application to a full-scale demonstration plant at Los Angeles	Slow regime, nonstationary process, gas and water phases with axial dispersion and convection	37 (1997)
Ozone mass transfer efficiency	Simulation results	Slow regime, gas and water phases in perfect mixing; two gas phases considered: bubbles and gas above the water level	13 (1997)
Ozone/UV radiation/ chlorophenols	264-L semibatch bubble column photoreactor, 254-nm low pressure Hg lamp (0.304 W), pH=2.5	Slow regime, gas and water phases in perfect mixing; intermediate, chloride, and hydrogen peroxide concentrations followed and simulated as well	38 (1998)
Ozone/UV radiation/ chlorophenols	264-L semibatch bubble column photoreactor, 254-nm low pressure Hg lamp (0.304 W), pH=9.5	Fast regime, gas and water phases in perfect mixing, balance of compounds in the bulk water and microscopic balance equations in the film layer	4 (1998)

TABLE 11.1 (continued)
Works on Kinetic Modeling of Ozonation Systems

Ozonation System	Reactor System	Kinetic Regime and Phase Flow Type	Reference No. (Year)
Ozone decomposition	Batch reactor, presence of natural organic carbon (NOM) and bromide	Homogeneous aqueous system, water phase in perfect mixing	39 (1998)
Ozone/H_2O_2/atrazine	4-L standard glass agitated reactor	Slow regime, water and gas phases in perfect mixing, following concentrations of intermediates	40 (1998)
Ozone/bromide	Different laboratory, pilot plant, and full size contactors	Tracer experiments, slow regime, determination of kinetic constant (laboratory batch reactors) and parameters of nonideal flow (dispersion number); predictions of bromate ion and ozone concentrations	41 (1998)
Ozone/p-chlorophenol	Semibatch stirred reactor, pH 2–8	Slow–fast regimes, water and gas phases in perfect mixing, two gas phases considered: bubbles and gas above the water level	6 (1999)
Ozone/H_2O_2/UV radiation/TCE, TCA	800-mL semibatch bubble photoreactor, 254-nm low pressure Hg lamp, 1.6×10^{-6} Einstein $L^{-1}s^{-1}$	Slow regimes, volatility coefficients used, gas and water phases in perfect mixing, evolution of TCA, TCE, and ozone (gas and water) concentrations	42 (1999)
Ozone/H_2O_2/UV radiation/fluorene, phenanthrene	4-L standard glass agitated reactor and 800-mL semibatch bubble photoreactor, 254-nm low pressure Hg lamp, 3.8×10^{-6} Einstein $L^{-1}s^{-1}$	Slow regimes, gas and water phases in perfect mixing, influence of intermediates and formation of hydrogen peroxide, mechanism and kinetic modeling	43 (1999)
Ozone/disinfection	Full-size contactor divided in 4 chambers (total length: 17 m, total height: 5 m)	Dispersion model in three spatial directions, the momentum equation is included; it predicts hydrodynamics of the ozone contactor with microorganism inactivation	44 (1999)

TABLE 11.1 (continued)
Works on Kinetic Modeling of Ozonation Systems

Ozonation System	Reactor System	Kinetic Regime and Phase Flow Type	Reference No. (Year)
Ozone/odorous compounds (geosmin and 2-MIB)	U-tube reactor: inner tube 7.5 cm diameter, outer tube 45.4 cm diameter; length 3.55 m	Plug flow through inner tube and N perfectly mixed tanks in series through the outer section; predictions of ozone and odorous compound concentrations; inner tube acts as an efficient ozone absorber while outer section acts as reactor to consume compounds	45 (1999)
Ozone decomposition in the presence of carbonates, hydrogen peroxide, and NOM	5-cm quartz cell magnetically stirred as batch reactor	Homogeneous aqueous system; water in perfect mixing conditions; comparison to experimental results and simulation in other conditions	46 (2000)
Ozone decomposition in water	Sequential stopped flow spectrophotometer, pH: 10.4–13.2	Homogeneous aqueous system; water in perfect mixing conditions; use of THG modified mechanism	47 (2000)
Ozone/ p-hydroxybenzoic acid	15-L stainless steel semibatch stirred reactor, pH 3 and 10	Slow regime, water and gas phases in perfect mixing conditions; intermediates considered in the model, THM formation potential	48 (2000)
Ozone/atrazine	Homogeneous batch reactors	Homogeneous kinetic model, influence hydroxyl radical reactions, effects of intermediates	49 (2000)
Ozone/mineral oil wastewater	Semibatch stirred reactor	Slow regime, water and gas phases in perfect mixing; two gas phases considered: bubbles and gas above the water level	50 (2000)
Ozone/biological oxidation/olive wastewater	1.5-L semibatch bubble column for ozonation, 3-L batch aerobic tank for biological oxidation	Slow regime, hydroxyl radical reactions considered, COD surrogate parameter, sequential pH cycle effects	51 (2000)
Ozone decomposition in natural river water	360-mL semibatch ozone bubble contactor	Slow regime, water and gas phases in perfect mixing conditions, NOM divided in humic and nonhumic substances	52 (2001)
Ozone mass transfer	Simulated results applied to water and wastewater treatment conditions	Slow regime, concentration of ozone in the gas phase as a function of position in column, one-phase axial dispersion model for the water phase	17 (2001)

TABLE 11.1 (continued)
Works on Kinetic Modeling of Ozonation Systems

Ozonation System	Reactor System	Kinetic Regime and Phase Flow Type	Reference No. (Year)
Ozone mass transfer, tracer study	Simulation of tracer experiments	Slow regime, gas phase in plug flow, transient back flow cell model for water phase	18 (2001)
Ozone/H_2O_2/MTBE	Batch homogeneous reactors	Slow regime, influence of hydroxyl radical oxidation, intermediates considered	53 (2001)
Ozone/pulp mill wastewater (750 mgL^{-1} COD)	Pilot plant impinging jet bubble column (venturi injectors)	One-phase axial dispersion model (1P-ADM), fast and moderate kinetic regimes	54 (2001)
Ozone mass transfer	Bubble columns (5.5 m high, 15 cm I.D.)	Absorption and desorption (with nitrogen runs), slow kinetic regime, gas phase in plug flow, water phase with axial dispersion (no convection term), nonstationary regime	55 (2001)
Ozone/dichlorophenol	5-L semibatch stirred reactor	Slow–fast regimes, water phase in perfect mixing, gas phase as three models: complete gas, plug flow, and perfect mixing models; mass flux at interface determined from film theory	5 (2001)
Ozone/domestic wine wastewaters	Bubble column for acid pH ozonation followed by standard agitated reactor for alkaline pH ozonation	Sequential pH ozonation (acid and alkaline pH cycles), evolution of COD and BOD, gas and water phase in perfect mixing conditions	56 (2001)
Ozone decomposition in sea water; removal of ammonia	Gas-lift type reactor: 30 cm long, 14 cm I.D., pH: 6.5–9	Slow regime, gas and water phase in perfect mixing conditions	57 (2002)
Ozone/phenols and swine manure slurry	1.5-L semibatch bubble reactor	Slow–moderate regimes, water phase in perfect mixing; mean value of ozone concentration in the gas between entrance and outlet concentrations, total mass balance of ozone instead of ozone gas balance	7 (2002)
Ozone/natural water and ozone/wastewater (theoretical studies)	Ozone bubble columns	Slow and fast regimes, comparison of axial dispersion and back flow cell models for the ozonation of natural and wastewaters (see Section 11.5)	58 (2002)

TABLE 11.1 (continued)
Works on Kinetic Modeling of Ozonation Systems

Ozonation System	Reactor System	Kinetic Regime and Phase Flow Type	Reference No. (Year)
Ozone/UV/natural water (TOC = 3 mgL^{-1})	Ozone bubble column plus annular photoreactor (80 cm length, 30 cm I.D.)	Slow regime, hydrodynamic model: application of mass and momentum of fluid and mass balance of species equations, profiles of UV intensity, ozone concentration and TOC	59 (2002)
Ozone/H$_2$O$_2$/simazine	Continuous nonsteady-state bubble column (30 cm high, 4 cm I.D.)	Slow regime, nonideal flow study: water phase perfectly mixed, gas phase with some dispersion, perfect mixing, plug flow, and axial dispersion were considered, intermediate products and direct and hydroxyl radical reactions also considered, deviations for high concentration of hydrogen peroxide	60 (2002)
Ozone/H$_2$O$_2$/alachlor in surface water	Continuous bubble column (2 m high, 4 cm I.D.)	Slow regime, nonideal flow study: water phase with some dispersion, gas phase perfectly mixed, application of axial dispersion and N perfectly mixed tanks in series models	61 (2002)
Ozone/ 2-chlorophenol in soil	Packed bed column (17.6 cm high, 3.125 cm I.D.)	Fast kinetic regime, gas–liquid–solid reacting system, dispersion model for the gas phase, nonstationary regime, ozone gas and 2-chlorophenol concentration profiles with position	62 (2002)
Ozone disinfection of wastewaters (2 secondary effluents and 1 tertiary effluent)	Bubble columns of different size (2.6 and 3.6 m high, 15 and 30 cm I.D.)	Slow regime, continuous and cuntercurrent operation, only water phase treated: N tanks in series model, inactivation of *E. coli*	63 (2002)
Ozone disinfection in drinking water plant	Pilot scale diffuser bubble column (2.74 m high, 15 cm I.D.)	Slow regime, cocurrent and countercurrent operation at steady regime, axial dispersion model applied to ozone (gas and water), natural organic matter, and microorganisms (*C. muris, C. parvus*)	64, 65 (2002)

TABLE 11.1 (continued)
Works on Kinetic Modeling of Ozonation Systems

Ozonation System	Reactor System	Kinetic Regime and Phase Flow Type	Reference No. (Year)
Ozone mass transfer	Packed (silica gel) bed column (20 to 50 cm high, 5 cm I.D.)	Slow regime, no decomposition of ozone on the solid bed is observed, plug flow operation for water and gas	66 (2003)

2. For ozone in the water phase:

$$-U_{L0}\frac{\partial C_{O3}}{\partial h} + N_{O3} + r_{O3} = \frac{\partial C_{O3}}{\partial t} \qquad (11.49)$$

with U_{L0} the actual water phase velocity at empty column conditions.

3. For any reacting nonvolatile species:

$$-U_{L0}\frac{\partial C_i}{\partial h} + r_i = \frac{\partial C_i}{\partial t} \qquad (11.50)$$

4. For any reacting volatile species in water:

$$-U_{L0}\frac{\partial C_{vi}}{\partial h} + N_{vi} + r_{vi} = \frac{\partial C_{vi}}{\partial t} \qquad (11.51)$$

5. For volatile species in gas:

$$U_{g0}\frac{\partial C_{vig}}{\partial h} + N_{vig}\frac{\beta}{1-\beta} = \frac{\partial C_{vig}}{\partial t} \qquad (11.52)$$

This is a very complex mathematical system, and, except for academic reasons, the model is solved for the case of steady-state operation ($dC_i/dt = dC_{O3}/dt = 0$). With this simplification, which better simulates a real situation, the mathematical model becomes a set of first-order nonlinear ordinary differential equations that can be solved numerically with the fourth-order Runge–Kutta method and a trial-and-error procedure as follows:

1. Assume a value for the concentrations of ozone (and volatile compound, if any) in the gas phase at the column outlet, i.e., for $z = 0$. These assumed values have to be lower than the corresponding ones at the column entrance, i.e., for $h = H$.
2. Solve the system of ODE with the Runge–Kutta method with the initial condition:

$$h = 0 \quad C_{O3g} = C_{O3gs} \quad C_{vig} = C_{vigs} \quad C_i = C_{i0} \quad C_{vi} = C_{vi0} \qquad (11.53)$$

3. Compare the calculated values of the concentration of ozone in the gas at the column inlet ($h = H$) with the actual concentration in the gas fed to the column (it is assumed that the ozone–air or ozone–oxygen does not carry any volatile species). If their difference in absolute value is lower than any low figure previously established, the model is solved. If not, return to step 1.

Note that for parallel flow operation starting from the top of the column ($h = 0$) all convection flow terms have a negative value as in Equation (11.49) to Equation (11.51) for countercurrent operation. Again, Table 11.1 lists a few instances where this model was used.

11.6.1.3 The Water Phase in Perfect Mixing Flow and the Gas Phase in Plug Flow

This is another possible case that occurs in bubble columns (laboratory or pilot plant size). A combination of equations, given in the two previous models, holds for this case. It is irrelevant whether the water and gas phases are fed countercurrently or in parallel, since the water phase is well-mixed. Equation (11.39) to Equation (11.41) apply for the water phase and Equation (11.47) and Equation (11.52) apply for the gas phase. However, compared to the other two ideal models, there is a significant difference in the mass transfer rate term included in the generation term in the ozone (and volatile species, if any) mass-balance equation. Thus, these terms are as follows:

$$N_{O3g} = -N_{O3} = -(k_L a)_{O3} \left[\frac{1}{H} \int_0^H (C_{O3}^* - C_{O3}) dh \right] \quad (11.54)$$

and

$$N_{vig} = -N_{vi} = -(k_L a)_{vi} \left[\frac{1}{H} \int_0^H (C_{vi} - C_{vi}^*) dh \right] \quad (11.55)$$

where the interface concentrations of ozone and volatile compounds, C_{O3}^* and C_{vi}^*, respectively, are expressed as a function of the concentrations in the gas at any position in the column, C_{O3g} and C_{vig}, respectively, with the Henry and perfect gas laws [see Equation (11.15) for the case of ozone]. The form of the mass transfer terms is due to the fact that, although the concentrations of species are uniform in the water, concentrations in the gas phase vary along the height of the column. Hence, an integrated form of the mass transfer rate term is required to determine its contribution in the mass-balance equations.

The mathematical model needs a numerical and trial-and-error method to reach the solution. If the numerical integration starts from the bottom of the column where the gas is fed ($h = 0$), convection rate terms have a negative value. The general conditions are

$$T = 0 \quad \text{any } h, \quad C_{O3} = 0 \quad C_i = C_{i0} \quad C_{vi} = C_{vi0}$$

$$h = 0 \quad C_{O3g} = C_{O3g} \quad \quad (11.56)$$

$$h = H \quad C_{O3g} = C_{O3gs} \quad C_{vg} = C_{vgs}$$

where C_{O3gi} is the concentration of ozone in the gas fed to the column, and C_{O3gs} and C_{vgs} are assumed values for the concentrations of ozone and volatile compounds, respectively, in the gas at the column outlet.

Similarly, in practical application, the system will work at steady state, so that the accumulation rate terms in the mass-balance equations are zero. For steady-state operation, a possible way to solve the mathematical model involves the following steps:

1. Assume a concentration profile for ozone (and volatile species, if any) in the gas with the position in the column [$C_{O3g} = f(h)$ and $C_{vg} = f(h)$].
2. Solve the set of nonlinear algebraic equations for the mass balances in the water phase. This will give the calculated concentration of species in the water phase, which are the same as in the water at the reactor outlet because of perfect mixing conditions.
3. Solve the set of differential equations in the gas phase (for ozone and volatile species, if any). This will give the concentration profiles in the gas along the column height.
4. Compare the calculated and assumed concentration profiles of ozone (and volatile species, if any) in the gas phase along the column height.
5. If acceptable concordance is achieved, the problem is solved. If not, return to step 1.

Table 11.1 presents ozonation examples where this model was followed.

11.6.1.4 The Water Phase as N Perfectly Mixed Tanks in Series and the Gas Phase in Plug Flow

This model is similar to the previous one but the flow of the water phase differs. In this case, the water phase flow is not ideal but it could be simulated with the water flow through N equal-sized perfectly mixed tanks in series. The value of N is deduced from the corresponding RTDF (see Appendix A3) and the residence time of the water phase in the actual column is the product of the residence time in one tank times the number, N, of tanks. Figure 11.1 depicts the situation assumed with this model. Equations of this model are, therefore, the same as those of the previous model except for step 3, which consists of the solution of the N set of mass-balance equations for the water phase. These equations have to be solved one after the other from one of the edges of the column (preferably from the water phase column inlet) to reach the concentrations at the column outlet. Equations for the k-th tank in the water phase are

Kinetic Modeling of Ozone Processes

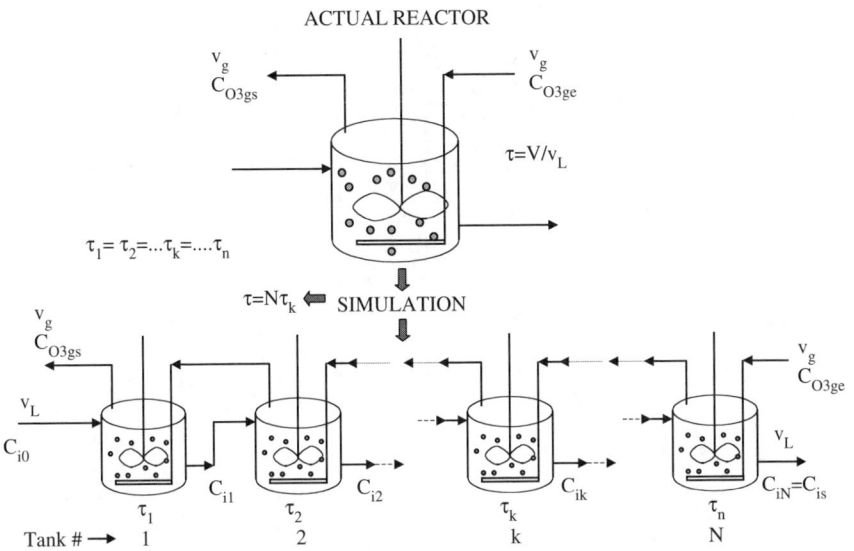

FIGURE 11.1 Water phase in a real contactor simulated with N perfectly mixed tanks in series.

1. For ozone:

$$\frac{1}{\tau_{Lk}}\left(C_{O3_{k-1}} - C_{O3_k}\right) + N_{O3_k} + r_{O3_k} = \frac{dC_{O3_k}}{dt} \quad (11.57)$$

2. For any reacting nonvolatile species i:

$$\frac{1}{\tau_{Lk}}\left(C_{i_{k-1}} - C_{i_k}\right) + r_{i_k} = \frac{dC_{i_k}}{dt} \quad (11.58)$$

3. For any reacting volatile compound:

$$\frac{1}{\tau_{Lk}}\left(C_{vi_{k-1}} - C_{vi_k}\right) + N_{vi_k} + r_{vi_k} = \frac{dC_{vi_k}}{dt} \quad (11.59)$$

where the subindex k refers to conditions at the outlet and inside of the k-th tank (see also Table 11.1 for examples of this model).

11.6.1.5 Both the Gas and Water Phases as N and N' Perfectly Mixed Tanks in Series

If the gas phase flow does not behave either as plug flow or perfectly mixed flow, then it could also be simulated with the flow in N' equal-sized perfectly mixed tank

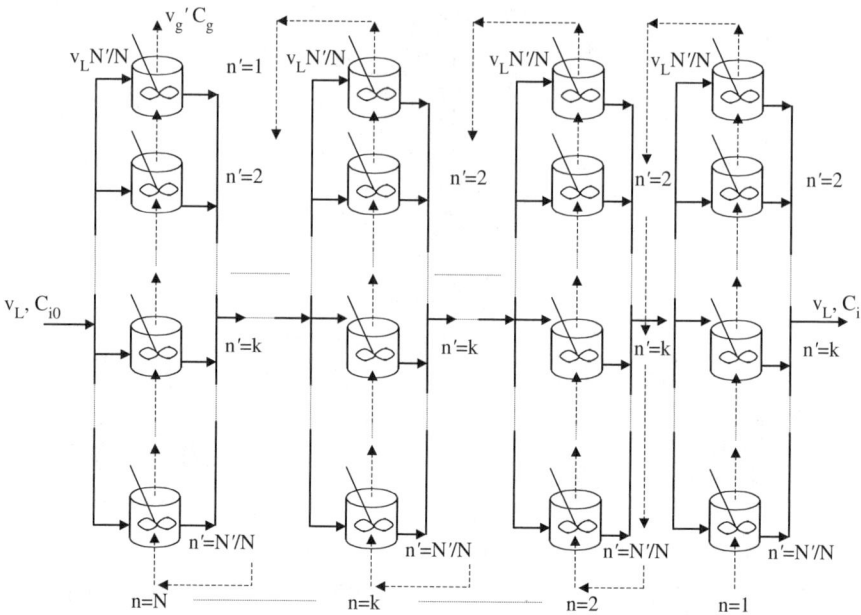

FIGURE 11.2 Water and gas phases in a real contactor simulated with N/N' perfectly mixed tanks in series (solid lines: wastewater flow; dotted lines: gas flow).

in series where the value of N' is obtained from the corresponding RTDF of the gas phase. Thus, for the water and gas phases, there will be N and N' zones, respectively, in the column with constant concentrations (i.e., in perfect mixing). Equations for this model are those corresponding to the perfect mixing model for the gas and water phases. They refer to any of the N and N' tanks that simulate the water and gas phases, respectively. This model, which apparently presents an easier mathematical solution, should be applied only in cases where the ratio N'/N is a natural nonfractional number. Note that in practical cases, N' is likely to be higher than N since the water phase flow is likely to be closer to perfect mixing conditions than the gas phase flow. The explanation given below will help us understand this condition.

If the ratio N'/N is a natural nonfractional number, then any one of the assumed tanks for the water phase will involve N'/N assumed tanks for the gas phase, and, given the property of perfect mixing, concentrations in each tank will be constant. In other words, the height of the column will be divided in N zones of equal height and the concentrations of species in the water that flows through these zones will be the same in each of the N zones. Also, in any of the N zones, there will be N'/N zones of equal height, and the concentrations of species in the gas phase that flows through them will also be the same for any N'/N zone. Figure 11.2 presents a scheme of the simulated processes. For the countercurrent operation at steady state, the possible steps to solve the model are as follows:

Kinetic Modeling of Ozone Processes

1. Starting conditions: concentrations at the top of the column. If unknown, they have to be assumed. For example, assume values for C_{O3g} (and C_{vig}, if any).
2. For the first tank of the water phase (concentration at the bottom of the column), also assume concentrations of species in water at the first tank inlet.
3. Solve the set of nonlinear algebraic mass-balance equations for species in the gas phase corresponding successively to the first tank, then the second, and so on, up to the N'/N-th tank. This will give the concentrations of species in the gas at the entrance of the first tank for the water phase.
4. Solve the set of nonlinear algebraic mass-balance equations for species in the water phase for the first tank and compare the results to the assumed values in step 2.
5. If agreement is achieved, repeat steps 2 to 4 for the second tank of the water phase. This has to be repeated for the N tanks representing the water phase. If there is no agreement, return to step 2 and repeat operations for the tank considered.
6. For the last tank of both water and gas phases, check agreement between concentrations of ozone in the gas phase at the column inlet. If agreement is achieved, the model is solved. If not, return to step 1 and repeat the procedure, assuming new values.

See Table 11.1 for some examples of this model.

11.6.1.6 Both the Gas and Water Phases with Axial Dispersion Flow

This is another possible scheme that can be followed when both the water and gas phase flows are nonideal. With this model, the flow through the column is not only due to convection but also due to axial diffusion. The parameter of this model is the dispersion number, N_D, or the inverse, the Peclet number, Pe (see Appendix A3):

$$N_D = \frac{D}{UH} = \frac{1}{Pe} \tag{11.60}$$

Both the above dimensionless numbers are also deduced from the corresponding RTDF, as presented in Appendix A3.

For countercurrent operation, i.e., the gas fed through the bottom of the column, the mass-balance equations in dimensionless form, derived from Equation (11.36), are as follows:

1. For ozone in the gas phase:

$$\frac{\partial^2 \Psi_{O3g}}{\partial \lambda^2} + (1-\beta)Pe_g \frac{\partial \Psi_{O3g}}{\partial \lambda} + \frac{H^2}{D_g C_{O3g}} \frac{\beta}{1-\beta} N_{O3g} = \frac{H^2}{D_g t_m} \frac{\partial \Psi_{O3g}}{\partial \theta} \tag{11.61}$$

2. For ozone in the water phase:

$$\frac{\partial^2 \Psi_{O3}}{\partial \lambda^2} - \beta Pe_L \frac{\partial \Psi_{O3}}{\partial \lambda} + \frac{H^2}{D_L} \frac{He}{RTC_{O3i}} (N_{O3} + r_{O3}) = \frac{H^2}{D_L t_m} \frac{\partial \Psi_{O3}}{\partial \theta} \quad (11.62)$$

3. For any nonvolatile species in water:

$$\frac{\partial^2 \Psi_i}{\partial \lambda^2} - \beta Pe_L \frac{\partial \Psi_i}{\partial \lambda} + \frac{H^2}{D_L} \frac{1}{C_{i0}} r_i = \frac{H^2}{D_L t_m} \frac{\partial \Psi_i}{\partial \theta} \quad (11.63)$$

4. In cases where the water contains ozone-reacting volatile species:
   a. For any volatile species, vi, in the gas phase:

$$\frac{\partial^2 \Psi_{vig}}{\partial \lambda^2} + (1-\beta) Pe_g \frac{\partial \Psi_{vig}}{\partial \lambda} + \frac{H^2}{D_g} \frac{RT}{C_{vi0} He} \frac{\beta}{1-\beta} N_{vig} = \frac{H^2}{D_g t_m} \frac{\partial \Psi_{vig}}{\partial \theta} \quad (11.64)$$

   b. For any volatile species, vi, in water:

$$\frac{\partial^2 \Psi_{vi}}{\partial \lambda^2} - \beta Pe_L \frac{\partial \Psi_{vi}}{\partial \lambda} + \frac{H^2}{D_L} \frac{1}{C_{vi0}} (N_{vi} + r_{vi}) = \frac{H^2}{D_L t_m} \frac{\partial \Psi_{vi}}{\partial \theta} \quad (11.65)$$

In Equation (11.61) to Equation (11.65) the dimensionless variables are defined as follows:

$$\Psi_{O3g} = \frac{C_{O3g}}{C_{O3gi}} \quad \Psi_{O3} = \frac{C_{O3}}{C_{O3}^*} = \frac{C_{O3}}{C_{O3gi}} \frac{He}{RT}$$

$$\Psi_i = \frac{C_i}{C_{i0}} \quad \Psi_{vi} = \frac{C_{vi}}{C_{vi0}} \quad \Psi_{vig} = \frac{C_{vig}}{C_{vgi}^*} = \frac{C_{vig}}{C_{vi0}} \frac{RT}{He_{vi}} \quad (11.66)$$

$$\lambda = \frac{h}{H} \quad \theta = \frac{t}{t_m}$$

where t_m is the hydraulic residence time of the water phase.

Note that Peclet numbers for the gas and water phases, Pe_L and Pe_g, respectively, are considered independent of chemical species [strictly speaking, the Peclet number depends on the axial dispersion coefficient of chemical species as observed from Equation (11.60)]. In other words, the diffusivity or dispersion coefficient, D, is considered independent of chemical species. A constant value, deduced from the tracer analysis (see Appendix A3), has been considered for any chemical species in the gas and water phases, D_L and D_g, respectively.

At steady state, when $\partial \psi / \partial \theta = 0$, one possible way to solve the mathematical model is:

Kinetic Modeling of Ozone Processes

1. Define new ϕ variables for each species in both phases:

$$\phi = \frac{\partial \Psi}{\partial \lambda} \quad \text{so that} \quad \frac{\partial \phi}{\partial \lambda} = \frac{\partial^2 \Psi}{\partial \lambda^2} \tag{11.67}$$

2. Assume values for the concentrations of the gas species (ozone and volatile compounds, if any) at the top of the column: $C_{O3g} = C_{O3gs}$ and $C_{vig} = C_{vigs}$.
3. Solve the mathematical model as a set of first-order ordinary differential equations with the following initial conditions:

$$\lambda = 0 \quad \Psi_{O3} = 0 \quad \Psi_i = 1 \quad \Psi_{vi} = 1 \tag{11.68}$$

$$\Psi_{vig} = \frac{RT}{He_{vi}} \frac{C_{vigs}}{C_{vi0}} \quad \Psi_{O3g} = \frac{He}{RT} \frac{C_{O3gs}}{C_{O3gi}} \tag{11.69}$$

$$\text{All} \quad \phi = 0$$

4. Apply some numerical method such as the fourth-order Runge–Kutta method (Appendix A5). This will give the dimensionless concentration profiles of species along the column height.
5. Check if calculated values of ψ_{O3g} and ψ_{vig}, at the bottom of the column, i.e., at $\lambda = 1$, are 1 and 0 (the feeding gas does not contain the ozone-reacting volatile species). If agreement is achieved, the process is finished. If not, return to step 2.

11.6.2 Fast Kinetic Regime

In the case where the kinetic regime of ozone reactions is fast (or instantaneous), there will not be dissolved ozone and the mass-balance equation of ozone in water is not needed. In the case of the reaction between ozone and a given compound B (this could be COD in a wastewater), the starting system of equations is the ozone mass balance in the gas [Equation (11.2)] and the mass-balance equation of B in the water [Equation (11.3)]. However, as indicated in Section 11.2, for the steady-state operation, this latter equation cannot be used since concentrations of ozone and B in the liquid film layer are unknown and vary with position within the film close to the gas–water interface. Instead, the equation for the total mass balance is recommended. The following equation can be applied to the whole reaction volume in cases where at least one of the phases flows is in perfect mixing condition (i.e., in an agitated tank):

$$v_g \left(C_{O3gi} - C_{O3gs} \right) = \frac{1}{z} v_L \left(C_{B0} - C_{Bs} \right) \tag{11.70}$$

The equation can be applied to a zone of the reaction volume (from the bottom or top of the column), in case both phases do not flow in perfect mixing (for example, pilot plant or actual bubble columns). In this case, the total balance is applied from one of the edges of the column (where most of the concentrations are known) and from a given position of height h. For example, if conditions at the bottom of the column are known, and the ozone concentration in the feeding gas C_{O3gi}, and the outlet concentration of B, C_{Bs}, the balance will be

$$v_g \left(C_{O3gi} - C_{O3g} \right) = \frac{1}{z} v_L \left(C_B - C_{Bs} \right) \tag{11.71}$$

Another important difference with respect to the slow kinetic regime is the generation term of ozone, G_{O3}, which is now given by Equation (11.18) as a function of the reaction factor, E. This term constitutes the main difficulty of this system because the reaction factor could be an implicit function of E, the Hatta number, Ha_2 [Equation (4.44)], and the instantaneous reaction factor, E_i [Equation (4.46)], unless the fast kinetic regime is of pseudo first order or instantaneous (see Section 4.2). In the two latter cases, E is a function of the concentrations of ozone and compound B. However, a second drawback to the fast kinetic regime arises when the reaction between ozone and B generates intermediate compounds that also react with ozone (as in the ozonation of phenol). Analytical solutions for equations to determine E, however, are available for one irreversible reaction between ozone and B but not for the situation of a series-parallel reacting system with more than three reactions (see Section 4.2). In that case, equations for E are known only for the ozonation of a given compound. Nonetheless, when compound B is present in a very high concentration, it can be assumed that this compound consumes most of the available ozone. Then, if these assumptions are considered, the kinetic model for a fast ozone reaction, which consists of the mass-balance equation of ozone in the gas and the total mass-balance equation, can be used to predict the degree of degradation of a given compound B that can be achieved with ozonation at steady state. In fact, this case also applies to the ozonation of wastewaters with high COD, as shown in Section 11.8.2.

The kinetic model, as mentioned in the preceding section, will also depend on the type of flow of the water and gas phases through the reactor. Hence, cases similar to those studied for the slow kinetic regime can be present. Here we discuss the following cases:

- Both the water and gas phases in perfect mixing
- The gas phase in plug flow and the water phase in perfect mixing
- Both the gas and water phases in plug flow

It is also assumed that the kinetic regime of ozonation is fast and of pseudo first-order, so that the Hatta number and E coincide (see Table 5.5):

$$E = Ha_2 = \frac{\sqrt{k D_{O3} C_B}}{k_L} \tag{11.72}$$

Solution for the case of instantaneous regime is similar, except that the reaction factor, E, coincides with E_i, which can be determined from Equation (4.46).

Kinetic Modeling of Ozone Processes

11.6.2.1 Both the Water and Gas Phases in Perfect Mixing

The kinetic model consists of the following equations:

1. Ozone mole balance in the gas phase:
   Equation (11.38) is used where N_g is given by Equation (11.18). In the case of fast, pseudo first-order kinetic regime, Equation (11.18) becomes:

$$N_g = -aC^*_{O3}\sqrt{kC_{Bs}D_{O3}} = -a\frac{C_{O3gs}}{HeRT}\sqrt{kC_{Bs}D_{O3}} \qquad (11.73)$$

   where the minus sign indicates that the ozone transfer goes from gas to water. Note, again, that for systems with three volume zones,[12] the ozone mass balance in the gas refers to the bubble gas and C^*_{O3} is expressed by Equation (11.45).

2. Total mass balance given by Equation (11.70):
   At steady-state conditions, the system is two algebraic equations where concentrations of ozone at the gas outlet and of compound B at the water outlet can be obtained. Note that for wastewater, COD represents the concentration of B.

A particular case often used in laboratory practice is the semibatch ozonation process where a nonsteady-state operation develops. In this case both mass balances of ozone in the gas phase and B in the water phase can be used. The first one is also given by Equation (11.38) with Equation (11.73), while in the second one, the generation rate term is also given by Equation (11.73), once the stoichiometric coefficient, z, has been accounted for:

$$-\frac{dC_B}{dt} = zN_g \qquad (11.74)$$

The mathematical system is, then, formed by two first-order differential equations, Equation (11.38) and Equation (11.74), which can be solved by numerical integration methods such as the Runge–Kutta method.

11.6.2.2 The Gas Phase in Plug Flow and the Water Phase in Perfect Mixing Flow

This is a situation commonly present in bubble columns, as indicated previously. The mathematical model is now:

1. Ozone mole balance in the gas phase [Equation (11.47)], where N_g is given by Equation (11.75):

$$N_g = -aC^*_{O3}\sqrt{kC_{B_s}D_{O3}} = -a\frac{C_{O3g}}{HeRT}\sqrt{kC_{Bs}D_{O3}} \qquad (11.75)$$

Compared to Equation (11.74), N_g in Equation (11.75) is now a function of the concentration of ozone in the gas, C_{O3g}, which varies with the height of the column.

2. Total mass balance [also given by Equation (11.70)]:
   Assuming the countercurrent flow of the water and gas phases at steady state, Equation (11.75) is substituted in Equation (11.47). After rearranging, variable separation, and integration, the concentration of ozone in the gas at the column outlet can be expressed as a function of concentrations at the bottom of the column:

$$C_{O3gs} = C_{O3gi} \exp\left(-a \frac{RT}{v_g He} \beta V \sqrt{kD_{O3} C_{Bs}}\right) \quad (11.76)$$

Then C_{O3gs} and C_{Bs} can be obtained using Equation (11.70) and Equation (11.76) by a simple trial-and-error procedure.

11.6.2.3 Both the Gas and Water Phases in Plug Flow

This model also applies in many cases for bubble columns. The equations are:

1. Mass balance of ozone in the gas phase: Equation (11.47)
2. Total mass balance: Equation (11.71)

Now, the ozone mass transfer rate at any point in the bubble column is given by Equation (11.77):

$$N_{O3g} = -aC_{O3}^* \sqrt{kC_B D_{O3}} = -a \frac{C_{O3g}}{HeRT} \sqrt{kC_B D_{O3}} \quad (11.77)$$

where, as a result of plug flow conditions, N_{O3g} is a variable function because of changing concentrations of ozone in the gas phase and B in the water phase along the height of the column. Solution for this model is accomplished by numerical integration and trial-and-error procedure. Again, if steady state and countercurrent operation are assumed, the method can be carried out as follows:

1. Integration is better commenced from the bottom of the column ($h = 0$). A value of the concentration of B in the water at the column outlet is assumed as C_{Bs}.
2. Take a known increment for h, Δh.
3. With Equation (11.47) in finite increments, once Equation (11.77) has been accounted for, calculate a value of C_{O3g} at position $h + \Delta h$ (i.e., at position Δh for the first iteration).
4. Using Equation (11.71), calculate the concentration of B in the water phase, C_B, at position $h + \Delta h$.
5. Repeat steps 3 to 4 until the top of the column is reached. At this position, compare the calculated value of C_B with the known value of C_{B0}. If

Kinetic Modeling of Ozone Processes

agreement is achieved, the modeling is terminated. If not, return to step 1 and repeat the process.

As deduced from the preceding paragraphs, when the flow of water and gas phases is simulated through the use of different nonideal models (i.e., tanks in series model or axial dispersion model, etc.), the procedure to follow is similar to that shown in Section 11.6.1 for the slow kinetic regime. In these cases, care should be taken with the expressions for the ozone mass transfer rate and total mass balance application.

11.6.3 THE MODERATE KINETIC REGIME: A GENERAL CASE

This kinetic regime presents the same characteristics of both fast and slow kinetic regime since the reaction zone develops in both the liquid bulk and the film layer. This means that part of B is consumed in the liquid film (where ozone simultaneously diffuses from the gas–water interface) and in the bulk liquid. Then, the generation rate term in Equation (11.1) must account for these contributions. Again, different models can be considered depending on the flow type of phases and the time regime. Here, the case of one irreversible reaction between ozone and B is considered as in the preceding section. Also, for informational purposes only, the case of both phases in perfect mixing conditions and nonsteady state operation is considered. The rest of the possible cases (plug flow, N tanks in series, etc.) follow a procedure similar to that presented in the previous sections for the slow or fast kinetic regimes.

The mathematical system consists of Equation (11.38) with N_g also given by Equation (11.18). For one irreversible reaction between ozone and a compound B, E can be determined from the combination of Equation (4.31) and Equation (4.25) with Ha_1 given by Equation (4.38). Since Ha_1 is also a function of E, solution to this model will imply a trial-and-error procedure with assumed values of E. The system is completed with the bulk mass-balance equations for ozone and B (or COD in the case of wastewater) expressed as follows:

- For ozone:

$$\frac{dC_{O3b}}{dt} = N^l_{O3b} a - \frac{1}{z} k_D C_{O3b} C_{Bb} \tag{11.78}$$

- For B:

$$\frac{dC_{Bb}}{dt} = -z \left[M_1 a N_{O30} - \frac{dC_{O3b}}{dt} \right] \tag{11.79}$$

In Equation (11.78), N^l_{O3b} represents the ozone flux rate from the liquid film to the bulk water, given by Equation (4.23). Again, the mathematical model of Equation (11.38), Equation (11.78), and Equation (11.79) can be solved numerically as in preceding cases.

Note that Equation (11.78) and Equation (11.79) are used for slow and fast kinetic regimes when the conditions of these regimes apply. Thus, for the slow regime, in Equation (11.78), N^l_{O3b} is $k_L(C^*_{O3} - C_{O3b})$, and in Equation (11.79), the expression between brackets becomes $k_D C_{O3b} C_{Bb}$. For the fast regime, Equation (11.78) is not needed ($C_{O3b} = 0$) and Equation (11.79) becomes Equation (11.74). Finally, in Equation (11.38) the only difference between the fast and slow regimes is the generation rate term that becomes $M_1 a C^*_{O3}$ and $k_L(C^*_{O3} - C_{O3b})$, respectively.

The approach presented above involves an ozonation system with just one irreversible reaction between ozone and another compound B. However, most ozonation systems consist of multiple reactions, with ozone reacting with the parent compound B and the intermediates formed. For these cases, when the moderate regime is developed, the procedure based on the depletion factor[8,10,11] (see Section 11.3) is an appropriate way to solve the kinetic modeling problem. Thus, for a system with water and gas phases in perfect mixing conditions, the mass-balance equations or bulk balance equations for ozone and compound B and, let us say, an intermediate compound of the ozonation of B also reacting with ozone, would be similar to Equation (11.38) to Equation (11.40) for the case of the slow kinetic regime but with the following differences:[10,11]

1. For ozone in the gas phase, Equation (11.18) and Equation (11.23) are substituted in Equation (11.38).
2. For ozone in bulk water, in Equation (11.39), N_{O3} is now

$$N_{O3} = N_{O3-bulk} = k_L a DF(C^*_{O3} - C_{O3}) \quad (11.80)$$

with DF given by Equation (11.22).

3. For any compound i reacting with ozone, in Equation (11.40), the molar rate of unreacted compound i leaving the bulk water and entering the film must be added. This molar rate of compound i is completely consumed by reaction with ozone in the film. Then, Equation (11.40) becomes:

$$\frac{1}{\tau_L}(C_{i0} - C_i) - \frac{1}{z_i} k_L a(E - DF)(C^*_{O3} - C_{O3}) + r_{O3} = \frac{dC_{O3}}{dt} \quad (11.81)$$

Solution of Equation (11.38), Equation (11.80), and Equation (11.81) involves determination of the reaction and depletion factors at different reaction times, i.e., determination of the first derivative of ozone concentrations with respect to the depth of liquid at both edges of the film layer ($x = 0$ for E, and $x = \lambda$ for DF), as shown in Figure 11.3. Obviously, this, in turn, involves the determination of the concentrations profiles of ozone and compounds i through the film layer. These profiles are obtained through simultaneous solutions of Equation (4.34) and Equation (4.35). Once the concentrations of ozone at both edges of the film layer are known, E and DF can be determined from Equation (11.22) and Equation (11.23), respectively. This kind of system has been solved only for the case of one irreversible second-order

Kinetic Modeling of Ozone Processes

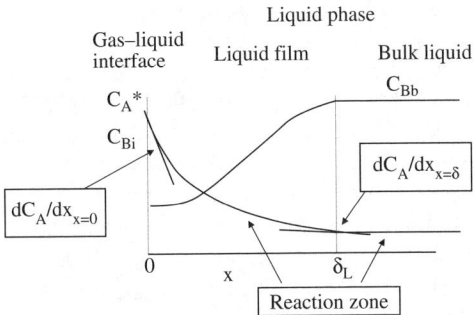

FIGURE 11.3 Concentration profiles through the film layer in an irreversible moderate second-order gas–liquid reaction. Determination of reaction and depletion factors.

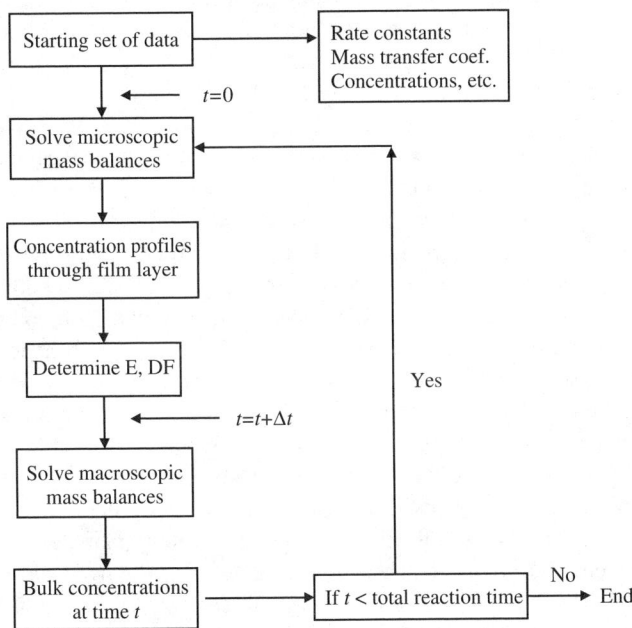

FIGURE 11.4 Flowchart to solve the kinetic model for moderate ozonation reactions.[10]

reaction.[9] For the more complicated ozone systems with more than one irreversible reaction, a numerical approach should be undertaken. For example, Benbelkacen[10] followed the algorithm shown in Figure 11.4 to solve the kinetic modeling of ozone with maleic or fumaric acid and three intermediate compounds formed (glyoxylic acid, formic acid, and oxalic acid). More details on this work are given in the next section.

11.7 EXAMPLES OF KINETIC MODELING FOR MODEL COMPOUNDS

Thus far, most of the information in the literature about kinetic modeling of ozone AOP for simple compounds considers the ozone reactions completed in the slow kinetic regime (see also Table 11.1). The lack of kinetic models valid for fast kinetic regimes is because ozone treatment is oriented mainly to drinking water, where compounds (water pollutants) are found at low concentrations (i.e., ppb or a few ppm). At these concentrations, the Hatta number of ozone reactions is lower than 0.3 and the kinetic regime is slow. Nonetheless, the literature also reports a few cases where fast and even moderate regimes of ozonation are treated.[4–7] Curiously enough, most of these cases deal with the ozonation of phenol compounds (especially chlorophenols). Ozonation was applied at neutral and basic pH in most cases so that a moderate or fast regime developed. An interesting case is the study of the kinetic modeling of maleic acid and fumaric acid ozonation carried out in the moderate kinetic regime.[10,11] The authors used the dimensionless numbers of the reaction and depletion factors to account for the mass flow of ozone that reacted in the film and in the bulk water. They used a bubble column where the water phase was perfectly mixed and the gas phase was in plug flow. For the concentration of ozone in the gas phase, however, an average value of the concentrations at the bottom and top of the column was used. This system is mainly used as a tool for the determination of the rate constants of the ozone reactions. The decomposition of ozone in free radicals is accounted for using an independent first-order kinetic term.

Studies on ozone AOP kinetic modeling, on the other hand, are classified as those analyzing the ozone decomposition or ozone mass transfer in both laboratory-prepared and natural water, and studies dealing with model compounds such as organochlorine compounds, herbicides, aromatic hydrocarbons, phenols, etc. (see Table 11.1). These studies follow the guidelines shown in the preceding sections regarding the mass-balance equations applied with some minor modifications to the value of the rate constant of the ozone decomposition reaction, volatility coefficients for volatile compounds, or influence of pressure drop in the ozonation in bubble columns of industrial size. Thus, Zhou et al.[29] used a specific value of the ozone decomposition rate constant in their kinetic model, while Laplanche et al.[31] applied an ozone concentration-dependent equation to determine the hydroxyl radical concentration and then the ozone decomposition rate constant that is a function of this concentration (see Chapter 7). This rate equation is deduced from the general mechanism of ozone decomposition. A similar approach has been adopted in other studies[20,40] where the contribution of the ozone decomposition reaction is also deduced from the mechanism of reactions, with the concentration of hydroxyl radicals given by Equation (7.12) or Equation (8.4). In other works,[49,53] the hydroxyl radical concentration is expressed as a function of the ozone concentration and the R_{CT} value of the water treated that was previously calculated, as shown in Section 7.4.3. With respect to the ozonation of volatile compounds in other studies,[42] a first-order volatility coefficient is proposed to account for the contribution of volatility to the general rate of compound disappearance. In these cases, no mass balance of volatile compound in the gas phase is used (in fact, there is no need to know the concentrations of these compounds in the gas phase). In other works, however, the Henry constant for the volatile compound is considered, and both mass-balance equations in the water and gas phases were part of the whole kinetic model.[22] Another aspect to be considered is related to the influence of the pressure drop on bubble columns of high or industrial size. This aspect, which was not dealt with in the previous section, is of great interest for industrial contactors. Thus, Zhou et al.[29] introduced the relationship between pressure and height of the column due to the hydrodynamics of the system. They used the following linear relationship:

$$P = P_T + \rho g(H-h)\beta \tag{11.82}$$

where P_T is the pressure at the top of the contactor and H is its height. Then, in both ozone mass-balance equations (gas and water phases), the ozone concentration in the gas is expressed as mole fraction times the total pressure given by Equation (11.82). Also, Cockx et al.[44] dealt with the effect of pressure by including in their model the momentum equation that was applied to model ozone mass transfer in a conventional ozonation tower, as in Figure 11.5. Another aspect not very often treated is the presence of intermediates of ozonation. This, of course, is considered in studies on the treatment of water containing nonvolatile complex compounds such as aromatic hydrocarbons, herbicides, or phenols. The presence of intermediates in the kinetic mechanism that yields the kinetic model equations are fundamental for a good fit between calculations and experimental results, as reported in the ozonation of nitrobenzene.[43] Nitrobenzene is a refractory compound in direct ozonation that is removed by hydroxyl radicals in ozonation processes. As a result of the ozonation, hydrogen peroxide is formed and the process becomes an ozone/hydrogen peroxide oxidizing system. The absence of the hydrogen peroxide mass balance results in a poor match between experimental and calculated concentrations. Consideration of intermediates, however, is not needed when the aromatic compound (as in the case of phenanthrene) is directly attacked by molecular ozone. Figures 11.6 and 11.7 show the mechanism proposed for the formation of hydrogen peroxide and intermediates (nitrophenol, etc.), and the verification of the kinetic model with and without considering reactions in Figure 11.6 in the basic mechanism as has been reported.[43]

As far as the type of gas and water flows is concerned, the literature reports examples such as the ones given in the preceding section. Perfectly mixed gas and water phases were considered in the pioneering work of Singer and Gurol[19] on the

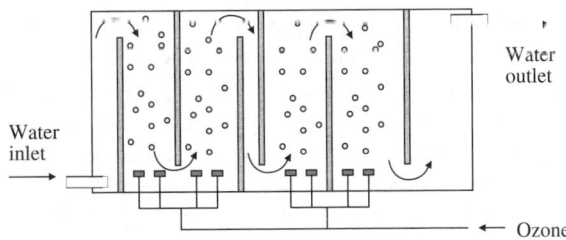

FIGURE 11.5 Conventional ozonation tower.

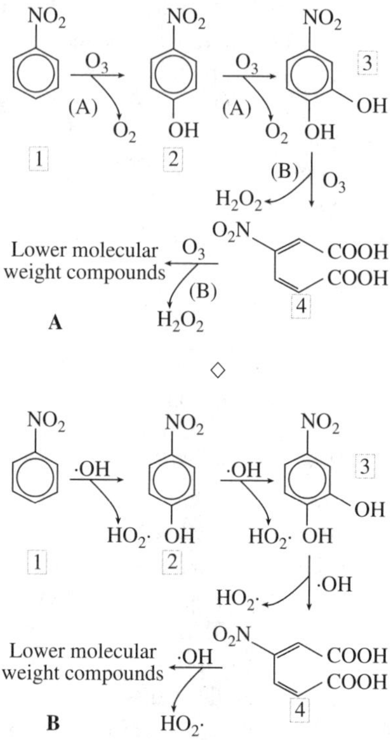

FIGURE 11.6 Proposed reaction mechanism for the ozonation of nitrobenzene through direct ozone reactions and hydroxyl radical free radical reactions: (A) electrophilic substitution, (B) dipolar cycloaddition: (1) nitrobenzene, (2) *p*-nitrophenol, (3) nitroresorcinol, (4) nitromuconic acid. (Reprinted with permission from Beltrán, F.J. et al., A kinetic model for advanced oxidation processes of aromatic hydrocarbons in water: application to nitrobenzene and phenanthrene, *Ind. Eng. Chem. Res.*, 38, 4189–4199, 1999. Copyright 1999, American Chemical Society.)

kinetic model of the ozonation of phenol or in the ozone AOP kinetic modeling of aromatic hydrocarbons.[43] In these studies a semicontinuous agitated tank was used. Plug flow for both the gas and water phases was also considered in other studies such as in the ozonation of toluene, a volatile compound,[22] or in the general mass-balance equations of Pedit et al.[37] Le Sauze et al.[26] used, among others, the N tanks in series model to simulate the water phase in the study of the ozone mass transfer in bubble column at steady state with the gas phase in plug flow. Zhou et al.[29] used the axial dispersion model for the water flow and the plug flow for the gas phase to simulate the experimental results obtained in other studies where the ozone mass transfer in water was treated. The work of Zhou et al.[29] is of particular interest because the authors checked experimental results of the ozone mass transfer in natural water of known TOC and alkalinity in a bubble column (4.2 m high, 15 cm I.D.) against that in an ozonation contactor consisting of one countercurrent flow chamber for ozone absorption and five subsequent reactive chambers of equal size.

Comparison between the performance of different flow models has also been the subject of kinetic modeling works. Beltrán et al.[60] compared the use of plug flow,

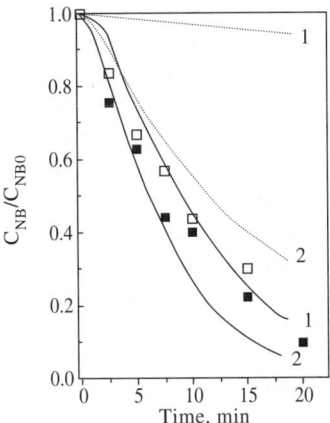

FIGURE 11.7 Verification of the kinetic model. Evolution of experimental and calculated dimensionless remaining concentration of nitrobenzene with time during (1) ozonation and (2) O_3/H_2O_2 oxidation. Symbols: experimental data: □ = ozonation alone; ■ = O_3/H_2O_2 oxidation; dotted curves: kinetic model without intermediate reactions; solid curves: kinetic model with intermediate reactions; conditions: standard reactor, $T=20°C$, pH 7, C_{TPH}: 10^{-3} M, $F_g = 50$ Lh^{-1}. For ozonation alone: inlet ozone partial pressure: 264 Pa, $C_{NBo} = 7.98 \times 10^{-5}$ M. For O_3/H_2O_2 oxidation: inlet ozone partial pressure: 304 Pa, $C_{NBo} = 1.52 \times 10^{-4}$ M, $C_{H2O2To} = 1.75 \times 10^{-4}$ M. (Reprinted with permission from Beltrán, F.J. et al., A kinetic model for advanced oxidation processes of aromatic hydrocarbons in water: application to nitrobenzene and phenanthrene, *Ind. Eng. Chem. Res.*, 38, 4189–4199, 1999. Copyright 1999, American Chemical Society.)

perfect mixing flow, and axial dispersion flow for the gas phase in the kinetic modeling of simazine ozonation in a laboratory bubble column in the slow kinetic regime. Tracer analysis of both the water and gas phases in this work showed that the water phase was perfectly mixed while some dispersion was present in the gas phase flow. Calculated results of the dissolved ozone concentration were best matched with the experimental ones when the kinetic model was solved by assuming the gas phase with axial dispersion in addition to the convection flow. In another work, Le Sauze et al.[26] studied the ozone mass transfer in a bubble column by assuming the gas phase in plug flow and four different flows for the water phase: perfect mixing, plug flow, axial dispersion, and three reactors with different degrees of axial dispersion. Two of these reactors corresponded to the top and bottom part of the column where some sort of backmixing took place due to the entrance and exit of gas and water phases, and the third reactor corresponded to the middle of the column where conditions were closer to the plug flow behavior. Tracer analysis led to the simulation of these three parts of the bubble column with different N tanks in series or different Peclet or dispersion numbers, assuming the axial dispersion model.

The main reactor types used in these studies were not only bubble columns and agitated tanks but packed bed reactors and deep U tubes as well.[22,45] The reactors were operated in continuous and steady-state regimes, as well as in semicontinuous and nonstationary regimes. In most of the cases, the kinetic model involved a heterogeneous reaction (with gas and water phases), but in a few cases the reaction was simply carried out in a homogeneous way in an ozonated aqueous solution, which are examples of studies on kinetic modeling of ozone decomposition reaction[23,36,39,46] as well as on the kinetic modeling of the ozonation of model compounds such as trichloroethene and perchloroethene.[34]

Finally, the kinetic models refer not only to the simple use of ozone alone but also to ozone combined with hydrogen peroxide or UV radiation.[21,34,42,60,61] The kinetic model of the ozonation of volatile organochlorine compounds[34,37,42] was used for the ozone/hydrogen peroxide systems, while the ozone/UV radiation system was studied for the ozone decomposition kinetics[33] or the ozonation kinetic modeling of aromatic hydrocarbons and organochlorine compounds.[25,42,43]

11.8 KINETIC MODELING OF WASTEWATER OZONATION

In wastewater the organic–inorganic matrix is too complex and individual concentrations of species cannot be used to study the ozonation kinetics. Instead, a general parameter such as COD better represents the concentration of ozone-reacting matter in the water. In wastewater ozonation, the ozone reactions usually develop in the fast kinetic regime, as the absence of dissolved ozone confirms. This is due to the high concentration of pollution in wastewater, which makes the Hatta number of ozone reactions higher than 3, especially at the start of ozonation. However, for some wastewater of low COD (see Chapter 6) ozone reactions could also be slow (e.g., domestic wastewater, food-derived wastewater, etc.).[56,67] Hence, for the ozonation of wastewater, kinetic modeling can be accomplished in the various kinetic regimes. Equations are also the mass balance of ozone in gas and wastewater (the latter only for slow reactions) and the mass balance of COD in the wastewater (for the slow kinetic regime) or the total mass balance, also referred to COD, in the reaction volume (for the fast kinetic regime). Here, kinetic modeling for slow and fast-to-moderate kinetic regimes for the cases of perfect mixing flow for both the wastewater and gas phases and perfect mixing flow for the wastewater phase with plug flow for the gas phase. In one case, the N perfectly mixed tanks in series flow model is used. Kinetic model equations for the remaining flow possibilities can easily be derived as in the preceding sections for individual compounds.

In the following two sections, a mathematical model of wastewater ozonation (both slow and fast-to-moderate kinetic regimes) in agitated tanks and bubble columns is presented. Kinetic models usually consider two possible reactions to remove the ozone-reacting matter from wastewater: on the one hand, a unique irreversible reaction that involves all possible direct and indirect ozone reactions with the matter present in wastewater:

$$O_3 + zCOD \xrightarrow{k} \text{Products} \qquad (11.83)$$

and on the other hand, in addition to Reaction (11.83), another reaction between hydroxyl radicals generated from the ozone decomposition and the matter in wastewater:

$$HO\bullet + COD \xrightarrow{k_{HOww}} COD\bullet \qquad (11.84)$$

11.8.1 CASE OF SLOW KINETIC REGIME: WASTEWATER WITH LOW COD

This case usually holds for COD lower than 300 mgL^{-1} although the exact value depends on the nature of wastewater components (see Chapter 6). In any case, the Hatta number of the surrogate wastewater ozonation reaction [Reaction (11.83)] must be lower than 0.3. Since the slow kinetic regime is characterized by the presence of dissolved ozone, detection of dissolved ozone constitutes the first check to confirm this kinetic regime, in the absence of rate constant data.

Kinetic modeling of slow wastewater ozonation follows the steps shown in Section 11.6.1 for model compounds. The mass-balance equations depend on the type of gas and wastewater phase flow through the contactor. For example, for the case of perfect mixing in both phases, Equation (11.38) to Equation (11.40) would constitute the mathematical model. The difference between this case and the previous one on model compounds is the expression of the reaction rate term for ozone, r_{O3}, and compounds, r_i. Paradoxically, the unknown nature of compounds present in wastewater simplifies the mathematics because only one equation of the type of Equation (11.40) is needed. In contrast to kinetic modeling of single compounds, few studies on this subject deal with wastewater (see Table 11.1). The examples given below illustrate the ozonation kinetic modeling of tomato wastewater in a semicontinuous operation in bubble columns of different sizes (but equal height/diameter ratio) and in a pilot plant bubble column.[30]

11.8.1.1 Kinetic Modeling of Wastewater Ozonation without Considering a Free Radical Mechanism

Preliminary experiments in a small bench scale bubble column allowed the rate and mass-transfer coefficients and stoichiometric parameters to be determined (see Chapter 6). For the slow kinetic regime, the dissolved ozone concentration constitutes another variable to be determined. The kinetic model consists of Equation (11.38) or Equation (11.41) (depending on the flow type of the gas phase), Equation (6.25), and Equation (6.26) (water is considered perfectly mixed), already used for the determination of the rate coefficient and stoichiometric ratio (see Section 6.6.3). Note that the system can also be solved from Equation (11.38) or Equation (11.47), and Equation (11.39) and Equation (11.40), as in the case of single compounds. The use of Equation (6.25) and Equation (6.26) in this case is a second possibility for solving the model following the method of rate constant determination in Chapter 6. The following assumptions are applied in solving the mathematical model:

- Ozone is consumed in only one irreversible second-order reaction with COD as a surrogate parameter to represent the concentration of the reacting matter present in the wastewater [see Equation (11.83)].
- Equations for the ozone absorption rate, N_{O3}, come from the application of film theory.
- The ozone reactor is a bubble column of height/diameter ratio equal to that of the laboratory column where experiments to determine the kinetic parameters were carried out (see Chapter 6).

- The bubble column is initially charged with a known volume of wastewater and is perfectly mixed because of the bubbling of the gas phase. The concentration values for ozone and COD remain constant regardless of the height of the column.
- The gas phase containing ozone is continuously fed to the column through a porous plate situated at the bottom. The gas phase is in plug flow so that the concentration of ozone in the gas or its partial pressure changes with the height of the column.
- At any time, the gas at the column outlet is in equilibrium with the wastewater, so that Henry's law is fulfilled.
- Loss of total gas pressure through the reactor is negligible.
- Parameters obtained in the small laboratory bubble column are valid for the large size column.

The objectives are to calculate the time profiles of COD, the ozone partial pressure or ozone concentration in the gas at the column outlet, and the dissolved ozone concentration.

The method of solving the model presents two trial-and-error procedures and starts by assuming a value for the concentration of dissolved ozone, C_{O3}. This value should be lower than C_{O3}^* calculated from the ozone partial pressure at the column outlet by applying Henry's law. Thus, the ozone gas concentration profile with the height of the column is determined from Equation (6.14) where h and C_g substitute for h_T and C_{go}, respectively. Then, numerical integration of the ozone mass balance in the bulk water [Equation (6.25)] with calculated values of C_g allows the determination of COD. Finally, the total mass balance [Equation (6.26)] is used to calculate the dissolved ozone concentration. For any time considered, the iterative procedure is complete when the assumed and calculated ozone concentrations coincide.[30] Some of the results obtained for the ozonation of tomato wastewater (COD < 100 mgL) are shown in Figure 11.8. The accuracy of the procedure can be established in terms of the oxidation and ozone efficiencies:[30]

$$E_{OX} = \frac{COD_0 - COD}{COD_0} \times 100 \qquad (11.85)$$

and

$$E_{O3} = \frac{C_{gi} - C_{go}}{C_{gi}} \times 100 \qquad (11.86)$$

For the particular cases mentioned here, the calculated oxidation and ozone efficiencies showed deviations within ±30 and ±6% of the experimental ones.[30]

11.8.1.2 Kinetic Modeling of Wastewater Ozonation Considering a Free Radical Mechanism

Since the kinetic regime is slow, direct ozone reactions represent only a partial contribution to the total removal rate of compounds in wastewater. In fact, in the

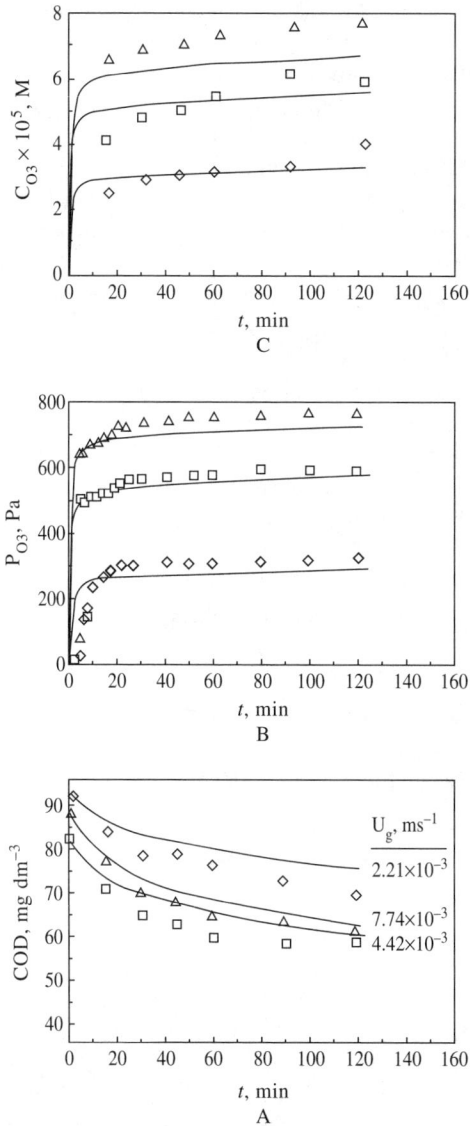

FIGURE 11.8 Tomato wastewater ozonation. Experimental and calculated profiles of (A) COD, (B) P_{O_3} (column outlet) and (C) C_{O_3}. Conditions: bubble column height: 1.05 m, $T=17°C$, inlet ozone partial pressure: 1026 Pa (average value), U_g, ms^{-1} as superficial velocity of gas phase: ◊ = 2.21, □ = 4.42, ∆ = 7.74. Curves represent calculated results. (Reprinted with permission from Beltrán, F.J., Encinar, J.M., and García-Araya, J.F., Modelling industrial wastewater ozonation in bubble contactors. 2. Scale-up from bench to pilot plant, *Ozone Sci. Eng.*, 1995, 17, 379–398, 1995. Copyright 1995, International Ozone Association.)

slow kinetic regime, part of the dissolved ozone decomposes in hydroxyl radicals that can act on the dissolved matter. If Reaction (11.84) is considered, a mechanism of reactions similar to that shown in Table 2.3 and Table 2.4 can be proposed to express the concentration of hydroxyl radicals as a function of known concentrations or to establish the corresponding mass-balance equations. This method for solving the kinetic model of wastewater ozonation in the slow kinetic regime has also been the subject of previous studies. For example, Andreozzi et al.[50] proposed the classic mechanism of Staehelin and Hoigné[68] to model the ozonation kinetics with and without hydrogen peroxide or UV radiation of a mineral-oil-contaminated wastewater in the slow kinetic regime. The mechanism also contained Reaction (11.83) to account for the direct ozone reactions. The mathematical model consisted of the mass-balance equations of ozone, COD, and hydrogen peroxide as the type of Equation (11.1) with some modifications, and perfect mixing was assumed to hold for the wastewater and gas phases. The modifications introduced were to the ozone mass balances because Andreozzi et al.[50] assumed the ozone mass balance in the bubble gas inside the reactor, the ozone mass balance in water, and the ozone mass balance in the freeboard of the reactor. In addition, the authors considered two possible cases depending on the stoichiometry of Reaction (11.84). In the first case, reaction between hydroxyl radicals and COD was as in Equation (11.84), but in the second case this reaction was assumed to yield initiators of the ozone decomposition, $I_j\bullet$:

$$HO\bullet + COD \xrightarrow{k_{HOww}} I_i \bullet \quad (11.87)$$

that subsequently react with ozone to propagate the radical chain:

$$O_3 + I_j \bullet \xrightarrow{k_j} HO\bullet + O_2 + products \quad (11.88)$$

Depending on the case it is evident that the expression for the concentration of hydroxyl radicals was different. This expression was then introduced in all mass-balance equations containing the concentration of hydroxyl radicals. Rate constants of reactions in this model were taken from previous studies except those of Reaction (11.83), Reaction (11.87), and Reaction (11.88), which were calculated from matching experimental results to the kinetic model. The authors obtained a good fit between experimental and calculated concentrations of ozone in the outlet gas and COD.

In another work, Rivas et al.[51] studied the ozonation kinetic modeling of wastewater derived from olive processing industries. They also started from the Staehelin and Hoigné mechanism[68] where wastewater and gas phases were considered perfectly mixed. Reaction (11.83), for the ozone direct reactions, and Reaction (11.84) were also included with some modifications. Thus, the latter was divided into further steps to account for a fraction of COD that directly led to mineralization (formation of CO_2 and water) and to propagate the radical chain in a way similar to what Andreozzi et al.[50] proposed. In addition, Reaction (11.83) was assumed to yield hydrogen peroxide because the wastewater studied contained significant concentration of phenol compounds. As is known, direct reaction of ozone with phenols develops through electrophilic substitution and cycloaddition reactions. The latter implies that the ring or carbon double bond breaks to yield hydrogen peroxide.[69] The COD-involving reactions assumed were

$$O_3 + COD \xrightarrow{k_D} H_2O_2 + Products \quad (11.89)$$

$$HO\bullet + \alpha COD \xrightarrow{k_{HOww}} H_2CO_3 \quad (11.90)$$

$$HO\bullet + (1-\alpha)COD \xrightarrow{k_{HOww}} Products \quad (11.91)$$

The kinetic model also considered changes in pH because the ozonation was carried out at different pH cycles with the aim of improving the ozone efficiency, as indicated in Chapter 6. To the system of first-order ordinary differential equations constituted by the mass-balance equations of ozone (in the gas and wastewater), COD, and hydrogen peroxide [Equation (11.40)], equations for total carbonate ion and free radicals were added. Figure 11.9 shows an example of the fit between experimental and calculated results.[51]

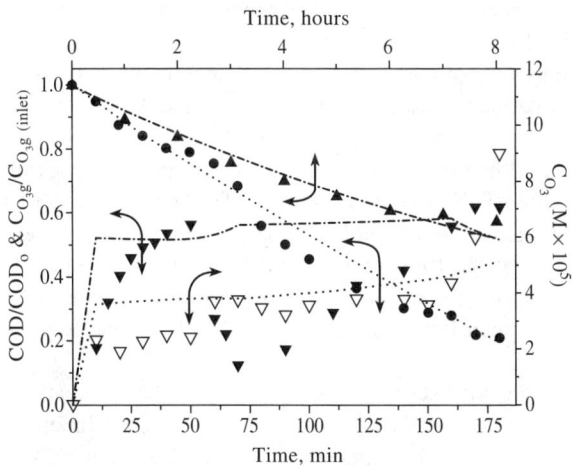

FIGURE 11.9 Experimental (symbols) and calculated (dotted lines) of the integrated sequential ozonation–aerobic biodegradation of olive oil wastewater. Conditions: 20°C, pH 7, gas flow rate: 20 Lh^{-1}, kLa = 0.02 s^{-1}; concentration of ozone fed: 45 mgL^{-1}; pH cycles: acid–basic–acid, COD0/VSS: 0.4, ● = COD profile from ozonation, ▲ = COD profiles from biodegradation, ∇ = dissolved ozone concentration profile, ▼ = ozone gas outlet concentration profile. (Reprinted with permission from Rivas, F.J. et al., Two-step wastewater treatment: sequential ozonation-aerobic biodegradation, *Ozone Sci. Eng.*, 22, 617–686, 2000. Copyright 2000, International Ozone Association.)

11.8.2 CASE OF FAST KINETIC REGIME: WASTEWATER WITH HIGH COD

When the COD of wastewater is higher than 300 mgL^{-1}, the wastewater usually contains different compounds at high concentrations. Then the Hatta number of ozone reactions is likely to be higher than 3 and the kinetic regime of ozonation becomes fast (see Chapter 6). However, as a result of ozonation the nature of compounds present in water changes with time. Thus, the resistant nature of compounds formed, toward ozone, increases with reaction time, and the kinetic regime also changes to become moderate or even slow. This change can be noted when the reaction factor E, as defined by Equation (4.31), goes from above to below unity. The mathematical model, however, can still be formed from the mass-balance equations of ozone in gas and the total mass-balance equation that refers to COD. In the case of gas and wastewater perfectly mixed and fed continuously to the ozone reactor, at steady state-conditions, the total mass balance is [see also Equation (11.70) for the case of model compounds]

$$v_L \left[C_{O3b} + z(COD_i - COD_s) \right] = v_g \left(C_{geb} - C_{gob} \right) \tag{11.92}$$

In this case, all the available ozone is consumed through fast or instantaneous reactions at the gas–wastewater interface (see Chapter 4) and there is no formation of free radicals (direct reactions predominate, see Chapter 7). Another point of practical interest concerns the use of other agents such as UV radiation or hydrogen peroxide combined with ozone. In some cases, when the fast ozone kinetic regime holds, these agents do not add any significant increment to the COD degradation rate compared to ozonation alone,[70] although this is not a general rule.[71]

The generation rate term in Equation (11.38) or Equation (11.47) is the main difficulty of the fast kinetic regime model. This term is the absorption rate of ozone, which is a function of the reaction factor E. This parameter depends on the fast kinetic regime. Bench scale or pilot plant experiments allow an estimation of the type of kinetic regime. The following paragraph details an example of the application of this kinetic model for the ozonation of a distillery wastewater in semicontinuous operation in a pilot plant bubble column (see conditions and hypothesis in Section 11.8.1.1).

The kinetic model consists only of Equation (11.47) and Equation (6.26) together with the ozone absorption rate law [general Equation (4.25)]. Integration of Equation (11.47) taking into account Equation (4.25) yields Equation (6.24), which relates the ozone concentration in the gas at the column outlet to the height of the column. On the other hand, from Equation (6.26), after rearranging for two successive moments of time, the following is obtained:[30]

$$COD_{t+\Delta t} = COD_t - \frac{U_g}{z \beta h_T} \left(C_{geb} - C_{gob} \right) \Delta t \tag{11.93}$$

From Equation (6.24) and Equation (11.93), C_{gbo} and COD can be determined at different times by using the physico-chemical parameters already known (ε_{O3}, $k_L a$, He, etc.; see Chapter 6). The method presented here was checked not only in two bubble columns of height/diameter ratio equal to 4 but also in a pilot plant bubble column of 14.4 height/diameter ratio (diameter: 0.08 m).[30] Bubble column sizes had a height/diameter ratio equal to that of another laboratory column used for kinetic data determinations (diameters were 0.04, 0.08, and 0.12 m). In the pilot plant bubble column, the water phase was recirculated at a rate of 70 L/h to assure perfect mixing conditions. However, the experimental COD profiles were the same with and without recirculation, so that the ideal mixing was assumed to hold. Figure 11.10 and Figure 11.11 present some of the reported results.[30] Beltrán et al.[30] presents detailed explanations on deviations observed.

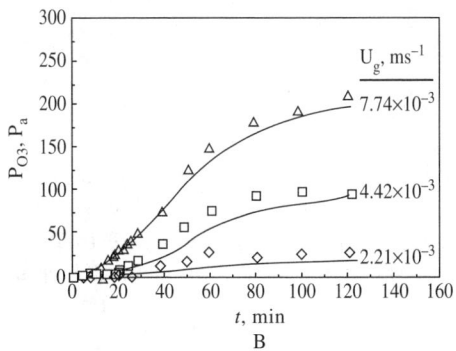

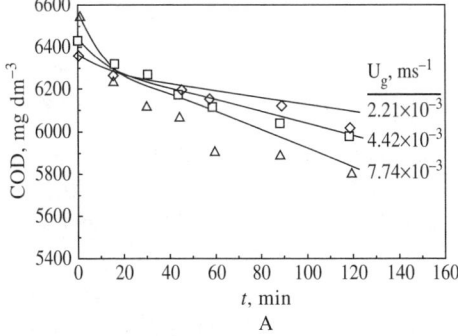

FIGURE 11.10 Experimental and calculated profiles of (a) COD and (b) P_{O3} (column outlet) during the ozonation of distillery wastewater. Conditions: bubble column height, 1.05 m; T = 17°C, inlet ozone partial pressure = 1038 Pa, U_g, ms^{-1} as superficial velocity of gas phase: ◇ = 2.21, □ = 4.42, △ = 7.74. Curves represent calculated results. (Reprinted with permission from Beltrán, F.J., Encinar, J.M., and García-Araya, J.F., Modelling industrial wastewater ozonation in bubble contactors. 2. Scale-up from bench to pilot plant, *Ozone Sci. Eng.*, 1995, 17, 379–398, 1995. Copyright 1995, International Ozone Association.)

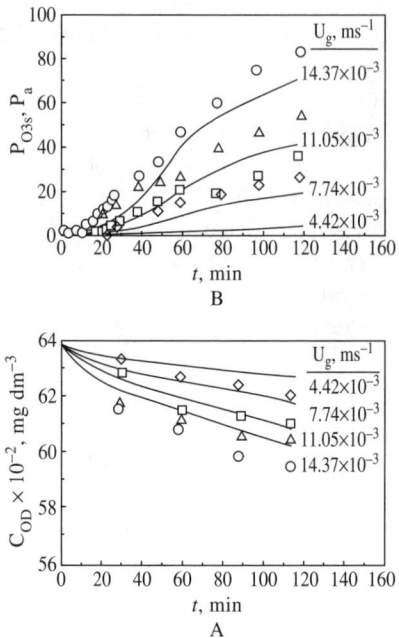

FIGURE 11.11 Experimental and calculated profiles of (a) COD and (b) P_{O3} (column outlet) during the ozonation of distillery wastewater. Conditions: Pilot plant bubble column 1.2 m high, $T = 18°C$, inlet ozone partial pressure = 921 Pa, U_g, ms^{-1} as superficial velocity of gas phase: ◊ = 4.42, □ = 7.74, △ = 11.05, ○ = 14.37. Curves represent calculated results. (Reprinted with permission from Beltrán, F.J., Encinar, J.M., and García-Araya, J.F., Modelling industrial wastewater ozonation in bubble contactors. 2. Scale-up from bench to pilot plant, *Ozone Sci. Eng.*, 1995, 17, 379–398, 1995. Copyright 1995, International Ozone Association.)

11.8.3 A General Case of Wastewater Ozonation Kinetic Model

For a general case, the mathematical model, however, can still be formed from equations used in the preceding section, with the reaction factor E given by the general Equation (4.25), Ha_1 by Equation (4.38), and the Hatta number for the second-order reaction expressed as a function of COD:

$$Ha_2 = \sqrt{\frac{\varepsilon_{O3} D_{O3} COD}{k_L}} \tag{11.94}$$

Note that using the general kinetic Equation (4.25) means that the ozone mass balance in water does not have to be used, since the concentration of ozone has been substituted as a function of the interfacial ozone concentration or ozone partial pressure as explained in Chapter 4. This kinetic model has also been reported for the ozonation of a domestic wastewater where, depending on the ozone dose applied, the regime changed from fast to slow.[56] Equations for this kinetic model were the

Kinetic Modeling of Ozone Processes

ozone mass balance in the gas phase and the total mass balance. In the latter equation the contribution of the mass flow of ozone in water at the column outlet was considered negligible and the kinetic system reduced to two nonlinear algebraic equations with COD and the ozone gas at the reactor outlet as unknowns to be determined. The kinetic model used the N–N' tanks in series model[56] because step input tracer experiments (see Appendix A3) indicated that the gas and wastewater phase flows in the column behaved as in a series of two and eight perfectly mixed reactors, respectively. Figure 11.2 shows the flow model proposed for the general kinetic modeling of wastewater ozonation. Thus, both the ozone gas mass balance and the total mass balance applied to the ith reactor out of the eight assumed reactors for the gas phase were as follows:[56]

Ozone in the gas phase:

$$\frac{1}{\tau_g}\left(C_{gb(i+1)} - C_{gb(i)}\right) + N_g \frac{\beta}{1-\beta} = 0 \qquad (11.95)$$

with N_g being G_{O3} given by Equation (11.18) and subindex i referring to the ith reactor of the gas phase.

Total mass balance:

$$\frac{v_L}{4}\left[z\left(COD_{inlet(j)} - COD_{(i)}\right)\right] = v_g\left(C_{gb(i+1)} - C_{gb(i)}\right) \qquad (11.96)$$

where $COD_{inlet(j)}$ represents the COD at the entrance of the jth reactor of the wastewater phase (j varied from 1 to 2). Note that the contribution of the ozone mass flow rate in the wastewater at the outlet of the ith gas reactor was suppressed due to the low values of experimental dissolved ozone concentrations [see also Equation (11.92)]. Also note that wastewater flow rate in the ith reactor is the fourth part ($v_L/4$) of the total flow rate because of the flow model assumed (see Figure 11.2).

The ozonation system worked continuously with the gas and wastewater phases circulated countercurrently through the bubble column. The kinetic model was solved from the known value of the COD of wastewater at the column inlet and by assuming the value of the ozone concentration in the gas leaving the column (top of the column). Equation (11.95) and Equation (11.96) were then simultaneously solved to find the values of the COD at the reactor outlet and the ozone gas concentration at the reactor inlet. E values were obtained from Equation (4.25), Equation (4.46), Equation (4.38), and Equation (11.94) by trial and error. The system was first solved by an iterative method from the first to the fourth reactors of the gas phase. The average value of the COD of the first four reactors was taken as the COD at the entrance to the second set of four gas phase reactors. Again, the new set of Equation (11.95) and Equation (11.96) were solved for each reactor of the second wastewater phase reactor (see Figure 11.2). Calculated results for COD were very close to the experimental ones. Major deviations were observed for the ozone gas concentration, especially at high ozone doses. The authors attributed these discrepancies to the absence of free radical reaction [Reaction (11.84)] contribution to the kinetic model.[56]

Table 11.1 summarizes the kinetic modeling studies on ozonation processes covering different reacting systems, flow types, and kinetic regimes that have been reported chronologically.

References

1. Danckwerts, P.S.V., *Gas–Liquid Reactions*, McGraw-Hill, New York, 1970.
2. Sotelo, J.L., Beltrán, F.J., and González, M., Kinetic regime changes in the ozonation of 1,3 cyclohexanedione in aqueous solutions, *Ozone Sci. Eng.*, 13, 397–419, 1991.
3. Beltrán, F.J. et al., Oxidation of mecoprop in water with ozone and ozone combined with hydrogen peroxide, *Ind. Eng. Chem. Res.*, 33, 125–136, 1994.
4. Hautaniemi, M. et al., Modelling of chlorophenol treatment in aqueous solutions. 2. Ozonation under basic conditions, *Ozone Sci. Eng.*, 20, 283–302, 1998.
5. Qiu, Y., Kuo, C., and Zappi, M.E., Performance and simulation of ozone absorption reactions in a stirred tank reactor, *Environ. Sci. Technol.*, 35, 209–215, 2001.
6. Andreozzi, R. and Marotta, R., Ozonation of *p*-chlorophenol in aqueous solution, *J. Haz. Mat.*, B69, 303–317, 1999.
7. Wu, J.J. and Masten, S.J., Oxidation kinetics of phenolic and indolic compounds by ozone: applications to synthetic and real swine manure slurry, *Water Res.*, 36, 1513–1526, 2002.
8. Benbelkacem, H. and Debellefontaine, H., Modeling of a gas–liquid reactor in batch conditions. Study of the intermediate regime when part of the reaction occurs within the film and part within the bulk, *Chem. Eng. Proc.*, 42, 723–732, 2003.
9. Schlüter, S. and Schulzke, T., Modeling mass transfer accompanied by fast chemical reactions in gas/liquid reactors, *Chem. Eng. Technol.*, 22, 742–746, 1999.
10. Benbelkacem, H., Modelisation du transfert de matiere couple avec une reaction chimique en reacteur fermé. Application au procede d'ozonation, Ph.D. thesis, University of Toulouse, France, 2002.
11. Benbelkacem, H. et al., Maleic acid ozonation: reactor modelling and rate constant determination, *Ozone Sci. Eng.*, 25, 13–24, 2003.
12. Anselmi, G. et al., Ozone mass transfer in stirred vessel, *Ozone Sci. Eng.*, 6, 17–28, 1984.
13. Chiu, C.Y. et al., A refined model for ozone mass transfer in a semibatch stirred vessel, *Ozone Sci. Eng.*, 19, 439–456, 1997.
14. Fogler, H.S., *Elements of Chemical Reaction Engineering*, 3rd ed., Prentice-Hall, Englewood Cliffs, NJ, 1999.
15. Beltrán, F.J. et al., Kinetic modelling of aqueous atrazine ozonation processes in a continuous flow bubble contactor, *J. Haz. Mater.*, B80, 189–206, 2000.
16. Laplanche, A. et al., Simulation of ozone transfer in water: comparison with a pilot unit, *Ozone Sci. Eng.*, 13, 535–558, 1991.
17. El-Din, M.G. and Smith, D.W., Designing ozone bubble columns: a spreadsheet approach to axial dispersion model, *Ozone Sci. Eng.*, 23, 369–384, 2001.
18. El-Din, M.G. and Smith, D.W., Development of transient back flow cell model (BFCM) for bubble columns, *Ozone Sci. Eng.*, 23, 313–326, 2001.
19. Gurol, M.D. and Singer, P.C., Dynamics of the ozonation of phenol. II. Mathematical simulation, *Water Res.*, 17, 1173–1181, 1983.

20. Glaze, W.H. and Kang, J.H., Advanced oxidation processes. Description of a kinetic model for the oxidation of hazardous materials in aqueous media with ozone and hydrogen peroxide in a semibatch reactor, *Ind. Eng. Chem. Res.*, 28, 1573–1580, 1989.
21. Glaze, W.H. and Kang, J.W., Advanced oxidation processes. Test of a kinetic model for the oxidation of organic compounds with ozone and hydrogen peroxide in a semibatch reactor, *Ind. Eng. Chem. Res.*, 28, 1580–1587, 1989.
22. Yurteri, C. and Gurol, M.D., Ozonation of trace organic compounds: model predictions versus experimental data, *Ozone Sci. Eng.*, 12, 217–229, 1990.
23. Chelkowska, K. et al., Numerical simulation of aqueous ozone decomposition, *Ozone Sci. Eng.*, 14, 33–49, 1992.
24. Le Sauze, N. et al., The residence time distribution of the liquid phase in a bubble column and its effect on ozone transfer, *Ozone Sci. Eng.*, 14, 245–262, 1992.
25. Hayashi, J. et al., Decomposition rate of volatile organochlorines by ozone and utilization efficiency of ozone with ultraviolet radiation in a bubble column contactor, *Water Res.*, 27, 1091–1097, 1993.
26. Le Sauze, N. et al., Modelling of ozone transfer in a bubble column, *Water Res.*, 27, 1071–1083, 1993.
27. Laplanche, A., Bastiment, R., and Boison, V., Modelisation de l'oxydation de l'atrazine et de ses sous-produits dans les colonnes d'ozonation á bulles, *Proceedings of the International Conference on Ozone*, Conference 29, Poitiers, France, September 1994.
28. von Gunten, U. and Hoigné, J., Bromate formation during ozonation of bromide-containing waters: interaction of ozone and hydroxyl radical reactions, *Environ. Sci. Technol.*, 28, 1234–1242, 1994.
29. Zhou, H., Smith, D.W., and Stanley, S.J., Modeling of dissolved ozone concentration profiles in bubble columns, *J. Env. Eng.*, 120, 821–840, 1995.
30. Beltrán, F.J., Encinar, J.M., and García-Araya, J.F., Modelling industrial wastewater ozonation in bubble contactors. 2. Scale-up from bench to pilot plant, *Ozone Sci. Eng.*, 1995, 17, 379–398, 1995.
31. Laplanche, A. et al., Modelisation of micropollutant removal in drinking water treatment by ozonation or advanced oxidation processes (O_3/H_2O_2), *Ozone Sci. Eng.*, 17, 97–117, 1995.
32. Roustan, M., Wang, R.Y., and Wolbert, D., Modeling hydrodynamics and mass transfer parameters in a continuous ozone bubble column, *Ozone Sci. Eng.*, 18, 99–115, 1996.
33. Gurol, M.D. and Akata, A., Kinetics of ozone photolysis in aqueous solution, *AIChE J.*, 42, 3283–3292, 1996.
34. Sunder, M. and Hempel, D., Oxidation of tri and perchloroethene in aqueous solution with ozone and hydrogen peroxide in a tube reactor, *Water Res.*, 31, 33–40, 1997.
35. Wright, P.C., Meeyoo, V., and Soh, W.K., A study of ozone mass transfer in a cocurrent downflow jet pump contactor, *Ozone Sci. Eng.*, 19, 17–33, 1997.
36. Westerhoff, P. et al., Application of ozone decomposition models, *Ozone Sci. Eng.*, 19, 55–73, 1997.
37. Pedit, J.A. et al., Development and application of a gas–liquid contactor model for simulating advanced oxidation processes, *Env. Sci. Technol.*, 31, 2791–2796, 1997.
38. Hautaniemi, M. et al., Modelling of chlorophenol treatment in aqueous solutions. 1. Ozonation and ozonation combined with UV radiation under acidic conditions, *Ozone Sci. Eng.*, 20, 259–282, 1998.
39. Westerhoff, P. et al., Numerical kinetic models for bromide oxidation to bromine and bromate, *Water Res.*, 32, 1687–1699, 1998.

40. Beltrán, F.J. et al., Aqueous degradation of atrazine and some of its main by-products with ozone/hydrogen peroxide, *J. Chem. Technol. Biotechnol.*, 71, 345–355, 1998.
41. Siddiqui, M. et al., Modeling disolved ozone and bromate ion formation in ozone contactors, *Water Air Soil Pollut.*, 108, 1–32, 1998.
42. Beltrán, F.J. et al., Aqueous degradation of VOCs in the ozone combined with hydrogen peroxide or UV radiation processes. 2. Kinetic modelling, *J. Env. Sci. Health*, A34, 673–693, 1999.
43. Beltrán, F.J. et al., A kinetic model for advanced oxidation processes of aromatic hydrocarbons in water: application to nitrobenzene and phenanthrene, *Ind. Eng. Chem. Res.*, 38, 4189–4199, 1999.
44. Cockx, A. et al., Use of computational fluid dynamics for simulating hydrodynamics and mass transfer in industrial ozonation towers, *Chem. Eng. Sci.*, 54, 5085–5090, 1999.
45. Muroyama, K. et al., Hydrodynamics and computer simulation of an ozone reactor for treating drinking water, *Chem. Eng. Sci.*, 54, 5285–5292, 1999.
46. Acero, J.L. and von Gunten, U., Influence of carbonate on the ozone/hydrogen peroxide based advanced oxidation process for drinking water treatment, *Ozone Sci. Eng.*, 22, 305–328, 2000.
47. Nemes, A., Fabian, I., and Gordon, G., Experimental aspects of mechanistic studies on aqueous ozone decomposition in alkaline solution, *Ozone Sci. Eng.*, 22, 287–304, 2000.
48. Ko, Y.W. et al., Kinetics of the reaction between ozone and p-hydroxybenzoic acid in a semibatch reactor, *Ind. Eng. Chem. Res.*, 39, 635–641, 2000.
49. Acero, J.L., Stemmler, K., and von Gunten, U., Degradation kinetics of atrazine and its degradation products with ozone and OH radicals: a predictive tool for drinking water treatment, *Environ. Sci. Technol.*, 34, 591–597, 2000.
50. Andreozzi, R. et al., Advanced oxidation processes for the treatment of mineral oil-contaminated wastewaters, *Water Res.*, 34, 620–628, 2000.
51. Rivas, F.J. et al., Two step wastewater treatment: sequential ozonation-aerobic biodegradation, *Ozone Sci. Eng.*, 22, 617–686, 2000.
52. Yavich, A.A., and Masten, S.J., Modeling the kinetics of the reaction of ozone with natural organic matter in Huron River water, *Ozone Sci. Eng.*, 23, 105–119, 2001.
53. Acero, J.L. et al., MTBE oxidation by conventional ozonation and the combination ozone/hydrogen peroxide: efficiency of the processes and bromate formation, *Environ. Sci. Technol.*, 35, 4252–4259, 2001.
54. El-Din, M.G. and Smith, D.W., Maximizing the enhanced oxidation of Kraft pulp mill effluents in an impinging-jet bubble column, *Ozone Sci. Eng.*, 23, 479–493, 2001.
55. Bin, A.K., Duczmal, B., and Machniewski, P., Hydrodynamics and ozone mass transfer in a tall bubble column, *Chem. Eng. Sci.*, 56, 6233–6240, 2001.
56. Beltrán, F.J., García-Araya, J.F., and Álvarez, P., Domestic wastewater ozonation: a kinetic model approach, *Ozone Sci. Eng.*, 23, 219–228, 2001.
57. Tanaka, J. and Matsumura, M., Kinetic studies of removal of ammonia from seawater by ozonation, *J. Chem. Technol. Biotechnol.*, 77, 649–656, 2002.
58. Smith, D.W. and El-Din, M.G., Theoretical analysis and experimental verification of ozone mass transfer in bubble columns, *Environ. Technol.*, 23, 135–147, 2002.
59. Kamimura, M., Furukawa, S., and Hirotsuji, J., Development of a simulator for ozone/UV reactor based on CFD analysis, *Water Sci. Technol.*, 46, 13–19, 2002.
60. Beltrán, F.J. et al., An attempt to model the ozonation of simazine in water, *Ind. Eng. Chem. Res.*, 41, 1723–1732, 2002.

61. Beltrán, F.J. et al., Use of the axial dispersion model to describe the O_3 and O_3/H_2O_2 advanced oxidation of alachlor in water, *J. Chem. Technol. Biotechnol.*, 77, 584–592, 2002.
62. Sung, M. and Huang, C.P., In situ removal of 2-chlorophenol from unsaturated soils by ozonation, *Environ. Sci. Technol.*, 36, 2911–2918, 2002.
63. Xu, P. et al., Wastewater disinfection by ozone: main parameters for process design, *Water Res.*, 36, 1043–1055, 2002.
64. Kim, J.H. et al., Inactivation of *Crystosporidium* oocysts in a pilot-scale ozone bubble-diffuser contactor. I. Model development, *J. Environ. Eng. ASCE*, 128, 514–521, 2002.
65. Kim, J.H. et al., Inactivation of *Crystosporidium* oocysts in a pilot-scale ozone bubble-diffuser contactor. II. Model validation and application, *J. Environ. Eng. ASCE*, 128, 522–532, 2002.
66. Farines, V. et al., Ozone transfer from gas to water in a co-current upflow packed bed reactor containing silica gel, *Chem. Eng. J.*, 91, 67–73, 2003.
67. Beltrán, F.J., Encinar, J.M., and García-Araya, J.F., Modelling industrial wastewater ozonation in bubble contactors. 1. Rate coefficient determination, *Ozone Sci. Eng.*, 17, 355–378, 1995.
68. Staehelin, S. and Hoigné, J., Decomposition of ozone in water the presence of organic solutes acting as promoters and inhibitors of radical chain reactions, *Environ. Sci. Technol.*, 19, 1206–1212, 1985.
69. Bailey, P.S., The reactions of ozone with organic compounds, *Chem. Rev.*, 58, 925–1010, 1958.
70. Beltrán, F.J. et al., Effects of single and combined ozonation with hydrogen peroxide or UV radiation on the chemical degradation and biodegradability of debittering table olive industrial wastewaters, *Water Res.*, 33, 723–732, 1999.
71. Beltrán, F.J., Encinar, J.M., and González, J.F., Industrial wastewater advanced oxidation. Part 2. Ozone combined with hydrogen peroxide or UV radiation, *Water Res.*, 31, 2415–2428, 1997.

Appendices

APPENDIX A1

IDEAL REACTOR TYPES: DESIGN EQUATIONS

Reactors are the vessels in which chemical reactions are carried out. For design purposes, reactors have been classified as ideal reactors based on how they function according to certain hypotheses regarding the level of mixture of reactants and type of flow through these reactors in the different phases (gases, liquids, or even solids). These hypotheses are fundamental for establishing the design equations that allow the size of reactors (i.e., the volume) to be determined from experimental results. The design equations are, in fact, the mass balances of species (reactants or reaction products) present in the phases circulating through the reactor (and the balance of energy, if needed). As a general rule, however, the design equation is the mass balance of the limiting reactant species. There are basically two ideal reactor types: the perfectly mixed reactor and the plug flow reactor. Both work continuously; that is, reactants and products are continuously fed to and withdrawn from reactors, respectively. The hypotheses concerning the mixture and flow of phases for these reactors are given below.

A1.1 Perfectly Mixed Reactor

These reactors are also called continuous stirred tank reactors (CSTR). They are mechanically agitated tanks (although small bubbles tanks or columns can belong to this reactor type) where the following hypotheses hold:

- The reactant mixture is complete.
- Concentrations of species (and temperature) in the phases flowing through the reactors are uniform and do not depend on position. At steady-state conditions, these intensive properties also do not depend on time.
- Concentrations of species (and temperature) of the effluent streams from the reactor are equal to those inside the reactor.

According to these hypotheses, for a continuous perfectly mixed reactor of volume V, the design equation or mass balance of a reacting species, A, in that volume is as follows:

$$F_{A0} - F_A + G_A V = \frac{dn_A}{dt} \quad (A1.1)$$

where (see Figure A1.1) F_{A0} and F_A are the molar rates of A at the reactor inlet and outlet, respectively, G_A is the generation rate term of A (i.e., the chemical reaction

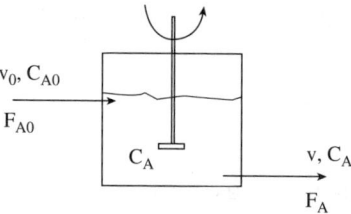

FIGURE A1.1 The continuously stirred tank reactor.

rate when A undergoes only homogeneous reactions inside the reactor), and n_A is the number of moles of A in the reactor with dn_A/dt the accumulation rate of A in the volume V of the reactor. In terms of concentrations, Equation (A1.1) becomes:

$$v_0 C_{A0} - v C_A + G_A V = \frac{d(VC_A)}{dt} \quad (A1.2)$$

where v_0 and v are the volumetric flow rates of inlet and outlet streams, respectively, that contain species A. For constant volume systems, for ozone reactions in water or wastewater, and once the hydraulic residence time is accounted for, Equation (A1.2) becomes

$$C_{A0} - C_A + \tau G_A = \tau \frac{dC_A}{dt} \quad (A1.3)$$

where τ, the hydraulic residence time, is defined as:

$$\tau = \frac{V}{v_0} \quad (A1.4)$$

In most practical situations, CSTRs work at steady state so that Equation (A1.3) reduces to Equation (A1.5):

$$C_{A0} - C_A + \tau G_A = 0 \quad (A1.5)$$

In some cases, mainly at laboratory scale, there are no inlet and outlet streams but the reacting species A is initially charged to the reactor. In these cases, the reactor becomes a batch reactor that is usually applied to kinetic studies (determination of rate constants of reactions, mass-transfer parameters if needed, etc.). Batch reactors work at nonsteady-state conditions so that Equation (A1.3) is reduced to:

$$G_A = \frac{dC_A}{dt} \quad (A1.6)$$

This equation is solved with the initial condition:

$$t = 0 \quad C_A = C_{A_0} \quad (A1.7)$$

Appendices

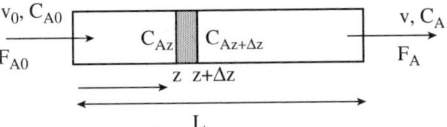

FIGURE A1.2 The plug flow reactor.

A1.2 Plug Flow Reactor

These are tubular reactors (i.e., bubble columns of large size in gas–liquid reactions) where the following hypotheses hold:

- There is no axial or radial mixture of fluid elements at any point inside the reactor.
- The concentration of species (and temperature for nonisothermal reactors) varies along the longitudinal axis of the reactor.
- Concentration (and temperature for nonisothermal reactors) then varies with position in the reactor.

As a result of these hypotheses, the reactor design equation is built from a mass balance of the reacting species A across a differential volume of reactor as shown in Figure A1.2. Thus, Equation (A1.8) is the starting design equation:

$$F_A\big|_V - F_A\big|_{V+\Delta V} + G_A \Delta V = \frac{\Delta n_A}{\Delta t} \qquad (A1.8)$$

where the terms of Equation (A1.8) have the same meaning as those in Equation (A1.1). Tubular reactors generally work at steady-state conditions so that the accumulation rate term (the right-side term) becomes zero. Dividing the resulting equation by ΔV, in the limit when this volume increment is zero, the following is obtained:

$$\lim \left(\frac{F_A\big|_V - F_A\big|_{V+\Delta V}}{\Delta V} \right)_{\Delta V \to 0} = -\frac{dF_A}{dV} = -G_A \qquad (A1.9)$$

If Equation (A1.9) is expressed as a function of concentration, it becomes:

$$\frac{dC_A}{d\tau} = G_A \qquad (A1.10)$$

As can be observed, Equation (A1.10) is mathematically similar to Equation (A1.6), which corresponds to a batch reactor, with the difference of time. For plug flow reactors the hydraulic residence time is used, while in batch reactors the real time is applied. Equation (A1.10) is commonly expressed as a function of the position within the tube (column), z. Equation (A1.10) then becomes:

$$\frac{dC_A}{dz} = \frac{S}{v_0} G_A = \frac{G_A}{U} \qquad (A1.11)$$

where S is the sectional area of the column and U the lineal velocity of the phase through the column. Equation (A1.10) or (A1.11) can be solved with the initial condition:

$$\tau = 0 \quad C_A = C_{A_0} \qquad (A1.12)$$

or

$$z = 0 \quad C_A = C_{A_0} \qquad (A1.13)$$

Note that the generation term, G_A, in the design equations takes different mathematical forms depending on the nature of the chemical reactions (homogeneous or heterogeneous), regardless of reactor type. For a homogeneous reaction, G_A is the chemical reaction rate term if a liquid phase is considered, while for a heterogeneous reaction the reaction type (gas–liquid, liquid–solid catalytic, etc.) and kinetic regime (mass-transfer control, chemical reaction control, etc.) are necessary to establish the correct mathematical expression of G_A. Details of expressions for the generation term can be found in specialized books.[1-3] In this book, expressions for G_A corresponding to gas–liquid and gas–liquid–solid catalytic reactions are also given in Chapter 4 and Chapter 10.

APPENDIX A2

Useful Mathematical Functions

As observed in this book, rate equations for heterogeneous reactions involve some mathematical functions such as hyperbolic and error functions. Here, only the definitions of these functions are given. More information can be obtained elsewhere.[4,5]

A2.1 Hyperbolic Functions

Definitions of hyperbolic functions are as follows:

- Hyperbolic sine of x:

$$\sinh x = \frac{\exp(x) - \exp(-x)}{2} \qquad (A2.1)$$

- Hyperbolic cosine of x:

$$\cosh x = \frac{\exp(x) + \exp(-x)}{2} \qquad (A2.2)$$

Appendices

- Hyperbolic tangent of x:

$$\tanh x = \frac{\sinh x}{\cosh x} = \frac{\exp(x) - \exp(-x)}{\exp(x) + \exp(-x)} \quad \text{(A2.3)}$$

As in Equation (A2.3) the remaining hyperbolic functions are defined as the corresponding trigonometric ones.

A2.2 The Error Function

Although the error function is not used as often as the hyperbolic function, it also appears in some rate expressions, mainly when nonsteady-state gas–liquid absorption theories are used. This function is defined as follows:

$$\operatorname{erf} x = \frac{2}{\sqrt{\pi}} \int_0^x \exp(-t^2) dt \quad \text{(A2.4)}$$

The error function can also be expressed as a series function:

$$\operatorname{erf} x = \frac{2}{\sqrt{\pi}} \left[x - \frac{x^3}{3} + \frac{1}{2!} \frac{x^5}{5} - \frac{1}{3!} \frac{x^7}{7} + \ldots \right] \quad \text{(A2.5)}$$

The complementary error function is defined as follows:

$$\operatorname{erfc} x = 1 - \operatorname{erf} x = \frac{2}{\sqrt{\pi}} \int_x^\infty \exp(-t^2) dt \quad \text{(A2.6)}$$

APPENDIX A3

THE INFLUENCE OF THE TYPE OF FLOW ON REACTOR PERFORMANCE

In Appendix A1 reactor design equations correspond to the ideal flow behavior of phases through the reactor. In practice, however, the real situation may be something different. In ozonation systems, both the wastewater and gas phases are usually continuously fed to the reactor and the fluid flow could be nonideal. Therefore, in order to establish the appropriate molar balance equations, and hence the design equations, we must know the type of flow the gas and water or wastewater present through the reactor.

A3.1 Nonideal Flow Study

Molar balances of species can only be established if the type of flow through the reactors is known. Thus, only cases for perfect mixing and plug flow are valid for these purposes. Note, however, that molar balances for laminar flow and axial dispersion flow can also be established in some cases.[1] The most common procedure to simulate the real flow through a reactor is to consider it equivalent to that in a series of N perfectly mixed tanks of equal size or that it follows the axial dispersion model. Application of these nonideal flow models requires some special experiments to be conducted. These are called tracer experiments and they determine the residence time distribution function (RTD) of the phase through the reactor.

A3.1.1 Fundamentals of RTD Function

The RTD function, generally called the E function, determines the amount of time molecules of a fluid spend flowing through any reactor. In fact, it is said that the product Edt represents the fraction of fluid that has spent a time dt inside the reactor. According to this definition, E is a normalized function of time and the following condition holds:

$$\int_0^\infty Edt = 1 \qquad (A3.1)$$

For flow characterization in actual reactors, it is evident that the E function is a key parameter to know because the extension of the chemical reactions depends on the time the substances spent inside the reactor. The fluid flow through any reactor has a given E function. For example, ideal reactors have specific E functions that can be compared to those of real reactors. This comparison, theoretically, could allow the fluid flow through any real reactor to be simulated with that present in one or more ideal reactors. This is the technique applied in the N perfectly mixed tanks in series model. However, the indicated comparison is not made with the corresponding E functions but from the variances of both distribution functions. Once the fluid flow has been characterized, mass-balance equations can be established through the real reactor and the mathematical kinetic model can be solved.

A3.1.1.1 Determination of the E Function

The RTD or E function is determined from tracer experiments carried out in the reactor studied. A tracer experiment consists of feeding a fluid containing the tracer through the reactor at the flow rate that will be used later when the process involves chemical reactions. The tracer should fulfill the following conditions:

- It has to be inert, that is, it should not react inside the reactor or be adsorbed onto the reactor walls.
- The concentration must be easily determined.
- If possible, the nature of the tracer and reacting species should be similar.

There are two types of tracer experiments: the pulse input and step experiments.

1. *Pulse input experiments*

In these experiments, a solvent fluid is first passed through the reactor at a known flow rate (i.e., just water in ozone reactors). Then, at a given time (which will be time zero for the experiment), a small amount of tracer is injected in the fluid, just

Appendices

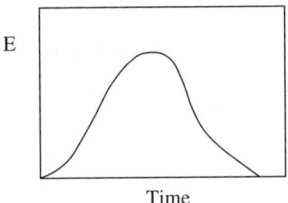

FIGURE A3.1 Typical form of the residence time distribution curve, E, against time from a pulse input tracer experiment.

at the reactor entrance. From this moment, the concentrations of the tracer are measured at the reactor outlet with time. The time–concentration profile of the tracer will allow the RTD function to be determined. In fact, it is easily shown that the E function can be expressed as follows:[1]

$$E = \frac{C_T}{\int_0^\infty C_T dt} \quad (A3.2)$$

where C_T is the tracer concentration at time t from the start of the experiment. Figure A3.1 represents a typical E function of a general reactor.

2. *Step experiments*

In these cases, the fluid flow of an inert solvent passing through the reactor is substituted, at a given time or time zero, for the same fluid containing a known concentration of the tracer, C_{T0}, at the same flow rate. From that time, the concentration of the tracer is also determined at the reactor outlet. The dimensionless tracer concentration, C_T/C_{T0}, is called the F function and it represents the fraction of fluid that has spent a time t inside the reactor. It is also shown that both E and F functions are related as:[1]

$$F = \int_0^t E dt \quad (A3.3)$$

From Equation (A3.3) E can be obtained from the first derivative of the F function with time. Figure A3.2 presents a typical F distribution curve.

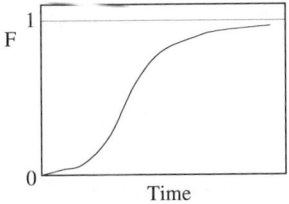

FIGURE A3.2 Typical form of the F curve against time from a step input tracer experiment.

A3.1.1.2 Moments of the RTD

Any distribution function, such as E in this case, has some parameters that allow its characterization. In fact E functions of different reactors are compared using these parameters, called the moments of the distribution. The main moments are the mean residence time, t_m, and the variance, σ^2:

$$t_m = \int_0^\infty tE\,dt \tag{A3.4}$$

$$\sigma^2 = \int_0^\infty (t_m - t)^2 E\,dt \tag{A3.5}$$

The variance represents the spread of the distribution, and it is the key moment to know in order to apply some fluid flow models such as the axial dispersion and N tanks in series models.

A3.1.2 RTD Functions of Ideal Flows through the Reactors

The ideal reactors or, more correctly, the ideal flows through reactors correspond to the situations called perfect mixing and plug flow. The E function of ideal fluid flows can be obtained from mass balances of the tracer in the corresponding reactors at nonsteady-state conditions. These functions are the following:

- For plug flow:

$$E = \delta(t - t_m) \tag{A3.6}$$

- For perfect mixing:

$$E = \frac{1}{t_m}\exp\left(-\frac{t}{t_m}\right) \tag{A3.7}$$

Note that $\delta(t_m - t)$ is called the Delta Dirac function. According to the properties of this function, $E \to \infty$ when $t = t_m$ and $E = 0$ for $t \neq t_m$. If we take into account the hypothesis of plug flow, the Delta Dirac function results in the correct way to express the RTD function or E in this kind of reactor. It can be shown that the mean residence time of the E functions for ideal flows coincides with the hydraulic residence time or spatial time:[2]

$$t_m = \tau = \frac{V}{v_0} \tag{A3.8}$$

Figure A3.3 shows the residence time distribution function of a plug flow and perfectly mixed reactors, respectively, with a hydraulic residence time of 3.7 min.[6]

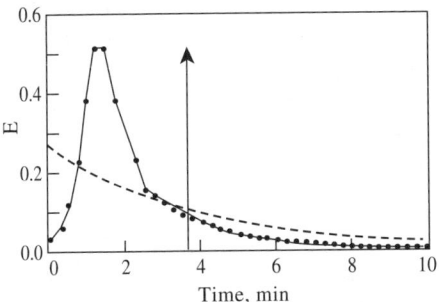

FIGURE A3.3 Comparison between actual residence time distribution function (RTD) determined with ozone as gas tracer and those corresponding to ideal patterns of perfect mixing (PM) and plug flow (PF) at the same experimental conditions of 20°C, pH 4, initial ozone gas concentration: 8.2 mgL^{-1}. Gas flow rate: 40 Lh^{-1}; water flow rate: 0.25 Lmin^{-1}. Flow patterns: ● = actual RTD (actual mean residence time: 2.3 min); dotted curve = perfectly mixed; arrow = plug flow (assumed mean residence time = hydraulic residence time = 3.7 min). (Reprinted with permission from Beltrán, F.J. et al., The use of ozone as a gas tracer for kinetic modeling of aqueous environmental processes, *J. Env. Sci. Health*, 135A, 681–699, 2000, p. 708. Copyright 2000, Marcel Dekker Inc.)

For fluid flow in a given reactor, deviations of E functions from those of equations (A3.6) and (A3.7) are due to phenomena such as bypass, dead zones in the volume of reaction, axial dispersion in the flow, etc.[1,2] In these cases, a fluid flow model is assumed to characterize the real one. This fluid flow model is usually the same as the model based on a combination of ideal flows. For example, the flow could be simulated as the same in a series of an agitated tank with perfect mixing plus a tubular reactor with plug flow or in a series of N equally sized agitated tanks with perfect mixing. Another possibility is the assumption of some type of dispersion if the fluid flows through tubular reactors. For more detailed explanations of this matter, the reader should consult other specialized works.[7] Once the fluid model is known, the mass-balance equations can be established for the real reactor.

As can be deduced from the above comments, both the real fluid flow and assumed fluid flow model must have the same E function. Because comparing E functions is a difficult task, their moments are used for that purpose.

A3.2 Some Fluid Flow Models

There are numerous fluid flow models that can be assumed. Only two of these models are presented here and used for the ozonation kinetic modeling (see also Chapter 11): the perfectly mixed tanks in series model and the axial dispersion model. The first type is useful for any kind of reactor while the second one is preferable for tubular reactors (i.e., bubble columns). Nonetheless, both models represent an intermediate situation between the two basic ideal fluid flow models: the perfect mixing and plug flow models.

A3.2.1 The Perfectly Mixed Tanks in Series Model

The fluid flow in the real reactor would be that observed in a series of perfectly mixed agitated tanks of equal size. The key parameter to be determined is the number of tanks. Another important requirement is that the tanks be the same size, and have the same mean residence time or hydraulic time:

$$\tau_i = \frac{\tau}{N} \qquad (A3.9)$$

where τ and τ_i are the residence times in the real reactor and in any one of the N reactors in series.

From mass balances of a tracer flowing through the assumed series of tanks, the E function corresponding to this model can be obtained:[1,2]

$$E = \frac{t^{N-1}}{(N-1)!\,\tau_i^N} \exp\left(-\frac{t}{\tau_i}\right) \qquad (A3.10)$$

From the definitions of mean residence time and variance [Equation (A3.4) and Equation (A3.5)], it is easily shown that the number of tanks, N, is the inverse of the dimensionless variance of the distribution:[1,2]

$$N = \frac{1}{\sigma_\theta^2} = \frac{t_m^2}{\sigma^2} \qquad (A3.11)$$

Once the number of tanks, N, has been determined from the E function obtained for the fluid flow through the real reactor, this reactor can be simulated with a series of N perfectly mixed tanks of equal size. Note that when N is one or infinite this model becomes an ideal flow in one reactor with perfect mixing or a plug flow, respectively. Finally, for kinetic modeling purposes related to the substances present in the fluid, the molar balances corresponding to N perfectly mixed tanks are used.

A3.2.2 The Axial Dispersion Model

This fluid flow model is more appropriate for tubular reactors. The assumption of the flow in this case is that the transport of mass is due to both convection and axial diffusion. Then, the mass flow rate for species A is

$$F_A = v_0 C_A - D_A \frac{S}{U} \frac{\partial C_A}{\partial t} \qquad (A3.12)$$

where D_A is the dispersion coefficient and U is the lineal velocity through the column. The axial dispersion model as the plug flow model also considers the species concentration varies with position in the axial position in the reactor. The design equation

Appendices

comes from the application of Equation (A1.8). If Equation (A3.12) is accounted for, after a series of transformation steps, the final design equation for the mass flow rate of A becomes:

$$-\frac{\partial C_i}{\partial \tau} + \frac{D_i}{U^2}\frac{\partial^2 C_i}{\partial \tau^2} + G_i = \frac{\partial C_i}{\partial t} \quad (A3.13)$$

or as a function of the contactor height, z:

$$-U\frac{\partial C_i}{\partial z} + D_i\frac{\partial^2 C_i}{\partial z^2} + G_i = \frac{\partial C_i}{\partial t} \quad (A3.14)$$

At steady state, in dimensionless form and considering a first-order reaction develops in the reactor ($G_A = -kC_A$), Equation (A3.14) is:

$$-\frac{d\psi_A}{d\lambda} + \frac{D_A}{UH}\frac{d^2\psi_A}{d\lambda^2} + \psi_A Dam = 0 \quad (A3.15)$$

where H is the length of the tube and the dimensionless variables are defined as follows:

$$\psi_A = \frac{C_A}{C_{A0}} \quad (A3.16)$$

$$\lambda = \frac{z}{H} \quad (A3.17)$$

$$Dam = k\tau \quad (A3.18)$$

Dam is also called the Damkohler number for a first-order reaction.

In Equation (A3.14) the key parameter is the dispersion number, $D/(UH)$, which represents the degree of axial dispersion. The higher this number, the closer the fluid flow to the perfectly mixed flow. The lower the dispersion number, the closer the fluid flow to the plug flow. Thus, the axial dispersion model also simulates an intermediate fluid flow between the two ideal ones.

The dispersion number can be obtained from different empirical correlations as a function of the Reynolds and Schmidt numbers[2] but it can also be determined from the E function and more specifically from the dimensionless variance. Equations that relate E with the dimensionless variance can be found in specialized texts.[1,2] If the dispersion number is known, the mass balances of substances present in the flowing phase are established by using Equation (A3.14) or Equation (A3.15).

A3.3 Ozone Gas as a Tracer

Tracer species for studies of fluid flow models are usually dyes and salts for the water phase and noble gases for the gas phase.[8] However, ozone can also be used to characterize the gas phase flow through reactors. Thus, when the E function of a gas phase flow (oxygen or air, for example) at empty column or reactor conditions is the objective, ozone fulfills all the conditions for a good tracer. It is easy to measure and it is the gas later used in ozonation processes. However, the ozone tracer experiment implies that both the water and gas phases simultaneously circulate through the reactor. When a water phase is also circulating through the reactor, one possible problem is that ozone is slightly absorbed. Thus, the measured ozone concentration at the reactor outlet would be due not only to the time it spends in the reactor but also to the ozone fraction absorbed in the water phase. This problem can be minimized if the water is free of reacting substances (for instance, it has been presaturated with ozone). Then ozone could be used in step experiments and its concentration at the reactor outlet would represent the tracer concentration. These types of experiments have been carried out in a previous work to characterize the gas fluid flow through bubble columns used in ozonation processes.[6] For example, Figure A3.3 shows the E function obtained in one step experiment using ozone as tracer corresponding to a gas flow (oxygen) through a bubble column with water circulating countercurrently to the gas. Also, Figure A3.3 shows for comparative purposes the E curves corresponding to the ideal flows of perfect mixing and plug flow at the same gas flow rate. For this case, the moments of E were as follows: $t_m = 2.3$ min, $\sigma^2 = 3.8$ min^2 with an actual gas and water flow rates of 40 Lh^{-1} and 0.125 Lmin^{-1}, respectively (spatial time for the gas phase: 3.7 min[6]). Parameter values corresponding to the N tanks in series and the axial dispersion models, N and the dispersion number, were found to be 1.3 and 1.11, respectively, which means that the gas phase flow could be simulated approximately to that in a perfectly mixed reactor with the same spatial time. Then the gas phase in the bubble column was perfectly mixed and Equation (A1.3) can be used to establish the mass balance of ozone in the gas phase during ozonation reactions.

APPENDIX A4

ACTINOMETRY

The intensity of the incident radiation, I_0, and the effective pathway of radiation L through a photoreactor are necessary parameters for studying the photolytic decomposition of any substance with the Lambert law model. These parameters depend on the radiation characteristics and geometry of the photoreactor. They have to be determined for each specific photoreactor and radiation source or lamp. The procedure is based on the photolytic decomposition kinetics of a substance called the actinometer. Actinometers are characterized by known quantum yields and molar absorptivity at different wavelengths of radiation. Furthermore, the quantum yield of an actinometer is usually constant in a wide range of wavelengths and higher than 0.1. In Table A4.1 some actinometer compounds, quantum yield values, and

Appendices

TABLE A4.1
Actinometer Substances and Quantum Yields[a]

Actinometer	Quantum yield (mol Einstein^{-1}) and wavelength range (nm) of application
Uranyl oxalate	$\lambda = 254–436$, $\Phi = 0.58–0.49$
Potasium ferroxyoxalate	$\lambda = 250–436$, $\Phi = 1.24$
o-Nitrobenzaldehyde	$\lambda = 254–300$, $\Phi = 0.5$
Chloroacetic acid	$\lambda = 254$, $\Phi = 0.31$
Decafluorobezophenone (in 2-propanol)	$\lambda = 290–370$, $\Phi = 0.6$
Reinecke salt	$\lambda = 316–750$, $\Phi = 0.27–0.32$
Benzophenone	$\lambda = 366$, $\Phi = 0.69$
trans-Azobenzene (in methanol)	$\lambda = 254–365$, $\Phi = 0.12$
Aberchrome 540	$\lambda = 310–370$, $\Phi = 0.2$
	$\lambda = 436–546$, $\Phi = 0.076–0.047$
Aberchrome 999P	$\lambda = 500$, $\Phi = 0.026$
	$\lambda = 540$, $\Phi = 0.058$
cis-2-Hexanone	$\lambda = 313$, $\Phi = 0.327$
Hydrogen peroxide[b]	$\lambda = 254$, $\Phi = 0.05$ (primary quantum yield)
	$\lambda = 254$, $\Phi = 1.0$
trans-Benzophenone/1,3-pentadiene	$\lambda = 366$, $\Phi = 0.44$
cis-Benzophenone/1,3-pentadiene	$\lambda = 366$, $\Phi = 0.55$

[a] From Guittonneau[9] unless otherwise indicated. [b] From Nicole.[10]

wavelength range of application are given. For 254 nm radiation, uranile oxalate is very often used as an actinometer but hydrogen peroxide is also used, especially for aqueous systems that are going to be used in UV/H$_2$O$_2$ or UV/O$_3$ advanced oxidations.[10,11]

From separate experiments of photolytic decomposition of actinometers, I_o and L can be determined, as explained below.

A4.1 Determination of Intensity of Incident Radiation

The intensity of incident radiation, I_0, can be determined from the photolytic decomposition of an actinometer substance B present in high concentration. Thus, in a batch photoreactor, from Equation (9.23), with $F_B = 1$ and $\mu L > 2$, the disappearance rate of the actinometer simplifies to yield:

$$-\frac{dC_B}{dt} = I_0 \Phi_B \tag{A4.1}$$

At these conditions, the photolytic decomposition follows zero-order kinetics for the actinometer, so that the plot of the actinometer concentration, C_B, with time should lead to a straight line of slope $I_0 \Phi_B$. Since the quantum yield is known, I_0 can be calculated. The literature reports many examples where this procedure has been

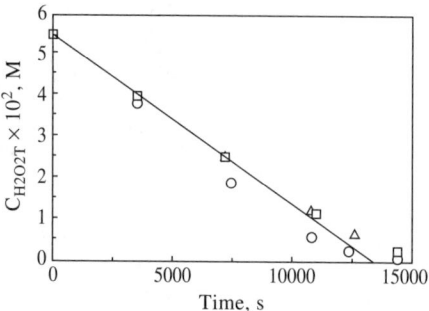

FIGURE A4.1 Determination of the intensity of incident radiation. Verification of zero-order kinetic Equation (A4.1). Conditions: pH 7, 20°C, initial concentration of hydrogen peroxide = 0.055 M. (Reprinted with permission from Beltrán, F.J. et al., Oxidation of polynuclear aromatic hydrocarbons in water. 2. UV radiation and ozonation in the presence of UV radiation, *Ind. Eng. Chem. Res.*, 34, 1607–1615, 1995. Copyright 1995, American Chemical Society.)

applied.[12,13] Figure A4.1 shows an example of this procedure with the use of hydrogen peroxide as actinometer.[12] Recall that for the determination of I_0, Φ_B is the total quantum yield of hydrogen peroxide (see also Table A4.1 and Section 9.2).

A4.2 Determination of the Effective Path of Radiation

The effective path of radiation [L in Equation (9.23)] can also be determined from the photolytic decomposition of an actinometer substance, provided I_0 is already known. In this case, the concentration of the actinometer should be low enough so that the exponential term in Equation (9.23), μL, < 0.4. In this case, the kinetics of photolytic decomposition is first order with respect to the actinometer, and the overall rate Equation (9.23) reduces to

$$-\frac{dC_B}{dt} = \mu L I_0 F_B \Phi_B = 2.303 I_0 \varepsilon_B L \Phi_B C_B \qquad (A4.2)$$

This simplification is based on the fact that the exponential term of Equation (9.23) can be expressed as the sum of an infinite series of terms:

$$\exp(-\mu L) = 1 - \frac{\mu L}{1!} + \frac{(\mu L)^2}{2!} - \frac{(\mu L)^3}{3!} + \ldots \qquad (A4.3)$$

For values of μL < 0.4 the series can be reduced to the first two members so that after substitution in Equation (9.23) the final kinetic equation is (A4.2). According to the integrated expression of Equation (A4.2) a plot of the logarithm of the concentration of the actinometer, $\ln C_B$, with reaction time will yield a straight line of slope $2.303 I_0 \varepsilon_B \Phi_B L$. From the value of this slope determined from least squares analysis of the experimental data, the effective path of radiation through the photoreactor can be obtained. Figure A4.2 shows the plot with hydrogen peroxide also used as an actinometer.[12]

Appendices

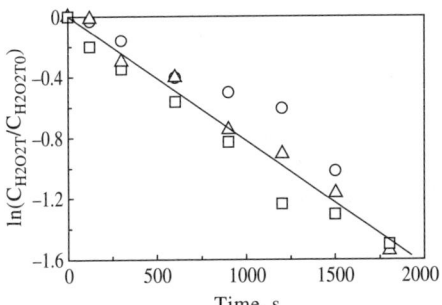

FIGURE A4.2 Determination of the effective path of radiation. Verification of first-order kinetic Equation (A4.2). Conditions pH 7, 20°C, initial concentration of hydrogen peroxide = 10^{-4} M. (Reprinted with permission from Beltrán, F.J. et al., Oxidation of polynuclear aromatic hydrocarbons in water. 2. UV radiation and ozonation in the presence of UV radiation, *Ind. Eng. Chem. Res.*, 34, 1607–1615, 1995. Copyright 1995, American Chemical Society.)

APPENDIX A5

SOME USEFUL NUMERICAL PROCEDURES

Kinetic modeling of reactors commonly involves the solution of simultaneous algebraic and differential nonlinear equations. For example, nonlinear algebraic equations are encountered to solve the kinetic model of ozonation processes in CSTRs while a system of differential equations of different order is found in the kinetic model of ozonation processes in tubular plug flow reactors or batch reactors. For the first problem, the extended Newton–Raphson method appropriate to solve roots of nonlinear algebraic equations is a powerful technique commonly used. For systems of nonlinear first-order differential equations, the Runge–Kutta methods are also applied. In view of their application to solve kinetic models of ozonation processes, a brief explanation of these methods is presented below. Again, detailed procedures and complementary variations of these methods can be found in specialized books.[14,15]

A5.1 The Newton–Raphson Method for a Set of Nonlinear Algebraic Equations

The method is based on the procedure of the same name applied to solve the roots of a nonlinear algebraic equation of the form $y = f(x)$. This equation can have multiple roots, x_i, such as $f(x_i) = 0$. The method starts by linearizing the problem function as a Taylor series function as follows:

$$f(x) = f(x_1) + f'(x_1)(x - x_1) + \frac{f''(x_1)(x - x_1)^2}{2!} + \ldots \qquad (A5.1)$$

The function is then truncated beginning with the second right-hand-side term:

$$x = x_1 - \frac{f(x_1)}{f'(x_1)} \tag{A5.2}$$

The process starts by assuming a value x_1 for the root x. If $f(x_1) = 0$, the problem is solved; if not, a new estimated value of x_1 is needed. Then the process is an iterative method with the general Equation (A5.3):

$$x_{i+1} = x_i - \frac{f(x_i)}{f'(x_i)} \tag{A.5.3}$$

until convergence, $f(x_{i+1}) = 0$, is achieved.

When instead of one single equation the mathematical system consists of a set of n nonlinear algebraic equations

$$f_1(x_1 \ldots x_n) = 0$$

$$\ldots\ldots\ldots\ldots \tag{A5.4}$$

$$f_n(x_1 \ldots x_n) = 0$$

the Newton–Raphson method can also be applied to transform Equations (A5.4) in a set of n linear algebraic equations. The steps to follow are as in the simple method with just one equation [$f(x)$]. First, functions are linearized as a Taylor series:

$$f_1(x_1 \ldots x_n) = f_1(x_{1_1} \ldots x_{n_1}) + \left.\frac{\partial f_1}{\partial x_1}\right|_1 (x_1 - x_{1_1}) + \ldots + \left.\frac{\partial f_1}{\partial x_n}\right|_1 (x_n - x_{n_1}) + \ldots$$

$$\ldots\ldots\ldots\ldots \tag{A5.5}$$

$$f_n(x_1 \ldots x_n) = f_n(x_{1_1} \ldots x_{n_1}) + \left.\frac{\partial f_n}{\partial x_1}\right|_1 (x_1 - x_{1_1}) + \ldots + \left.\frac{\partial f_n}{\partial x_n}\right|_1 (x_n - x_{n_1}) + \ldots$$

where $x_{11}..x_{n1}$ are initial estimated values with the second subindex representing the iteration number. If the Taylor series is truncated in the second derivatives and the left side of Equations (A5.5) are set to zero, the following equations are obtained:

$$\left.\frac{\partial f_1}{\partial x_1}\right|_1 \Delta_{1_1} + \ldots + \left.\frac{\partial f_1}{\partial x_n}\right|_1 \Delta_{n_1} = -f_1(x_{1_1} \ldots x_{n_1})$$

$$\ldots\ldots\ldots\ldots \tag{A5.6}$$

$$\left.\frac{\partial f_n}{\partial x_1}\right|_1 \Delta_{1_1} + \ldots + \left.\frac{\partial f_n}{\partial x_n}\right|_1 \Delta_{n_1} = -f_n(x_{1_1} \ldots x_{n_1})$$

Appendices

where $\Delta_{1_1}...\Delta_{n_1}$ represent the variable increment factors that relate the unknown variables to the estimated values:

$$\Delta_{1_1} = x_1 - x_{1_1}$$

$$\ldots\ldots\ldots \quad (A5.7)$$

$$\Delta_{n_1} = x_n - x_{n_1}$$

Equation (A5.6) constitute a system of n linear algebraic equations with n unknowns, the increment factors, with the first derivatives their corresponding known coefficients and the values of the right side the independent coefficients:

$$d_{11}\Delta_{1_1} + d_{12}\Delta_{2_1} + \ldots + d_{1n}\Delta_{n_1} = c_{1_1}$$
$$d_{21}\Delta_{1_1} + d_{22}\Delta_{2_1} + \ldots + d_{2n}\Delta_{n_1} = c_{2_1}$$
$$\ldots\ldots\ldots \quad (A5.8)$$
$$d_{n1}\Delta_{1_1} + d_{n2}\Delta_{2_1} + \ldots + d_{nn}\Delta_{n_1} = c_{n_1}$$

For low values of n (2 or 3) unknowns, Δ_{i1} can be determined with Cramér's rule.[14] For this rule, the system of equations can be expressed in matrix form as follows:

$$\begin{bmatrix} d_{11} d_{12} \ldots d_{1n} \\ d_{21} d_{22} \ldots d_{2n} \\ \ldots\ldots\ldots\ldots \\ d_{n1} d_{n2} \ldots d_{nn} \end{bmatrix} \begin{bmatrix} \Delta_{1_1} \\ \Delta_{2_1} \\ \ldots \\ \Delta_{n_1} \end{bmatrix} = \begin{bmatrix} c_{1_1} \\ c_{2_1} \\ \ldots \\ c_{n_1} \end{bmatrix} \quad (A5.9)$$

With this method for the j unknown, Δ_{j1}, the solution is the ratio between the determinants of the matrix of coefficients [the first matrix on the left side of Equation (A5.8)] and that resulting after substitution of d_{1j} to d_{nj} coefficients in this matrix by the independent coefficients, c_1 to c_n:

$$|D_j| = \begin{bmatrix} d_{11} d_{12} \ldots c_{1_1} \ldots d_{1n} \\ d_{21} d_{22} \ldots c_{2_1} \ldots d_{2n} \\ \ldots\ldots\ldots\ldots\ldots \\ d_{n1} d_{n2} \ldots c_{n_1} \ldots d_{nn} \end{bmatrix} \quad (A5.10)$$

thus:

$$\Delta_{j_1} = \frac{|D_j|}{|D|} \qquad (A5.11)$$

However, Cramér's rule is a tedious method when n is greater than just 3. For these cases the Gauss methods (Gauss elimination or Gauss–Jordan reduction methods) are recommended.[14,15]

Note that the increment factors Δ_{j_1} thus far calculated correspond to the first iteration to find the roots of $f(x_1,\ldots,x_n)$ functions. Thus, new estimated values of x_1 to x_n are calculated with Equations (A5.7) from the initial estimated values x_{j_1} and the calculated increment factors, Δ_{j_1}. These values are then used to check Equations (A5.4). If convergence is not achieved, then the new values are used as the estimate in the following iteration. Thus, the general formula is

$$x_{j_{n+1}} = x_{j_n} + \Delta_{j_n} \qquad (A5.12)$$

A5.2 The Runge–Kutta Method for a Set of Nonlinear First-Order Differential Equations

In kinetic modeling solving a system of nonlinear first-order differential equations is a common problem. The Runge–Kutta method is one of the most powerful tools for solving this kind of problem when the mathematical system presents an initial condition. For example, this condition can be used to determine the concentrations of species charged in a batch reactor at the start of the process. The Runge–Kutta method is classified as a function of its complexity and accuracy as second, third, fourth order, etc. For a system of n nonlinear first-order differential equations:

$$\frac{dy_1}{dx} = f_1(x, y_1, y_2, \ldots y_n)$$

....   (A5.13)

$$\frac{dy_n}{dx} = f_n(x, y_1, y_2, \ldots y_n)$$

The procedure starts by linearizing the solution of Equation (A5.13) as an infinite series function:

$$y_{j,i+1} = y_{j,i} + \Delta y'_{j,i} + \frac{\Delta y''_{j,i}}{2!} + \frac{\Delta y'''_{j,i}}{3!} + \ldots \qquad (A5.14)$$

where $y_{j,i+1}$ and $y_{j,i}$ are the values of the j variable at two consecutive values of the independent variable so that $x_{i+1} = x_i + \Delta$ and y'_j, y''_j and y'''_j refer to the first, second and third derivatives of y_j. Thus, $y'_j = f_j(x, y_1 \ldots y_n)$. The final solution is of the type:

$$y_{j,i+1} = y_{j,i} + \sum_{p=1}^{q} w_p k_{j,p} \quad (A5.15)$$

where q is the order of the derivative from which Equation (A5.14) is truncated. For the fourth-order Runge–Kutta method, $q = 4$, Equation (A5.14) has five members, with w_p and k_p as follows:[14]

$$w_1 = w_4 = \frac{1}{6}$$
$$w_2 = w_3 = \frac{1}{3} \quad (A5.16)$$

and

$$k_{j,1} = \Delta f_j \left(x_i, y_{1,i} \ldots y_{n,i} \right) \quad (A5.17)$$

$$k_{j,2} = \Delta f_j \left[x_i + \frac{\Delta}{2}, y_{1,i} + \frac{k_{11}}{2}, \ldots, y_{n,i} + \frac{k_{n1}}{2} \right] \quad (A5.18)$$

$$k_{j,3} = \Delta f_j \left[x_i + \frac{\Delta}{2}, y_{1,i} + \frac{k_{12}}{2}, \ldots, y_{n,i} + \frac{k_{n2}}{2} \right] \quad (A5.19)$$

$$k_{j,4} = \Delta f_j \left[x_i + \Delta, y_{1,i} + k_{13}, \ldots, y_{n,i} + k_{n3} \right] \quad (A5.20)$$

This method can easily be solved using a computer program. More details about the fundamentals of the Runge–Kutta method can be found elsewhere.[14,15]

References

1. Fogler, H.S., *Elements of Chemical Reaction Engineering*, 3rd ed., Prentice-Hall, Englewood Cliffs, NJ, 1999
2. Levenspiel, O., *Chemical Reaction Engineering*, 3rd ed., McGraw-Hill, New York, 1999.
3. Froment, G.F. and Bishoff, K.B., *Chemical Reactor Analysis and Design*, John Wiley & Sons, New York, 1979.
4. Beyer, W.H., *Standard Mathematical Tables*, 27th ed., CRC Press, Boca Raton, FL, 1981.
5. Spiegel, M.R., *Manual de Fórmulas y Tablas Matemáticas*, McGraw-Hill Latinoamericana, México D.F., 1981.
6. Beltrán, F.J. et al., The use of ozone as a gas tracer for kinetic modeling of aqueous environmental processes, *J. Env. Sci. Health*, 135A, 681–699, 2000.

7. Levenspiel, O. and Bischoff, K.B., Patterns of flow in chemical process vessels, *Adv. Chem. Eng. Series*, 4, 95, 1963.
8. Shah, Y.T., *Gas–Liquid–Solid Reactor Design*, McGraw-Hill, New York, 1979.
9. Guittonneau, S., Contribution á l'etude de la photooxydation de quelques micropolluants organichlorés en solution aqueuse en presence de peroxyde d'hydrogène — Comparaison des systèmes oxydants: H2O2/UV, O3/UV et O3/H2O2, Ph.D. thesis, University of Poitiers, France, 1989.
10. Nicole, I. et al., Utilization du rayonnement ultraviolet dans le treatement des eaux. Measure du flux photonique par actinometrie chimique au peroxyde d'hydrogene, *Water Res.*, 24, 157–168, 1990.
11. Leighton, W.G. and Forbes, G.S., Precision actinometry with uranyl oxalate, *J. Am. Chem. Soc.*, 52, 31–39, 1930.
12. Beltrán, F.J. et al., Oxidation of polynuclear aromatic hydrocarbons in water. 2. UV radiation and ozonation in the presence of UV radiation, *Ind. Eng. Chem. Res.*, 34, 1607–1615, 1995.
13. Beltrán, F.J. et al., Application of photochemical reactor models to UV radiation of trichloroethylene in water, *Chemosphere*, 31, 2873–2885, 1995.
14. Constantinides, A., *Applied Numerical Methods with Personal Computers*, McGraw-Hill, New York, 1987.
15. Carnahan, B., Luther, H.A., and Wilkes, J.O., *Applied Numerical Methods*, John Wiley & Sons, New York, 1969.

INDEX

A

Abnormal ozonolysis, 9
Absorptivity at 254 nm, 124
Acenaphthene
  ozonation kinetics, 101
  see also Table 5.6
Acenaphthylene
  UV photolysis and UV/H_2O_2 oxidation, see Table 9.2
Actinometry, 342
  examples of actinometers, 343
  parameter determination, 343, 344
Activated carbon as catalyst, 253, 257
Activity coefficients, 73
Advanced oxidation processes, 1, 14, 21, 151, 162, 175, 193
Agitated cells, 84
Agitated tanks, 77, 84, 302, 332
Alachlor ozonation kinetics, see Table 5.6
Aldicarb ozonation kinetics, 101
Alkalinity, 159
Ammonia, kinetic regime of ozonation, see Table 6.3
Anatase TiO_2, 206
Apparent Henry's law constant, 73, 74
Aquaculture wastewater ozonation, 113, see also Table 6.1
Aromatic compounds, 11
Atrazine
  as scavenger of hydroxyl radicals, 83, 103, 106
  oxidation by-products, see Table 9.2
  ozonation kinetics, 153, 185
  UV photolysis and UV/H_2O_2 oxidation kinetics, see Table 9.2
Attenuation coefficient, 195
Axial dispersion model, 303, 339, 340

B

Back flow cell model, 288
Band gap energy, 265
Beer–Lambert law, 194
Benzene
  as scavenger of hydroxyl radicals, 34, 40
  kinetic regimes of ozonation of, see Table 6.3

Benzoic acid
  kinetic regimes of ozonation of, see Table 6.3
Biodegradability
  biological oxidation, 2, 113
  effects of ozone on, 130
  in wastewater, 2, 123, 129
  measured as BOD/COD, 123
Biological oxygen demand (BOD), 123
  interference of nitrification in analysis of, 123
Boiled water in power plants
  ozone treatment, 113, see also Table 6.2
Bromamine
  reactions with ozone, 18
Bromate, 1, 18
Bromide, 1, 18
  reactions with ozone, 18
BTX, see Table 6.3
Bubble column, 77, 87
Bunsen coefficient, see Ozone solubility

C

Carbon adsorption, 2, 257
Carbonate ion radical, 18, 159
Carbonates
  as scavengers of hydroxyl radicals, 159, 164, 177, 180, 181
Carboxylic acids
  rate constants of ozone direct reactions, see also Table 5.6
Catalysts, 227
Catalytic ozonation, 227
  examples, 227, see also Tables 10.1 and 10.2
Cavitation processes, 175
Chemical absorption kinetics, 50
Chemical biological processes, 129
  chemical oxidation influence on, 129
Chemical oxygen demand (COD), 121, 122, 129
  effects on wastewater ozonation, 121
  interference of chloride ion on analysis of, 123
  interference of hydrogen peroxide on analysis of, 122
Chemical potential, 72
Chloramines, 1
Chlorination, 1
Chlorine, 1

Chloro-alkali wastewater ozonation, 113, *see also* Table 6.1
Chlorophenol
  kinetic regime of ozonation, *see* Table 6.3
  ozonation kinetics, 185, 186
$Co^{(2+)}$ catalysts, 227
Coagulation, 2, 113
Coke plant wastewater ozonation, *see* Table 6.1
Competition of direct and indirect ozone reactions, 103
Complementary error function, 335
Conduction band, *see* TiO_2 semiconductor photocatalysis
p-Cresol, kinetic regime of ozonation, *see* Table 6.3
Criegge mechanism, 9
Crotonic acid
  ozonation kinetics, 84, 99
Cyanide wastewater
  kinetic regime of ozonation, *see* Table 6.3
1,3-Cyclohexanedione ozonation kinetics, 92, 108

D

Danckwerts theory, 62
Deactivating groups, *see* Electrophilic substitution reactions
Debittering table olive wastewater
  kinetic parameters of UV/H_2O_2 oxidation, *see* Table 9.2
Degree of dissociation, 41, 185
Deisopropylatrazine, *see* Atrazine
Density flux of radiation, *see* Table 9.1
Dental surgery wastewater
  ozone treatment, *see* Table 6.2
Depletion factor, 284, 310
Dethylatrazine, *see* Atrazine
Dibromochloropropane, DBCP
  kinetic parameters of UV/H2O2 oxidation, *see* Table 9.2
Diffusion time, 67
Diffusional kinetic regime, 105
  determination of volumetric mass transfer coefficient, 105
Diffusivity, 69
  of compounds in water, 71
  of ozone in water, 69, 70
1,4-Dioxane contaminated water ozonation, 113, *see also* Table 6.3
Dipolar cycloaddition reactions, 9
Direct and decomposition reactions of ozone, 151
  competition between, 151
  diffusion and reaction times on, 152
  pH effects, 151
Direct photolysis, *see* UV radiation

Direct photolysis and hydroxyl radical reactions (from UV/H_2O_2) of compounds, 209
  comparison between, 209
  hydrogen peroxide concentration effect, 210
Direct reactions and direct photolysis of ozone, 211
  comparison between, 211
  kinetic regimes of, 211
  reaction and diffusion times, 211
  reaction rates of, 214
    effects of intensity of UV radiation on, 215, 216
Disinfection, 1, 21
  by-products of, 1, 2
Dissociating compounds, 41
Dissolved organic carbon, 123
Distillery wastewater
  kinetic regimes, *see* Table 6.3
  ozonation, *see* Table 6.2
Drinking water, 1, 151
Dyes
  in wastewater, *see* Table 6.2
  kinetic regimes of ozonation, *see* Table 6.3
  rate constant of ozone direct reactions, *see* Table 5.6

E

E function, *see* Residence time distribution function
EDTA, kinetic regime of ozonation, *see* Table 6.3
  kinetic regimes, *see* Table 6.3
  ozonation, *see* Table 6.1
Effective diffusivity, 247
  determination of, 264
Effective path of radiation, *see* Actinometry
Einstein unit definition, 195
Electric power plant wastewater ozonation, *see* Table 6.1
Electrophilic substitution reactions, 11
  activating groups, 12
  deactivating groups, 12
  mechanism for reactions in aromatic compounds, 13
Electroplating wastewater, *see* Table 6.2
Energy of radiation, 194
Energy of resonance, 11
EPA limits, 1
Error function, 335
Explosives in wastewater
  kinetic regimes of ozonation, *see* Table 6.3
External diffusion, *see* Kinetics of gas–liquid–solid catalytic reactions
External mass transfer kinetic regime, 245, 261

INDEX

F

F function, 337
Fast kinetic regime, 54, 92, 120, 121, 136, 141, 245, 305, 322
Fast ozone demand, *see* Ozone decomposition reaction
Fenton system, 175, 239
Fick's law, 49, 178
Film theory, 48, 51, 107
Flocculation, 2, 113
Fluorene ozonation kinetics, 104
  rate constant of direct ozone reaction, *see* Table 5.6
Fluorescence, 194
Food and kindred products wastewater ozonation of, *see* Table 6.1
Fruit cannery effluents, ozonation of, *see* Table 6.2
Fugacity, 71

G

Gas absorption theories, 47
Gas–liquid absorption reaction kinetics, 31, 47
Gas–liquid absorption reactions, *see* Chemical absorption kinetics
Gas–liquid solid catalytic reactions, *see* Kinetics of gas–liquid–solid catalytic reactions
Gas phase mass-transfer coefficient, 66
Gas phase ozone catalytic decomposition, 251
Gas phase resistance, 65
Gasoline tank leaking, Kinetic regimes of ozonation, *see* Table 6.3
Gibbs free energy, 71
Global effectiveness factor, 249

H

Hatta number, 52, 56, 107, 176, 186, 187, 324
  in wastewater ozonation, 118
Haynuk–Laudi equation, *see* Diffusivity
Haynuk–Minhas equation, *see* Diffusivity
Henry's law, 65, 71, 87
  constant values for the ozone–wastewater system, 136
  constant values for the ozone–water system, 79
Heptachlor, kinetic regime of ozonation, *see* Table 6.3
Herbicide manufacturing wastewater, Kinetic regimes of ozonation, *see* Table 6.3
Heterogeneous catalytic ozonation, 227
Heterogeneous direct ozonation, 43
  kinetics, 43
  stoichiometry determination, 44
Heterogeneous ozone decomposition reaction, 15
Homogeneous catalytic ozonation, 227
Homogeneous direct ozonation, 31, 33
  batch reactor kinetics, 33
  effect of pH, 40
  flow reactor kinetics, 39
  inhibition of indirect reactions, 34
  reaction rate constant determination, 35
    absolute method, 35
    competitive method, 36
  stoichiometry determination, 42
Hospital wastewater ozonation, *see* Table 6.1
Humic substances, 1, 124, 163
Hydraulic residence time, 278
Hydrogen peroxide, 15, 17, 151, 154, 175, *see also* Ozone/hydrogen peroxide system and UV/hydrogen peroxide system
Hydrogen peroxide photolysis, 193, 206
Hydroperoxide ion, 7, 17
Hydroxide ion, 7, 17
Hydroxyl radical reactions, 15, 20
  kinetics, 24
  pH influence on ozone decomposition to yield, 21
  rate constant values or equations, 22, 34
Hydroxyl radicals, 1, 8, 14, 151, 154, 158, 162, 164, 169, 177, 179, 183, 186, 190, 193, 206, 207, 217, 227, 254, 270
Hyperbolic functions, 334
Hypobromous acid, 18

I

Ideal flow reactor types, 331
  perfectly mixed reactor, 338
  plug flow reactor, 333, 338
Indigo disulphonate instantaneous ozonation kinetics, 91
Indirect ozone reactions, 7, 14, 164, 190
  with compounds in the UV/H_2O_2 system, 219
Individual mass transfer coefficient, 49, 50, 88
Inhibitors of ozone decomposition, 18, 34, 83, 157, 161
Initiators of ozone decomposition, 15
Instantaneous kinetic regime, 57, 89
Instantaneous reaction factor, 57, 64
Intensity of incident radiation, 194, *see also* Actinometry
Internal conversion in photochemical processes, 194
Internal diffusion kinetic regime, *see* Kinetics of gas–liquid–solid catalytic reactions

Internal diffusion mass transfer, *see* Kinetics of gas–liquid–solid catalytic reactions
Internal effectiveness factor, 248, 265
Intersystem crossing in photochemical processes, 194
Ionic strength, 73
Iron and steel wastewater ozonation, *see* Table 6.1

J

Johnson and Davis equation, *see* Diffusivity

K

Kinetic model types, 288
    comparison between, 314
    in fast kinetic regime, 305
    in moderate kinetic regime, 309
    in slow kinetic regime, 289
    one phase axial dispersion model, 288
    ozone mass transfer and hydrodynamics, 312, 313
    perfect mixing flow, 289, 299, 307, 318
    plug flow, 291, 299, 300, 307, 318
    tanks in series, 300, 325, 340
Kinetic modeling of ozone processes, 277
    flow type effect, 286
    gas volumetric flow rate, 278
    generation rate term, 277, 279, 282, 289
    liquid volumetric flow rate, 278
    stationary and nonstationary processes, 285
Kinetic modeling of wastewater ozonation, 316
    examples, 317, 318, 322, 324
    fast kinetic regime, 322
    slow kinetic regime, 317
Kinetic regimes of direct ozone reactions, 69, 83, 107
    conditions for, 84, *see also* Table 5.5
    influence of secondary reactions, 84
Kinetic regimes of ozone decomposition reaction, 80
    pH effect, 80
    reaction and diffusion times, 80
Kinetics of direct photolysis, 195
    flux density of radiation, 196
    local rate of absorbed radiation, 195
Kinetics of gas–liquid reactions, 50, 51
    irreversible first-order reactions, 51, 62
    irreversible second-order reactions, 54, 62
    series-parallel reactions, 58, 65
Kinetics of gas–liquid–solid catalytic reactions, 241
    external diffusion effects, 241
    fast kinetic regime, 245
    internal diffusion effects, 241
    internal diffusion kinetic regime, 246
    kinetic regimes, 242
    slow kinetic regime, 242
    surface reaction effects, 242
Kinetics of heterogeneous catalytic ozonation of compounds in water, 258
    external mass transfer kinetic regime, 261
    internal diffusion kinetic regime, 263
    slow kinetic regime, 259
Kinetics of heterogeneous catalytic ozone decomposition, 258
    agitation speed effect, 253
    liquid–solid catalytic reaction, 253
    mechanism of reactions, 256
    particle size effect, 253
    pH effect, 255
Kinetics of indirect ozone reactions, 151
Kinetics of ozone direct reactions, 31
    heterogeneous ozonation, 43
    homogeneous ozonation, 31, 33
Kinetics of ozone/UV oxidation, 193
Kinetics of photocatalytic ozonation, 269
Kinetics of physical absorption, 47
Kinetics of semiconductor photocatalysis, 265
Kinetics of UV/hydrogen peroxide oxidation, 206
Kinetics of wastewater ozonation, *see* Wastewater ozonation

L

Lambert law model, *see* Photoreactor models
Langmuir–Hinshelwood mechanism, 243, 256, 260
Leachate ozonation, 113, *see also* Table 6.2
Linear source with emission in parallel planes model, *see* Photoreactor models
Liquid holdup, 88

M

Maleic acid ozonation, 102
Marine aquaria wastewater ozonation, 113
Mass-transfer coefficient in gas–liquid–solid catalytic reactions, 262, 263
Matrozov et al. equation, *see* Diffusivity
Maximum chemical reaction rate in bulk water, 52, 57
Maximum chemical reaction rate in film layer, 52, 57

INDEX

Maximum diffusion rates of compounds in film layer, 57
Maximum physical absorption rate in film layer, 52, 57, 179, 188
MCPA ozonation, *see* Rate constants for hydroxyl radical reactions
Mean oxidation number of carbon, 124
    determination of, 124
    interference in the analysis of, 125
Mears criterion, 250
Mecoprop ozonation, 188
Methanol as hydroxyl radical scavenger, 16
Moderate kinetic regime, 101
Molar absorptivity, 194
MTBE, *see* Rate constant for hydroxyl radical reactions
Muconic acid, *see* Rate constant for hydroxyl radical reactions
Municipal wastewater, *see* Wastewater ozonation

N

Natural organic matter, 18
    characterization, 103
    hydroxyl radical initiating and scavenging character, 157
Natural substances, 1, 181
Nitrification, 123
Nitrites, 8
Nitroaromatics
    kinetic regime of ozonation, *see* Table 6.3
    rate constants for hydroxyl radical reaction, *see* Table 9.2
p-Nitrophenol ozonation kinetics, 90, 98, 105
Nitrotoluenes, kinetic regime of ozonation, *see* Table 6.3
Nonideal flow studies, 335
Nonlinear algebraic equations, 345
    Newton–Raphson method, 345
Nonlinear first order differential equations, 348
Nonpurgeable organic carbon, 124
Nucleophilic substitution reactions, 13

O

Oils shale wastewater ozonation, *see* Table 6.2
Olive oil wastewater ozonation, *see* Table 6.2
Organochlorine compounds, 1
Overall quantum yield, *see* Kinetics of UV/hydrogen peroxide oxidation
Oxalic acid ozonation, 179, 279
Oxidant positive hole, *see* Kinetics of semiconductor photocatalysis

Oxidation–competition values, 164
    for batch and plug flow reactors, 167
    for continuous perfectly mixed reactors, 169
Oxidation–reduction reactions, 7
Oxygen transfer reactions, 8
Ozone
    adsorption, 15, 273, 276
    decomposition mechanism, 15, 16, 19, 151, 163, 175, 251, 256, *see also* STB and TFG mechanisms for ozone decomposition
    formation, 2
    molecule of, 3
    origin, 2
    photocatalytic decomposition, 239
    photolysis, 15, 193, *see also* Ozone/UV radiation
    physical-chemical properties, 4
    quantum yield, 201 *see also* Quantum yield determination
    reactor types for reactions of, 33 *see also* Ideal flow reactor types
    resonance forms, 3, 12, 13
    solubility, equilibrium conditions, 71, 87
    determination of solubility of, 75
Ozone catalytic system, 175, 227
Ozone–hydrogen peroxide system, 175, 176, 182, 183, 185
    critical hydrogen peroxide concentration, 178
    diffusion and reaction times, 176
    fast kinetic regime, 177
    pH effect, 175
    slow kinetic regime, 177
    volatile organochlorine compounds in, 182
Ozone–UV radiation system, 175, 193
Ozonides, 9, 10

P

Paint and varnish wastewater ozonation, *see* Table 6.1
Parathion, *see* Rate constant of hydroxyl radical reactions from UV/H_2O_2 oxidation
Particle porosity, 247
Peclet number, 303
Pellet catalysts, *see* Kinetics of heterogeneous catalytic ozonation of compounds in water
Perfect gas law, 87
Pesticide manufacturing wastewater ozonation, 113
Petroleum refinery wastewater ozonation, *see* Table 6.1
pH sequential ozonation, 127

Pharmaceutical wastewater ozonation, *see* Tables 6.1, 6.2 and 6.3
Phenol ozonation, 11, 41, *see also* Table 5.6
Phenol wastewater ozonation, 113, *see also* Tables 6.1, 6.2 and 6.3
Phloroglucinol ozonation kinetics, 92, 108
Phosphorescence, 194
Photocatalytic processes, 175, 239
Photofenton processes, 175, 239
Photoinduced oxidation, *see* Kinetics of semiconductor photocatalysis
Photoinduced reduction, *see* Kinetics of semiconductor photocatalysis
Photoprocessing wastewater ozonation, *see* Table 6.1
Photoreaction, 265
Photoreactor models, 197
Photosensitization, 194
Physical absorption kinetics, *see* Kinetics of physical absorption
Plastic and resins wastewater ozonation, *see* Table 6.1
Point source with spherical emission model, *see* Photoreactor models
Polynuclear aromatic hydrocarbons, *see* Tables 5.6 and 6.3, *see also* Rate constants for hydroxyl radical reactions from UV/H_2O_2 oxidation
Powdered catalysts, *see* Kinetics of heterogeneous catalytic ozonation of compounds in water
Precursors, *see* Trihalomethanes
Pressure loss in ozonation kinetic modeling, 313
Primary quantum yield, *see* Kinetics of UV/hydrogen peroxide oxidation
Promoters of ozone decomposition, 157, 160, 161, 177
Pulp and paper wastewater ozonation, *see* Tables 6.1, 6.2, and 6.3
Pulse input experiments, *see* Tracer studies for nonideal flow
Purgeable organic carbon, 124

Q

Quantum theory, 194
Quantum yield, 194, 196, 199, *see also* Actinometry
  absolute method determination, 199
  competition method determination, 201
  values of, 201, *see also* Table 9.2

R

Radiant energy of the lamp, *see* Table 9.1
Rate constants of catalytic ozone reactions, 264
Rate constants of direct ozone reactions, 31
  determination from heterogeneous ozonation
    in fast kinetic regime, 92
    in moderate kinetic regime, 101
    in slow kinetic regime, 102
  determination from homogeneous ozonation, 35
    absolute and competitive methods, 35, 37
  values of, *see* Table 5.6
Rate constants of hydroxyl radical reactions, 160
  determination from single ozonation, 161, 162
    values of, 162
  determination from ozone/H_2O_2 oxidation, 180, 181
    natural substance effect on, 181
  determination from UV/H_2O_2 oxidation, 206
    natural substance effect on, 208
  values of, *see* Table 9.2
R_{CT} concept, 34, 170, 312
  in batch and plug flow reactors, 170
  in continuous flow perfectly mixed reactors, 170
Reaction factor, 54, 103, 107, 134, 281, 284
  effect of COD, 134
Reaction rate coefficient for wastewater ozonation, 134, 136, 140, 141
Reaction time, 67
Reactions of compounds with hydroxyl radicals and direct photolysis, 221
  comparison between, 221
  hydrogen peroxide concentration effect, 222
  intensity of incident radiation effect, 222
  scavenging of hydroxyl radicals, 222
Reactions of compounds with ozone and hydroxyl radicals from ozonation and ozone/hydrogen peroxide oxidation, 154
  competition between, 155
  kinetic regimes on the, 186
  pH and hydrogen peroxide effects on reaction rates, 151, 187, 190
Reactions of compounds with ozone and hydroxyl radicals from ozone/UV oxidation, 216
  comparison between, 217
  hydrogen peroxide concentration effect, 217
  reaction rates of, 218
Reactions of ozone with compounds and hydrogen peroxide, 183
  competition between, 183
  diffusion and reaction times on, 184
  kinetic regimes of ozonation, 183
  pH effects, 184

INDEX

Reactor design equations for ideal reactors, *see* Ideal flow reactor types
Reactor design equations for nonideal reactors, 33
Residence time distribution function, 86, 336
  mean residence time, 338
  variance of the distribution, 338
Resorcinol ozonation kinetics, 92, 108, *see also* Table 5.6
Rinse water ozonation, *see* Table 6.2
Runge–Kutta methods, 348

S

Schmidt number, 88
Schumpe equation, *see* Ozone solubility
Sechenov equation, *see* Ozone solubility
Sedimentation, 2, 113, *see also* Wastewater ozonation
Semiconductors, *see* Kinetics of semiconductor photocatalysis
Sensitized reaction, *see* Kinetics of semiconductor photocatalysis
Sewage water ozonation, *see* Table 6.2
SHB mechanism for ozone decomposition, 15, *see also* Table 2.3
Slow kinetic regime, 54
Sludge production, *see* Wastewater ozonation
Sludge reduction, *see* Wastewater ozonation
Sludge settling, *see* Wastewater ozonation
Sludge volumetric index, *see* Wastewater ozonation
Soaps and detergent wastewater ozonation, *see* Tables 6.1 and 6.2
Solubility ratio, *see* Ozone solubility
Specific external surface area per mass of catalyst, 249
Specific internal surface area per mass of catalyst, 248
Standard redox potential, 7, 265
Stoichiometry, 29, 42, 44
Supercritical wet air oxidation, 175
Superoxide ion radical, 8, 16, 17
Surface reaction, *see* Kinetics of gas–liquid–solid catalytic reactions
Surface renewal theories, 50, 62, 108, 152
Suspended organic carbon, 124
Swine marine wastewater ozonation, *see* Tables 6.2 and 6.3

T

Table olive wastewater ozonation, *see* Tables 6.2 and 6.3
Textile wastewater ozonation, *see* Tables 6.1 to 6.3
TFG mechanism of ozone decomposition, 15, *see also* Table 2.4

Theoretical oxygen demand, 122
Thiele number, 248
TiO_2 semiconductor, 239, 260, 266
Toluene ozonation kinetics, *see* Table 6.3
Tomato wastewater ozonation, *see* Tables 6.3 to 6.5 and 9.2
Tortuosity factor, 247
  determination of, 264
Total mass-balance equation, 305, 308, 322
Total organic carbon, 123
Tracer studies for nonideal flow, 336, 342
  ozone as a tracer, 342
Trichloroethylene ozonation kinetics, 181, *see also* Table 9.2
Trihalomethanes, 1, 124

U

Underground water, 1
UV radiation, 15, 151, *see also* Kinetics of UV/hydrogen peroxide oxidation, Kinetics of ozone/UV radiation oxidation, *and* Kinetics of semiconductor photocatalysis

V

Valence band, *see* TiO_2 semiconductor photocatalysis
Van Krevelen and Hoftijzer equation, *see* Ozone solubility
Very slow kinetic regime, 54
Vibrational relaxation, 194
Volatile aromatic compounds, 1, 182
Volatile organochlorine compound ozonation in ozone/hydrogen peroxide oxidation, 182
Volatility coefficients, 183
Volumetric mass transfer coefficient, 77
  from instantaneous ozonation kinetic regime, 90
  from slow diffusional ozonation kinetic regime, 105
  from wastewater ozonation, 136

W

Wastewater, 1
  characterization, 121
Wastewater ozonation, 113, *see also* Chemical biological processes
  Hatta number, 118
  ozone diffusivity, 135
  ozone solubility, 136

pH effect, 125, *see also* pH sequential ozonation
scavenger effect, 126
Wastewater ozonation kinetics, 113, *see also* Kinetic modeling of wastewater ozonation
   fast kinetic regime, 135
   reactivity, 135
   slow kinetic regime, 143

Weiz–Prater criterion, 250
Wet-air oxidation, 175
Wilke–Chang equation, *see* Ozone diffusivity
Wood chips contaminated wastewater, 113

Z

Zwitterion, 9